MW00843744

HISTORY OF INDUSTRIAL GASES

HISTORY OF INDUSTRIAL GASES

Ebbe Almqvist
Ebbe Almqvist Consulting
Sweden

Kluwer Academic / Plenum Publishers
New York, Boston, Dordrecht, London, Moscow

Library of Congress Cataloging-in-Publication Data

Almqvist, Ebbe.
History of industrial gases/by Ebbe Almqvist.
p. cm.
Includes bibliographical references and index.
ISBN 0-306-47277-5
1. Gases—History. I. Title.

TP242 .A38 2002
665.7′09—dc21

2002066906

ISBN 0-306-47277-5

233 Spring Street, New York, New York 10013

http://www.wkap.nl/

10 9 8 7 6 5 4 3 2

A C.I.P. record for this book is available from the Library of Congress

Printed in the United States of America

PREFACE

All history goes to show that the progress of society has invariably been on lines quite different from those laid down in advance, and generally by reasons of inventions and discoveries which few or none had expected.

J. H. Beadle, 1893

The history of industrial gases is a story that starts at the dawn of science and continues to today's huge global industry, which plays a role in almost every area of life.

This book gives an overview of the history of gases, from the ancient theories of mysterious "airs", to the discovery and development of different gases, to their present production and application procedures. Such an understanding is essential for people involved in the gas industry and should also be of interest for anyone curious about how our environment came to be known, investigated, and developed.

The historical development of industrial gas technology and business and its impact on society during the last century is treated in particular detail. Current production, distribution, and applications of gases are based on discoveries and developments in different areas of science and technology. This aspect of the book may be especially valuable for engineering and marketing staff in industrial gas companies and for their customers. The uses of gases range from those of steelworks, which use thousands of tons of oxygen per day, to those of semiconductor fabrication plants, which may handle kilogram quantities of high-purity specialty gases.

Just as interesting as the applications are the ways in which the gases are manufactured. Even to people with some familiarity with industrial gas production methods, it is remarkable that liquid oxygen and nitrogen in low quantities can be sold for around the price of milk.

The present industrial gas business may be divided into the "on-site market", in which gas is made for an individual user and produced on or adjacent to the customer's operation site and delivered by pipeline for immediate consumption, and the "merchant gases market," the bulk and cylinder business, in which the products are made at a central location and transported in pipelines, tankers, cylinders, and insulated flasks to thousands of applications in many customers' facilities.

On-site production of industrial gases is now feasible even for smaller consumption due to new production methods. This reduces costs for customers because of the reduction of transportation costs. Such plants are owned and run by industrial gas companies and customers do not have to worry about the capital cost of equipment or personnel to operate the plant. Today, the on-site market makes up at least 20 percent of the worldwide industrial gas business. It is the fastest growing segment and is expected to continue to expand with a very high growth rate in the next decades.

The driving forces for the present growth of the business are:

- *Environmental protection.* The stricter emissions regulations in more and more countries have made environmental protection an especially important issue. Environmental applications and technology have been extremely important since improved processes of water treatment or purification of waste air have been developed. Furthermore, environmental pressures in developed countries have forced petrochemical plants to shift to the substitution of oxygen for air.
- *New geographic areas.* Additional opportunities are being afforded due to the industrialization of the developing world, particularly Asia and Latin America, where basic steelmaking and chemical capacity is being built.
- *Higher purity, quality, and efficiency in industrial processes.* Greater purity of gas is required in such areas as electronics and special optical manufacturing. Quality foods require particularly high product purity. For example, rapid freezing using liquid nitrogen preserves the fresh taste of shrimp. Food handling, such as transporting, freezing, packaging, and storing, is also a growth area for gases. New and expanding areas for gases include control of the ripening of fruits and high-pressure extraction applications in areas such as fruit juice production and decaffeination of coffee.

Acknowledgements

The compilation of a broad history like this one would not have been possible without assistance from specialists in the different subjects treated. Many colleagues in the major industrial gas companies have influenced, encouraged, and otherwise contributed positively to the writing of this book.

Because much company archive material was destroyed during the world wars, there is a lack of much historical evidence of when and how the industrial gas business started and developed in different countries (see Chapter 6). I have made every effort to provide the most accurate information available, but the record is not complete. I have, however, received much information on how the global business was created through interviews, discussions, and e-mails.

Without singling out any individuals, I would like to express my gratitude for their support to all who contributed their time and expertise.

I would, however, like to specifically thank my former AGA colleague John Campbell and his staff at JR Campbell & Associates, Lexington, Massachusetts, for helping me turn my original manuscript into a book.

Ebbe Almqvist
October 2001

FOREWORD: HISTORY OF INDUSTRIAL GASES

This marvelous book tracks the history of industrial gases from some of the very earliest theories of gases by Aristotle, through the observation and discovery of gases by some of the great scientists of the past 500 years, through discovery of oxygen by Scheele and Priestley in the 1770s, and further to current and anticipated developments in related gas products, production technologies, market uses and unique applications. A great deal of attention is paid by author Ebbe Almqvist to the major players who have built up this industry which produces and distributes and provides many unique customer services – like supplying liquid oxygen, liquid nitrogen, liquid hydrogen and liquid helium to NASA's Space Shuttle, and keeping the super sophisticated medical diagnostic system, MRI, topped off with liquid helium.

This book has taken Ebbe a decade of painstaking research gathering of some of the little known facts of our industry's development, and working with a large number of people to provide some wonderful drawings and photos of aspects of our industry and its development.

This is the very first work of its kind about our industry and provides great background reading for newcomers to our industry and to old hands who continue to wonder how this fascinating business came to be so much a part of all of our lives, in industry, medicine, science and commerce. It is particularly appropriate at this time as several of the major companies which have built this specialized industry are celebrating being at or over the century mark in business.

J. R. Campbell
Publisher, Cryogas International

CONTENTS

Short Stories

1

HISTORY OF GASES

Introduction[1]

Discoveries in England by Priestley, Black, Cavendish, and Rutherford

The parishioners of Mill Hill Chapel in Leeds, England, were disturbed. Their new minister, Joseph Priestley, was spending all his spare time at Jake and Nell's brewery. Had the devil drink taken command of their preacher? Speculation was rife and rumor spread like wildfire, even reaching the ears of the cause of their worry. Yet he continued in his investigation of the contents of the brewer's vat. The year was 1767, and his studies were related to a recently discovered gas, then known as fixed air and today known as carbon dioxide.

Priestley discovered that the gas was easily soluble in water and that it gave a "light, pleasantly acid taste." This discovery was later to give rise to carbonated mineral waters and soda water. It was also the discovery that, toward the end of his life, Priestley described as having given him the most enjoyment, despite having brought him the least scientific merit.

A misadventure prevented further research at the brewery. During a determination of the solubility of carbon dioxide in ether, a quantity of the solution ran into the vat, thus ruining the whole brew. Priestley was cast out from the natural carbonic acid environment of the brewery. Building some laboratory equipment in his kitchen, he initially used beer glasses and bowls, but in time used more refined equipment. His research into gases continued and was intensified when in 1772, acting on a recommendation by his friend Benjamin Franklin, he resigned his ministry to become librarian and literary companion to Lord Shelburne.

In time, Priestly discovered additional gases. The stimulus from his attempts at the brewery led to his later discovery and isolation of oxygen simultaneously with, but independently of, other scientists. Ammonia, hydrogen chloride, nitrous oxide, hydrogen sulfide, nitrogen oxide, nitrogen dioxide, silicon tetrafluoride, and carbon monoxide were all isolated during a short period. Priestley was also the first to study the gas absorption process by plants, photosynthesis.

Some years previously, several scientists had been engaged in examining different types of "air," that is, gases. Joseph Black, a professor of chemistry at Glasgow, had shown that gases were chemically identifiable substances. He produced isolated carbon dioxide in 1756. He was able to detect its presence in human exhalation by allowing slaked lime to be exposed to the gas that flowed from the ventilator in the roof of a church in which a congregation of 1500 were taking part in a 10-hour-long religious ceremony.

Around this time there lived in London an eccentric, rich nobleman, Henry Cavendish. He was a recluse who engaged in chemical experiments of great importance. In 1766 he produced and analyzed hydrogen gas ("inflammable air"). Later, in 1783, he repudiated Aristotle's theory that water was a basic substance and determined its composition. He also predicted the existence of further gases in air, such as inert gases. These were first isolated during the 1890s.

Daniel Rutherford, one of Black's students, reported on the production of nitrogen in 1772 in his master's dissertation.

Sweden and Scheele

It was not only in England that experiments with gases were taking place. An apothecary in Uppsala, Sweden, Karl Wilhelm Scheele, had already made similar experiments concerning the composition of air prior to those of Priestley and Rutherford, but publication had been delayed. At the same time Scheele's friend and colleague Torben Bergman, a professor of chemistry, was making efforts to produce artificial mineral water by analyzing waters from different springs and reproducing them using carbonic acid and minerals. Bergman suffered from illhealth and, like many of that period, believed in the healing effect of drinking spring water, although he was not able to stay at a spa for long. The commercial production of different forms of artificial mineral water was carried out on Bergman's initiative from 1776.

It All Begins with Ancient Theories and Alchemy

Commencing in 1756, it took less than a quarter of a century to discover and identify the most important of the naturally occurring gases. Why then had not the components of air and the effect of oxygen on combustion and other life processes been discovered earlier? The fact that air contained two components had already been pointed out by da Vinci in the 16th century and by Robert Hooke and John Mayow in the 17th century. Their results, however, had not been complete and had therefore fallen into oblivion.

The answer goes back to the time of Aristotle and earlier. The major theory of materials of that time, based on ideas propounded by the Sicilian/Greek philosopher Empedocles in the fifth century BC, was that the four primal elements—earth, water, air, and fire (Aristotle added a fifth, the "aether")—were the basis of all materials. Other chemical substances contained them in varying proportions. According to the principles of alchemy as they developed around the year 200 BC, substances could be transmuted from one into another.

Alchemy first appeared in Hellenic Alexandria. Its origin is to be found in a combination of the philosophy of Aristotle and the practice of metallurgy from Ancient Egypt, where the fraudulent art of casting metals that resembled gold was prevalent. As alchemy developed, practical chemical methods were mixed with newer religions and a richly developed store of sorcery and astrology. The alchemists created an almost inaccessible wisdom that they were intent on keeping to themselves. One much-sought-after goal was the "philosopher's stone," which, besides having the ability to transmute simple elements into gold, in the Middle Ages was also believed to contain the medicinal qualities of the universe. The stone was an epitome of the constant striving toward wealth and immortality, the same endeavor that drives today's stock exchanges and health care apparatus. Regardless of what one might think about it today, it did act as an incentive to engaged researchers for 1000 years.

Paracelsus and Hydrogen Gas

Interest in how to make medicines gained in importance, in the 16th century, which has been called the "atrochemical period" (*atros* is the Greek word for physician).

The year after Columbus's first voyage to the American continent, the physician–chemist Theophrastus Bombastus von Hohenheim, better known as Paracelsus (1493–1541), was born. His practical approaches to both medicine and chemistry during his relatively short life span had a far-reaching effect on modern science. Still under the influence of alchemy, he believed that all substances were composed of mercury, sulfur, and salt. He called these the three *principles* of matter because he still considered the elements to be air, earth, and water. Paracelsus extended "air" to include all aerial matter and gaseous emanations. He observed that when iron is dissolved in sulfuric acid, "air rises and breaks out like wind"—hydrogen.

Two generations after Paracelsus the Flemish chemist Johann Baptista van Helmont (1577–1644) applied the term "chaos" to what we now call "gas". He was the first to realize that gaseous substances other than air exist. He differentiated between gases and vapors; the latter were formed from water (liquid) by the effect of heat and could condense to give water again on cooling, whereas the former were dry, airlike substances which cannot be changed into liquid. Van Helmont was also able to separate different gases, and found that "*spiritus sylvestre*," or carbon dioxide, arose from the fermentation of alcohol and the burning of coal. However, van Helmont's convenient term "gas" did not find ready acceptance, as men like Robert Boyle (1627–1691) and Joseph Priestley (1733–1804) preferred terms such as "artificial air" or "factitious air" or even "different kinds of air."

Our modern idea as to what constitutes an element was first expressed by Robert Boyle in his treatise *The Sceptical Chymist* published in 1661. He also developed the first chemical gas theory. Boyle's colleague, Robert Hooke, stated in 1665 in his theory of combustion that air contains a substance which, upon being heated, has the ability to dissolve all combustible substances. The same substance is found in large quantities in saltpeter.

In 1669, John Mayow tried to burn camphor in air in which a candle had burnt until it went out. Because that was not possible, he concluded that air consisted of two components, one of which is capable of supporting combustion. Had his observations been followed up they could have laid the foundation for modern chemistry and the theory of combustion a century earlier than what in fact occurred.

The Defense of Alchemy

Alchemy was deeply rooted, however, and was defended by those in power at that time. It made all the more difficult the acceptance of Boyle's conceptions of materials and the newly won knowledge of the nature of the gases and combustion. Instead, the next century, from 1691 on, came to be dominated by an erroneous, yet systematically and thoroughly developed science of combustion, the phlogiston theory. Hovering over this theory like lost souls were the five elements and three principles. A German chemist, Georg Ernst Stahl, proposed the theory of the phlogiston, thought of as existing in all combustible substances, which disappeared when that substance was burnt. That the weight of this substance was sometimes positive and sometimes negative was of little consequence to the true phlogistonian.

It was not until the discovery of oxygen in 1772 by Scheele and 1774 by Priestley that phlogiston could be abolished and replaced by a more modern theory of combustion, which allowed the advancement of modern chemistry. The fact that this took place as early as 1774 was perhaps quite by chance. In Paris the brilliant Antoine Lavoisier had been researching the nature of combustion for some years. Scheele wrote him a letter in which he mentioned in passing his discovery of oxygen. In the autumn of 1774, Lavoisier was visited by Priestley, who was on his only official journey abroad accompanying his employer, Lord Shelburne. Based on these unplanned contacts between researchers, Lavoisier was able to put together the pieces of the puzzle that were to overturn the phlogiston theory. In a thesis of 1783 he reported that by careful weight determination he had determined that oxygen united with other substances during combustion, and that matter was neither created nor disappeared.

It is remarkable that both Priestley and Scheele died devotees to the phlogiston theory.

What Happened to the Newly Discovered Gases and Their Discoverers?

What Became of the Discoverers?

In 1780 Priestley returned to the ministry and moved to Birmingham. Over and above his preaching, he took part in political and scientific debates. Birmingham was home to the Lunar Society, whose members were prominent scientists and cultural personalities. They met on the evening of each Monday before the full moon to exchange opinions, many of which were to have a strong effect upon future scientific and cultural development. Among the members were James Watt, the inventor of the steam engine among other things; William Herschel, the astronomer; Josiah Wedgewood, the pottery king; and Erasmus Darwin, the philosopher. Priestley took the side of the fighters for American independence and sympathized with the French Revolution. For 3 years he was hounded by both the local populace and the authorities, and in 1794 he immigrated to America, where he again served as a minister, unaffected by offers of honorable professorships. He died in 1804.

The only physical contact Cavendish had with the outside world was through meetings and dinners at the Royal Society. His letters and writings describe in detail his great achievements in gas chemistry.

Black was a great teacher who, over a period of 30 years, shared his knowledge with and inspired several generations of students, among them Daniel Rutherford and James Watt. He published very few of his findings.

Scheele and Bergman made pioneering contributions despite their lack of funds. Their unselfish and painstaking research under conditions of poor heating and ventilation resulted in premature death, neither reaching the age of 50.

Lavoisier contributed to many areas of development. In order to finance his scientific work, from 1768 he held many public offices including a commissary of the treasury in 1791. He and his associates were falsely accused of fraud by the revolutionary tribunal and Lavoisier was guillotined in 1794, only 54 years old.

What Happened to the Gases That Were Discovered?

The answer is: “Not much.” The newly discovered gases could not be directly utilized in part because industrial methods of production were lacking. It was to be another 100 years before the cryotechnical air gas process made it economically feasible to produce pure oxygen and nitrogen. As time went by a large-scale demand was created, primarily as a result of developing chemical and scientific knowledge.

Certain unusual applications were, however, created at an early stage: Hydrogen gas was used from 1783 on for manned balloons, a development that was not terminated until 1937 with the catastrophe of the *Hindenburg* airship. Pure carbon dioxide was used in mineral water from about 1780. Oxygen found use in medicine from about 1800 and, from 1825 until the advent of electric lighting in 1890, as a source of lighting in theaters. Nitrous oxide, or laughing gas, was “sniffed” frequently at European aristocratic feasts during the first half of the 19th century. Prince Oscar of Sweden, later King Oscar I, often invited the chemist Berzelius together with the mood-elevating gas to his parties. During the second half of the century, nitrous oxide was used as an anesthetic. The use of nitrogen in agriculture, primarily saltpeter imported from Chile, doubled production per unit of area during the 19th century. This development paved the way for the use of aerial nitrogen during the 20th century.

The Liquefaction of Gas

The exploitation of industrial gases had to wait some decades for progress in physical and chemical science. Many new discoveries then happened simultaneously, apparently independently, in different parts of the world. This was mainly due to the broad interest in gas physics, gas theory, thermodynamics, and cooling technology generated by early industrialization.

The first gas liquefaction was announced on Christmas Eve 1877 and marked the start of a new technological era, with major practical applications after the turn of the century. Each gas has its own history. The aim of this book is to describe the history of the industrial gases and how they have revolutionized technical development, our lives, and our welfare.

Reference

1. Almqvist, E. (1989). Chaos was gas. *AGA Gasetten* **5**: 8–11.

2

FROM ARISTOTLE TO THE BIRTH OF MODERN CHEMISTRY

2.1. FROM ARISTOTLE TO THE BIRTH OF PNEUMATIC CHEMISTRY

Ancient Theories[1]

Philosophers in ancient Greece believed that everything in the universe could be reduced to one simple elementary substance. Thales (640–546 BC) concluded that this substance was water. Anaximenes (560–500 BC) believed it was air, Heraclitus (536–470 BC) thought that it was fire, and Xenophanes (570–480 BC) guessed that earth might be the fundamental element of the universe. Empedocles (490–430 BC) suggested that there were four basic elements: earth, water, air, and fire. To these Aristotle (384–322 BC) (Fig. 2.1.1) added a fifth, aether, the element of the heavens. For centuries these were accepted as the elements that made up the universe. In our day we recognize them as states of matter: solid, liquid, gas, and plasma.

The Chinese recognized five chemical elements: wood, metal, water, earth, and fire; ancient Hindus described the world in terms of nine elements: water, earth, fire, air, ether, time, space, soul, and sense.

An interesting aspect of the notions of the ancients is the importance given to fire. This concept was developed much later into complex theories in which fire played an immensely important part, such as the phlogiston theory.

Some of the substances we now classify as elements were known to the ancients, but were not recognized as elements. Copper, lead, gold, and silver were known to the Egyptians before 3500 BC. They also used bronze (an alloy of copper and tin) and tools made of iron. Mercury seems to have been discovered around 1500 BC. The only nonmetallic elements known to the ancients were carbon and sulfur, or "brimstone," as it is called in *Genesis*. All of these elements either occur in nature in the free state or can be separated from their ores at fairly low temperatures.

Alchemy (300 BC to 1600 AD)

The philosophical tradition of ancient Greece and the metallurgical tradition of ancient Egypt met in Alexandria, Egypt, and alchemy was the result. The early alchemists used Egyptian techniques for the handling of materials to investigate

Figure 2.1.1. Aristotle. Roman copy of a Greek statue. From *Bonniers Konversationslexikon* (1922).

theories concerned with the nature of matter. The philosophical content of alchemy incorporated elements of astrology and mysticism into the theories of the earlier Greeks (Fig. 2.1.2). A dominant theme was the transmutation of base metals, such as iron and lead, into the noble metal, gold. They believed that a metal could be changed by changing its qualities (particularly its color) and that such changes occurred in nature as metals strove for the perfection represented by gold.

Furthermore, the alchemists believed that these changes could be brought about by means of a very small amount of a powerful transmuting agent, later called the philosopher's stone.

In the seventh century AD the Arabs conquered the centers of Hellenistic civilization and alchemy passed into their hands. Greek texts were translated into Arabic and served as the foundation for the work of Arab chemists. The Arabs called the philosopher's stone *aliksir* (which was later corrupted into *elixir*). Arab alchemists believed that this substance not only could ennoble metals by transmuting them into gold, but also could ennoble life by curing all diseases.

From early alchemy only one type of gas, namely air, was known; although it was studied by natural philosophers, it was not considered to be in the practical chemist's sphere of interest. The gaseous products that were formed during various processes (for instance, fermentation or putrefaction) were considered by scientists to be varieties of air. Many other gases apart from air could be found in nature.

Figure 2.1.2. Alchemical view of matter: 7 metals, 12 zodiac constellations, water, air, fire, earth. From Valentinus (1651).

Eruptions of natural gas were observed from very early times and the dangers of fire-damp in mines were soon realized. Pliny the Elder (23–79 AD) refers to air which cannot be breathed and which is inflammable.

Alchemists often observed the evolution of gases during their experiments, and referred to them as *spiritus*, a term which also included acids, and the meaning of which was not quite clear. They supposed that the evolved gases were merely air, so they paid very little attention to them and did not carry out experiments to determine their nature.

From great antiquity it had been known that air played an essential role in combustion, but its actual function in the process of burning was not addressed by the philosophers of chemistry before the 16th century (Fig. 2.1.3).

The first European to state that air is not an element was the versatile artist–scientist Leonardo da Vinci (1452–1519). Leonardo noticed that air is consumed in respiration and combustion, but that it is not completely consumed. He recognized that the atmosphere contains at least two constituents.

Iatrochemistry[2]

Paracelsus and Hydrogen Gas

Alchemy came to the Western world during the crusades of the 12th century. The European variant changed direction during the Middle Ages. The making of gold

Figure 2.1.3. Alchemical assay laboratory. From L. Ercker, *Treatise on Ores and Assaying* (1574).

became less important than the task of creating preparations and drugs against human ailments and sickness. Thus was founded iatrochemistry, the use of chemical medicines, which contributed to the decline of the old material theories. Most renowned during this transitional period was the Swiss physician Paracelsus (Theophrastus von Hohenheim; 1493–1541) (Fig. 2.1.4).

Paracelsus forcefully disputed the teachings deriving from Aristotle, and for this was dismissed from his post as lecturer at the University of Basel. He advocated the increased use of chemical medicines and conducted numerous experiments in which he tested the old methods. He declared a new theory of materials in which all existing substances were composed of three principles: sulfur, mercury, and salt. The efficacy of Paracelsus as a doctor became widely known, but his ideas were strongly contested by the learned of the time. His death in 1541 was possibly caused by his own medication. One his far-reaching theories concerned the effect of air on combustion and in the processes of life. He considered life to be a slowly burning fire. By experiment he produced "inflammable air," probably hydrogen, which burned with a blue flame.

Chaos Becomes Gas

Johann Baptista van Helmont (1577–1644) (Fig. 2.1.5), a Flemish, physician, furthered the teachings of Paracelsus and developed a mystical, alchemically colored natural philosophy combined with scientific theories. He proposed an "*archeus*," a kind of spirit that by fermentation could unite all forms of material particles. He

Figure 2.1.4. Paracelsus. From Aberle (1891).

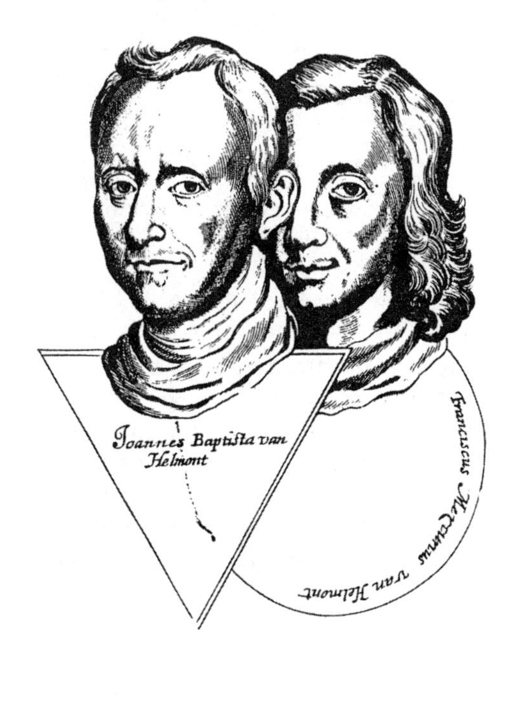

Figure 2.1.5. Johann Baptista van Helmont and his son Franciscus Mercurius, who published his father's works in 1648. From J. van Helmont, *Oriatrike or Physik refined* (1662).

believed that water was the prime element and the source of everything. Evidence for this was that he had been able to create a large willow tree purely by adding only water. Van Helmont believed that the growth was due to a fermenting agent in the seed, which, affected by the water, gave off a smell that attracted the *archeus*.

The *archeus* was to be found in scientific conceptions up until the 19th century as a bearer of a special vitality. Organic substances contained this force and could not be produced synthetically. It was not until 1828, when the German chemist Friedrich Wöhler synthesized urea, that this myth was dispelled.

In one of his experiments van Helmont burnt 62 pounds (30 kilograms) of wood and obtained only 1 pound of ash. What was the rest of the wood transformed into? Into a "wood spirit" (*spiritus silvestre*), the scientist believed. He called this previously unknown "spirit" by a new name, "gas," which he derived from the Greek word *chaos*. Van Helmont observed what we now call carbon dioxide and gave other instances of the formation of this gas, remarking that it occurs in caves, cellars, and mineral waters. He also noted that other gases were formed from the combustion of sulfur and saltpeter. Van Helmont did not collect gases because he thought they were uncontrollable and could not be contained because they possessed an untamable, wild spirit, which would burst a sealed vessel in its efforts to escape. So he did not examine the gases any further and saw in the new gases only a variety of air. Thus van Helmont did not understand his discoveries and it took 100 years for carbon dioxide to be once again defined and produced, by the English physicist Joseph Black.

In spite of the work of van Helmont the general view in the 18th century was that all gases were composed of air. Black's discovery that although carbon dioxide is present in air, it is a quite different gaseous substance, provided the breakthrough for the discovery of the gaseous elements.

Van Helmont's convenient term "gas" did not find acceptance until the great French chemist, Pierre Joseph Macquer (1718–1784), who published encyclopedic works on theoretical and practical chemistry, reintroduced the word into the literature. This usage was also followed by Antoine Laurent Lavoisier, the "father of modern chemistry," in his work between 1770 and 1790.

The New Theory of Elements

The discovery of the vacuum pump and the famous Magdeburg demonstration by Otto von Guericke (Figs. 2.1.6 and 2.1.7) in 1654 also showed the necessity of air for life and combustion.

The next important stage of development in the history of gases came around 1660 when the Anglo-Irish scientist Robert Boyle (1627–1691) developed his chemical gas theory. In his publication *The Sceptical Chemist* (1661) (Fig. 2.1.8) he criticized Paracelsus's principles and theories of transmutation. He stated that all substances were made up of components, "corpuscles," which he called elements. He defined an element as a basic substance that could be combined with other elements to form compounds, but could not itself be broken down into any simpler substance. This definition is remarkable because Boyle and his contemporaries still thought that gold could be made from other metals, and Boyle's list of elements included such

Figure 2.1.6. Otto von Geuericke's demonstration of the power of a vacuum, performed in Magdeburg, 1654. Sixteen horses could not pull apart the two halves of the evacuated sphere. From Ålund (1875).

materials as salt, water, and air. It was not until a century later that scientists actually began to identify which substances could be broken down into simpler substances and which could not.

Boyle also observed the formation of other gases; he described the formation of hydrogen and found a way to contain the gas. He placed some iron nails in a narrow-necked flask filled with sulfuric acid and then covered the neck of the flask with a vessel filled with water, and collected the bubbles that were evolved. In spite of these experiments Boyle did not believe that the various kinds of gases were essentially different from air.

Ideas about respiration had not advanced much from the theories of Aristotle: Breathing was thought to cool and ventilate the blood. Robert Boyle had noticed analogies with combustion (e.g., neither life nor flame survives in an enclosed volume of air) and speculated that both processes might be due to the presence of a "nitrous" substance in the air.

Inspired by von Guericke's vacuum pump experiments of the 1650s, Robert Hooke built a single-barreled air pump (Fig. 2.1.9). In 1665 in his book *Micrographia* he presented a complete theory of combustion. Hooke thought that air contains a

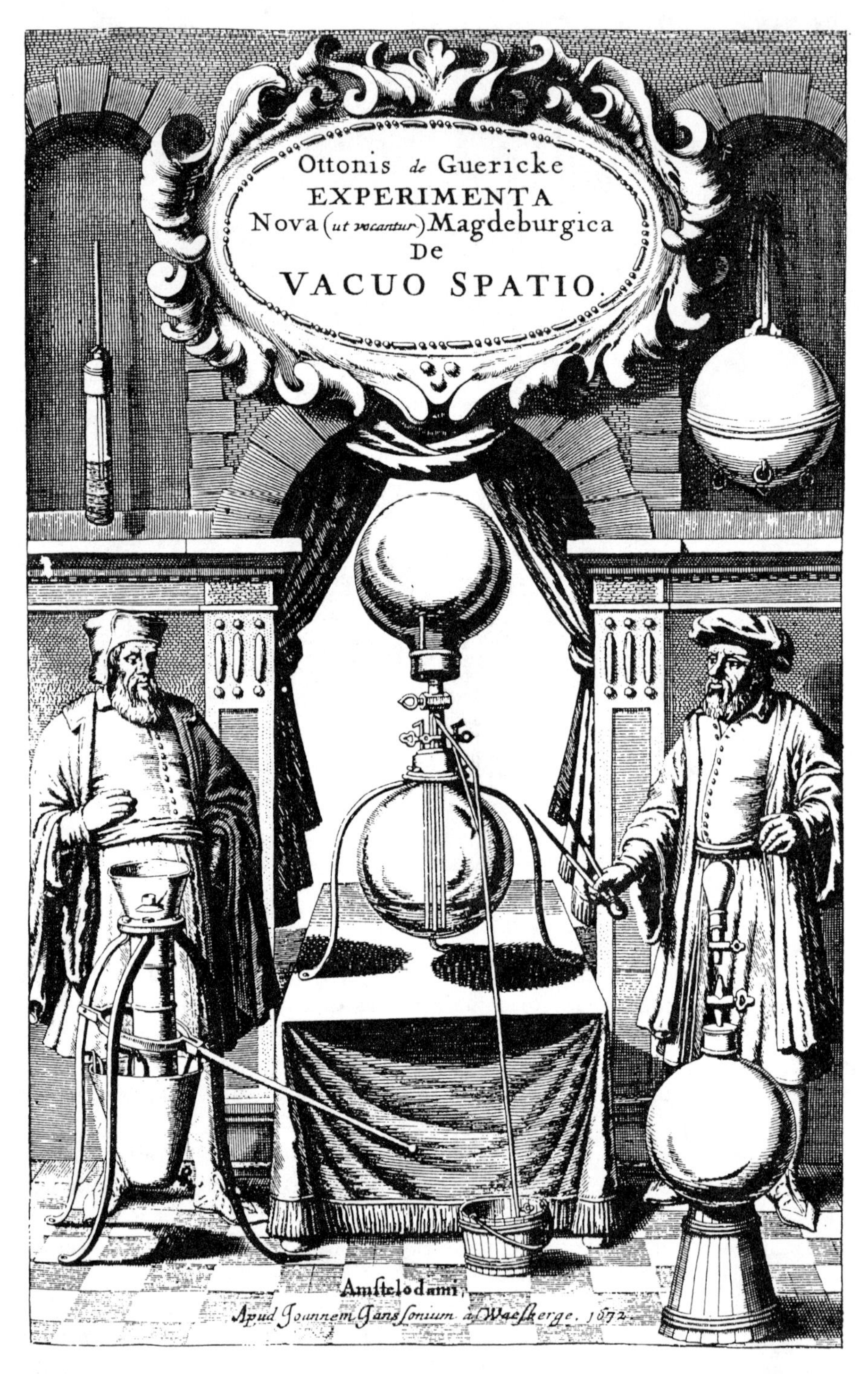

Figure 2.1.7. Title page of von Guericke's book, *Experimenta Nova Magdeburgica* (1672).

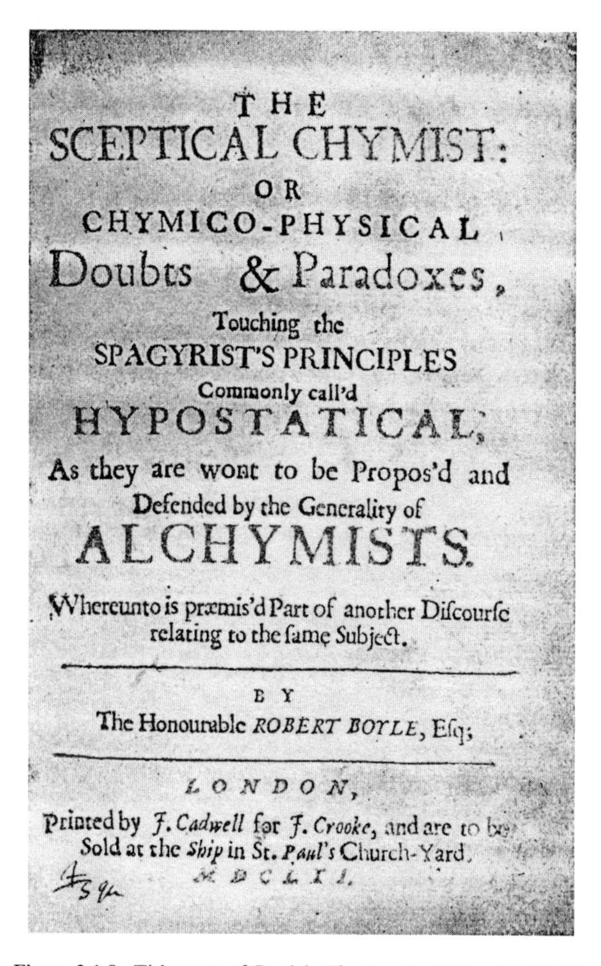
THE
SCEPTICAL CHYMIST:
OR
CHYMICO-PHYSICAL
Doubts & Paradoxes,
Touching the
SPAGYRIST'S PRINCIPLES
Commonly call'd
HYPOSTATICAL;
As they are wont to be Propos'd and
Defended by the Generality of
ALCHYMISTS.
Whereunto is præmis'd Part of another Discourse
relating to the same Subject.

BY
The Honourable *ROBERT BOYLE*, Esq;

LONDON,
Printed by *J. Cadwell* for *J. Crooke*, and are to be
Sold at the *Ship* in St. *Paul's* Church-Yard.
M DC LXI.

Figure 2.1.8. Title page of Boyle's *The Sceptical Chymist* (1661).

substance that exists in solid form in saltpeter and a larger quantity of an inert substance.

In 1669, John Mayow found that camphor could not burn in air in which a candle had burnt until it went out. He constructed an apparatus for collecting gases similar to the apparatus used by Boyle. Mayow explained combustion by saying that air contains a "*spiritus nitroaereus*" (oxygen), a gas that is consumed in respiration and burning, with the result that substances no longer burn in the air that is left (Fig. 2.1.10). He thought that his *spiritus* was present in saltpeter, and stated that it existed, not in the alkaline part of the salt, but in the acid part. According to Mayow, all acids contain the *spiritus*, and all animals absorb it into their blood as they breathe.

Pneumatic Chemistry

In the last part of the 17th century many chemists were performing experiments on gases. Techniques and equipment for experiments had been refined since the efforts

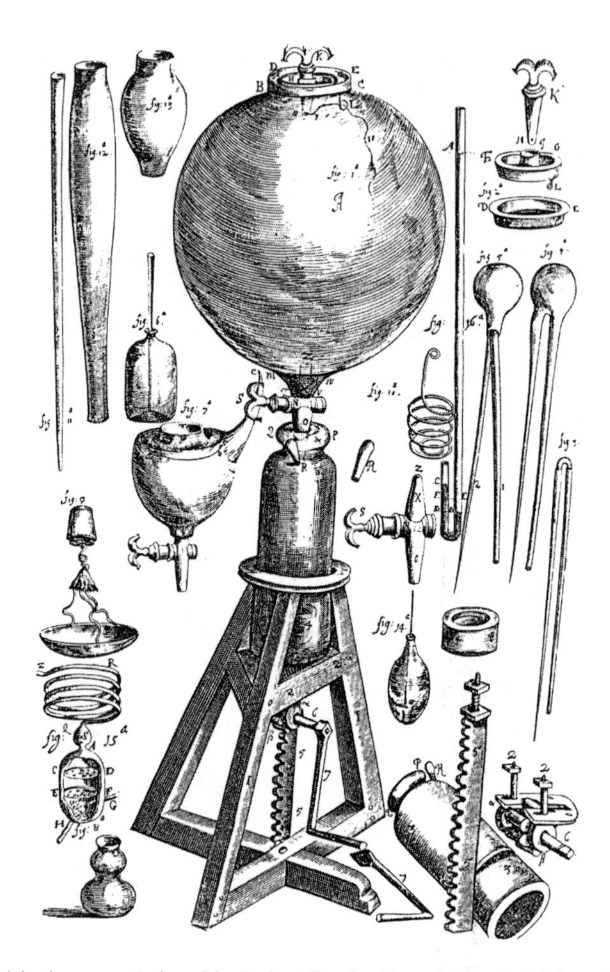

Figure 2.1.9. Boyle's air pump, designed by Robert Hooke. From R. Boyle, *Philosophical Works* (1725).

of van Helmont. Although quantitative results were obtained, less attention was paid to qualitative experiments.

The new interest in the study of gases contributed to the development of pneumatic chemistry, the scientific discipline that laid the foundation for the discovery of the gases in our environment. The study of gases had been difficult due to the absence of adequate methods for their preparation and collection and procedures for analysis of their properties. Bladders of animals were almost the only experimental vessels for collecting and weighing liberated gases, and it was much more difficult to study gases than solids or liquids.

The separation and identification of different gases was greatly facilitated by the construction of a "pneumatic trough" by the English botanist Stephen Hales (1677–1761) (Fig. 2.1.11). This, the earliest form of gas-collection apparatus, is still in use today. Previously, gases had been generated and collected in the same vessel,

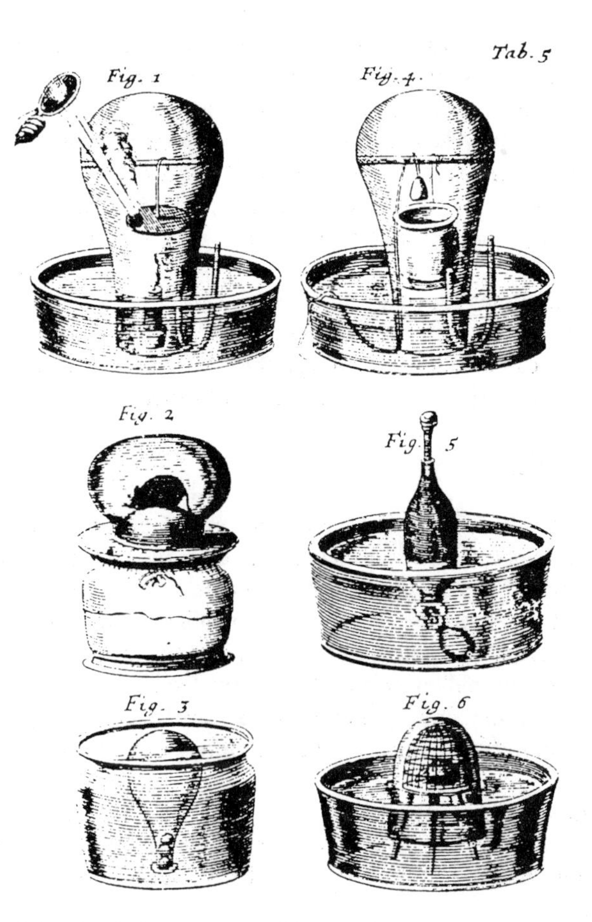

Figure 2.1.10. Some of the chemical apparatus used by John Mayow. From J. Mayow, *Tractatus quinque medico-physici* (1674).

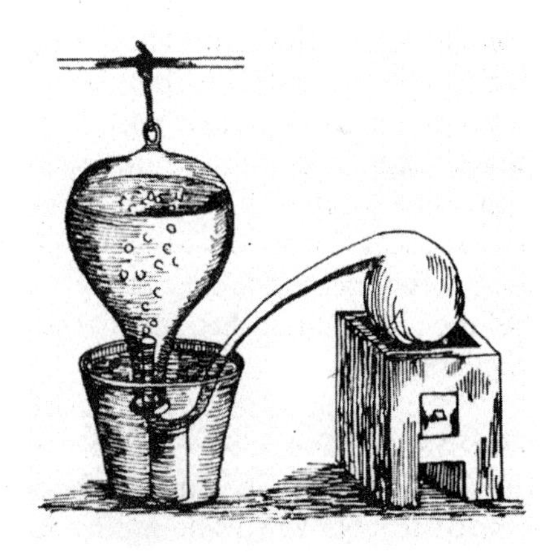

Figure 2.1.11. Pneumatic trough designed by Stephan Hales. From S. Hales, *Statical Essays Containing Vegetable Staticks* (1727).

but in Hales's apparatus two separate vessels were used, connected by a tube. The advantage of this apparatus was that gases evolved by the action of heat on a substance could be isolated. Hales made tests on wood, tobacco, oils, sulfides, limestone, fat, etc., and showed that many substances evolved "airs" when they were heated. He also measured the amount of gas obtained from a certain weight of substance, using water as the confining liquid. Although Hales succeeded in isolating many gases, he did not identify any of them, believing them all to be air.

Pneumatic chemists found in the latter part of the 18th century that there were many different varieties of air. Among these "airs," carbon dioxide, hydrogen, oxygen, and nitrogen were discovered and interpreted. All chemists in those days, however, thought of these "airs" in terms of the phlogiston theory.

The Phlogiston Theory

In the 14th century Geber in Spain mentioned that certain metals gained weight when heated or calcined. Eck von Sulzbach, a German chemist, made (circa 1490) the first recorded experiments on the oxidative weight gain of a metal.

In the early history of pneumatic chemistry Johann Joachim Becher (1635–1682) came up with a theory of the constitution of matter in which he treated heat as a material substance later called *caloric*. His work, *Physica Subterranea*, was published in 1681. He renamed the three principles of Paracelsus the vitreous, fatty, and fluid earths, which respectively corresponded to the properties of body, combustibility, and density.

In the minds of chemists in the 17th century, air was a nonreactive medium that was usually present, but acted only as a physical agent. It was not regarded as a constituent of matter. Thus Becher concentrated his attention on solids as the sole constituents of matter.

The ideas of Becher were taken up by his pupil, Georg Ernst Stahl (1659–1734), who is credited with founding the phlogiston theory and the explanation of the phenomenon of combustion. In a treatise in 1717 he gave special significance to the fatty earth described by Becher. Stahl called the fatty earth "phlogiston" (from Greek "*phlogistos*" = combustible) and upon it he erected the first great generalization of chemistry. Phlogiston to him was the fire principle. Observers had long felt as they watched the flame from a burning body that something was departing. Stahl believed that all substances are a compound of a combustible and an incombustible component. The combustible part is identical in all substances and is as phlogiston. Combustion was simply the loss of this mysterious component. The greater the amount of phlogiston, the more readily is the substance combustible. Phlogiston was also believed to be lost when metals were combusted, leaving metal-lime or oxide or calx remains.

Corrosion of metals was also believed to be a slow loss of phlogiston from the metal. Stahl never solved the problem of the gain in weight which metals exhibit when they corrode. Some chemists suggested that phlogiston had a negative weight, but this caused trouble when attempting to explain why, for example, a piece of wood lost weight when it was burned.

Stahl's ideas were quite logical when the object burning was an organic substance because in this case only a small residue was left behind. It was also the first comprehensive hypothesis to account for what we know as oxidation and reduction, but unfortunately the mechanisms of the phlogiston theory were exact the reverse of those now accepted.

The phlogiston theory was the first real attempt to explain different chemical phenomena from a unified point of view after the theoretical basis of alchemy had been rejected. The simplicity and clarity of the theory made it immediately acceptable and by 1740 it was the dominant theory of chemistry. It remained so until the last decades of the 18th century. The fact that phlogiston could not be separated or isolated from any substance and was thought to have a negative weight did not stop the discoveries of gases. It did, however, confuse many of the investigators.

References

1. Riedel, E., and Schram, J. (1994). Gas Antik: Episoden aus der Geschichte der Gase. *Gas Aktuell* **50**:30–38.
2. Ganot, A. (1873). *Elementary Treatise on Physics Experimental and Applied* (trans. E. Atkinson). London: Longman's Green.

Further Reading

Partington, J. R. (1962). *A History of Chemistry*, Vol. 3. London: Macmillan.
Strube, W. (1976). *Der historische Weg der Chemie*, Vols. 1 und 2. Leipzig: VEB Deutscher Verlag für Grundstoffindustrie.
Szabadváry, F. (1996). *History of Analytical Chemistry.* Oxford: Pergamon Press.

2.2. THE DISCOVERY OF GASES AND THE TRUE NATURE OF COMBUSTION[1–3]

The latter part of the 18th century saw a series of investigations in pneumatic chemistry. This started with the work of Joseph Black and continued with the discoveries of other gases by Henry Cavendish, Joseph Priestley, David Rutherford, Carl Wilhelm Scheele, Antoine Laurent Lavoisier, and others. These studies were made possible by new forms of apparatus, including the pneumatic trough, the gas jar, and the beehive shelf. Chemical science was now strongly influenced by technological demands and developments of the Industrial Revolution (Fig. 2.2.1). The collaboration and interchange of ideas between chemists and industrial entrepreneurs, for example, the correspondence between James Watt and Joseph Black, was very important.

Carbon Dioxide

The Dutch teacher of chemistry Hermann Boerhaave (1668–1738) began to suspect that air could take part in chemical reactions. The first evidence of that and the proof that gases have chemical identities of their own was submitted in a thesis

Figure 2.2.1. The chemist's workroom, between the alchemist's kitchen and the modern laboratory. From D. Diderot and J. d'Alembert, *Encyclopédie* (1752).

by a student for the medical degree in Edinburgh in 1754, Joseph Black (1728–1799). He was of Scotts origin, but was born in Bordeaux, France.

In 1754 Black discovered carbon dioxide. He demonstrated the uptake of carbon dioxide by quicklime (calcium oxide) to form chalk (calcium carbonate), and the reverse of this process when chalk was heated. Because of the radically new idea that a gas could be present in a solid, he called carbon dioxide "fixed air." His work was very important for the discovery of other gases mainly because scientists now began to consider carbon dioxide to be, not a variety of air, but an independent substance different from air and contained in many solids. Black also discovered the latent heats of fusion and vaporization and measured the specific heats of many substances.

Black used a balance to weigh the reactants and products at every stage of his gas experiments. Because the mass of the product formed exceeded that of the initial product in oxidation reactions with carbon, the main principle of the phlogiston theory was undermined. It took a long time, however, before the significance of this fact was recognized.

Black also demonstrated the role of carbon dioxide in respiration on a massive scale. His colleague John Robison wrote about an experiment which Black performed in Glasgow in the winter of 1764–1765. A considerable quantity of caustic alkali was collected in an apparatus placed above one of the spiracles in the ceiling of a church in which a congregation of more than 1500 persons been sat for nearly 10 hours. This may say more about the church-going habits of 18th-century Scots than it does about Black's experimental techniques, but it is an example of Black's unconventional thinking and creativity.

Hydrogen

Hydrochloric, sulfuric, and nitric acids as well as iron and zinc were already known by the alchemists, which means that scientists had in their possession the components for reactions that could give rise to hydrogen. Literature of the 16th through the 18th centuries reports many occasions when chemists observed how the pouring of, for instance, sulfuric acid on iron shavings produced bubbles of a gas, which they believed to be an inflammable variety of air.

The Russian scientist Mikhail Vasilievich Lomonosov (1711–1765) made important contributions in many scientific fields. Besides the principle of the indestructibility of matter, he also investigated the phenomenon of combustion, preceding the work of Lavoisier in these subjects.

In 1745 he wrote "*On Metallic Lustre*," in which he described observations of "inflammable vapor" when he dissolved some base metal, especially iron, in acids: "vapor shoots out from the opening of the flask." Thus Lomonosov observed hydrogen, but thought, as all chemists of that time in terms of generating phlogiston. Because metal dissolved in the acid liberated "inflammable vapor," it was very convenient to assume that dissolving metal releases phlogiston.

Henry Cavendish (1731–1810) was a near contemporary and pupil of Joseph Black. As a gentleman of means (being related to the Duke of Devonshire), he did not have to work for his living, but devoted most of his time to scientific research. Like Black, he published little of his work. Cavendish characterized "inflammable air," or hydrogen, in his first paper read to the Royal Society, on 29 May 1766. Cavendish described the processes of preparation and the properties of inflammable air in greater detail than his predecessors. He obtained it by the same technique—the action of sulfuric and hydrochloric acids on iron, zinc, and tin—but he was the first to obtain definite proof that the same type of air was formed in all cases. Cavendish believed that the gas came from the metal rather than the acid and that different metals contain different proportions of inflammable air. As a follower of the phlogiston theory, he could give only one interpretation of the nature of the substance: Like Lomonosov, he identified inflammable air as phlogiston.

The elementary nature of inflammable air was still not understood because chemistry had not yet matured enough for such an insight. Cavendish is generally regarded the discoverer of hydrogen because he was the first to establish that it is a separate gas quite different from air.

A fortnight after Cavendish's paper on inflammable air was read to the Royal Society in 1766, Joseph Priestley, who at this time was in the early stages of self-education in science, was elected a Fellow. In 1767 Cavendish and Priestley set out to study the action of electric discharges on various gases and exchanged information on their experiments on gas properties. There were only a few known gases at that time: ordinary air, fixed air, and inflammable air. Although the experiments could not produce definite results, it was shown later that electric discharge in humid air yields nitric acid.

Cavendish was the first to measure the density of gases by filling bladders with the gas and weighing them. The most important of his many articles was "*The Experiments on Air*," published in 1784.

Many years had to pass before hydrogen finally got its name and its proper place in chemistry. Its Latin name, *hydrogenium*, stems from the Greek *hydr* and *gennao*, which together mean "producing water"; the name was proposed in 1779 by Lavoisier after the composition of water had been established.

Nitrogen

Very early in history, humans used nitrogen compounds, for instance, saltpeter and nitric acid, and would have been able to see the generation of brown vapors of nitrogen dioxide. It would have been impossible, however, to discover nitrogen by decomposing its inorganic compounds. Tasteless, colorless, odorless, and chemically rather inactive, nitrogen could not be detected in those early days.

In 1772 Cavendish discovered that if atmospheric air is passed several times over red-hot coal and then passed through a solution of potassium hydroxide, an airlike substance remains which is lighter than air and will not support combustion. In a private letter to Priestley, he wrote that he had succeeded in preparing a new variety of air named by him "asphyxiating" or "mephitic" air. At the time Cavendish did not study it thoroughly, and only reported the fact to Priestley. He returned to study "mephitic air" much later.

When Priestley received the letter from Cavendish he was busy with important experiments and read it without due attention. Priestley burned various inflammable compounds in a given volume of air and calcinated metals; the fixed air formed during these processes was removed with the aid of lime water. He noticed that the volume of air decreased considerably. Priestley had at that time no idea of the existence of such a gas as oxygen (which he discovered 2 years later) and turned to phlogiston to explain the observed phenomenon. He believed that the result of metal calcination was due exclusively to the action of phlogiston. The remaining air is saturated with phlogiston and consequently was named "*phlogisticated air*" because it sustained neither respiration nor combustion. Thus, Priestley was in possession of a gas which subsequently became known as nitrogen and Cavendish was one of the first scientists to believe that phlogisticated air is a component of ordinary air.

Neither Cavendish nor Priestley could understand the real nature of the new gas. The understanding came later when oxygen appeared on the scene of chemistry. In September 1772 the English physician David Rutherford, a pupil of Black, published "*On the So-Called Fixed and Mephitic Air*," in which he described the properties of nitrogen. This gas, according to Rutherford, was absorbed neither by lime water nor by alkali and was unsuitable for respiration; he named it "corrupted air." Rutherford is considered to be the discoverer of nitrogen together with the Swede Carl Scheele, but in fact did nothing new compared with his famous colleagues.

Not properly discovered or understood as a gaseous chemical element, nitrogen in the 1770s had three names: "phlogisticated," "mephitic," and "corrupted" air. The name "azote" was proposed in 1787 by Lavoisier and other French scientists, who developed the principles of a new chemical nomenclature. They derived the word from the Greek negative prefix *a* and the Greek word *zoe* meaning "life." Lifeless, not supporting respiration and combustion, that was how the chemists saw the main property of nitrogen. Later this view turned out to be erroneous and nitrogen took its name from the Latin *nitrogenium*, which means "saltpeter-forming," when it was found that this element is vitally important for plants.

Oxygen

In 1774 the French pharmacist Pierre Bayen wrote a paper (1774) in *Journal de Physique* in which he discussed the causes for an increase in the mass of metals (mercury) during calcination. He stated that when mercury is calcined it does not lose phlogiston, but combines with the gas. He believed that a peculiar variety of air—a fluid heavier than ordinary air—was added to a metal in the process of calcination. He obtained this fluid by thermal decomposition of mercury compounds. Conversely, acting on metallic mercury, the fluid transformed it into a red compound. Bayen, however, neglected to make a thorough study of the properties of the gas and failed to recognize it as a new substance.

Joseph Priestley experimented (Fig. 2.2.2) in the same year with the same compound. Shortly before, he had discovered that in the presence of green plants, "fixed air," not supporting respiration, turned into ordinary air suitable for respiration by living organisms. This fact was extremely important not only for chemistry, but for biology as well. Priestley proved for the first time that plants release oxygen.

On August 1, 1774, he placed red mercury oxide into a sealed vessel and focused sunbeams on it with a big lens. The compound began to decompose, yielding bright metallic mercury and a gas. Unlike nitrogen, it was not initially isolated from the atmosphere, but extracted from a solid. The new gas proved to be suitable for respiration. A candle burnt much more brightly in the atmosphere of this gas than in ordinary air. Nothing was observed when the new gas was mixed with air, but upon being mixed with saltpeter, the gas yielded brownish vapors (now known to be NO_2 formed from NO). Priestley called the newly discovered gas "dephlogisticated air."

After the discovery Priestley joined his patron, Lord Shelburne, for a business trip to Paris in October 1774, where he met Lavoisier and other French scientists and told them at a dinner about his experiments. Lavoisier appreciated at once the importance of the experiments of his English colleague, and had a much clearer idea about them than Priestley. Priestley kept talking about dephlogisticated air, unable to see the greatness of his own discovery; he considered "dephlogisticated air" to be a complex substance. Only in 1786, influenced by the ideas of Lavoisier, did he begin to view it as an elemental gas.

The Swedish chemist Carl Scheele published a book, *Chemical Treatise on Air and Fire*, 1775 (it had been written 2 years earlier, but the publisher's problems delayed its appearance). Near the beginning of his book, Scheele defines air: "It is that fluid invisible substance which we continually breathe; which surrounds the whole surface of the earth, is very elastic, and possesses weight." After defining the properties characteristic of air, Scheele demonstrated that air must consist of at least two elastic fluids, "one which does not manifest the least attraction for phlogiston, while the other is peculiarly disposed to such attraction."

Scheele described his discovery of oxygen in his laboratory notes (Fig. 2.2.3) as early as 1772 and his description was much more detailed and accurate than that of Bayen and Priestley. Scheele obtained oxygen, "fiery air," in various ways, by decomposing five different inorganic compounds and believed that fiery air was a component of ordinary air. Mixing it with "mephitic air" (nitrogen), which Rutherford

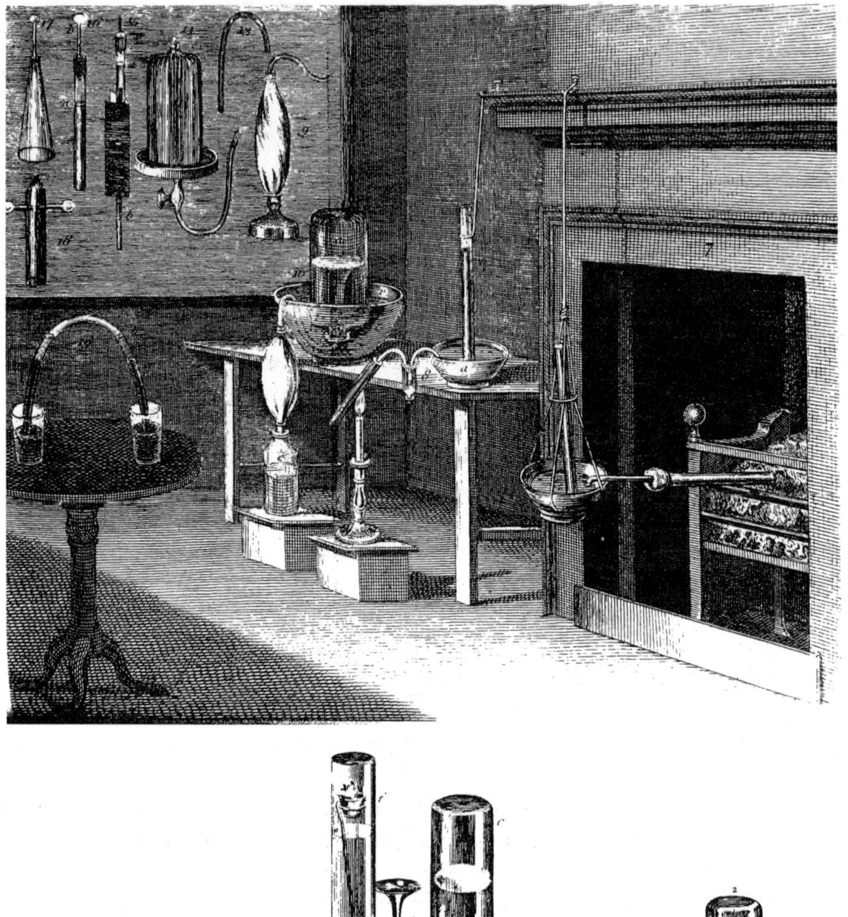

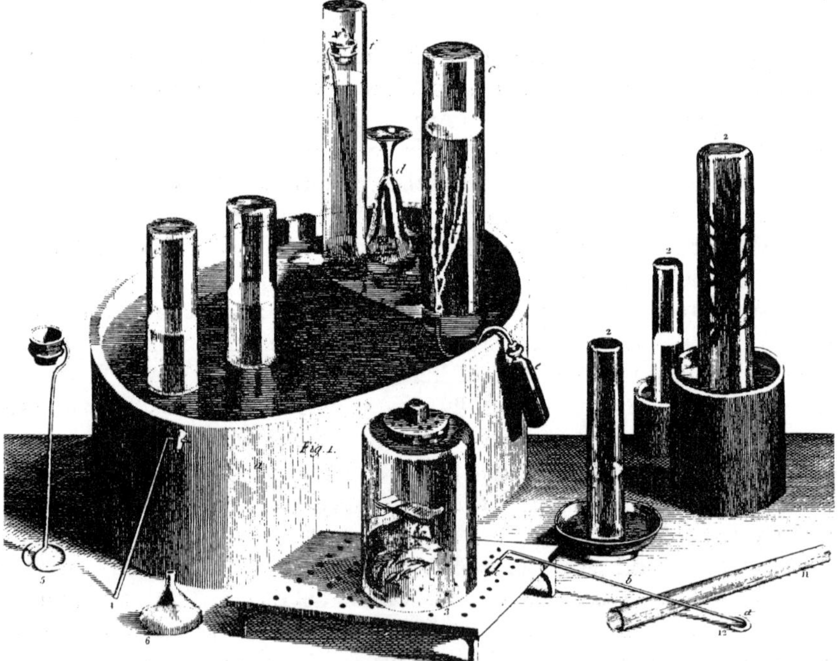

Figure 2.2.2. Part of Priestley's research laboratory and laboratory equipment for gas experiments. From J. Priestley, *Experiments and Observations on Different Kinds of Air* (1790).

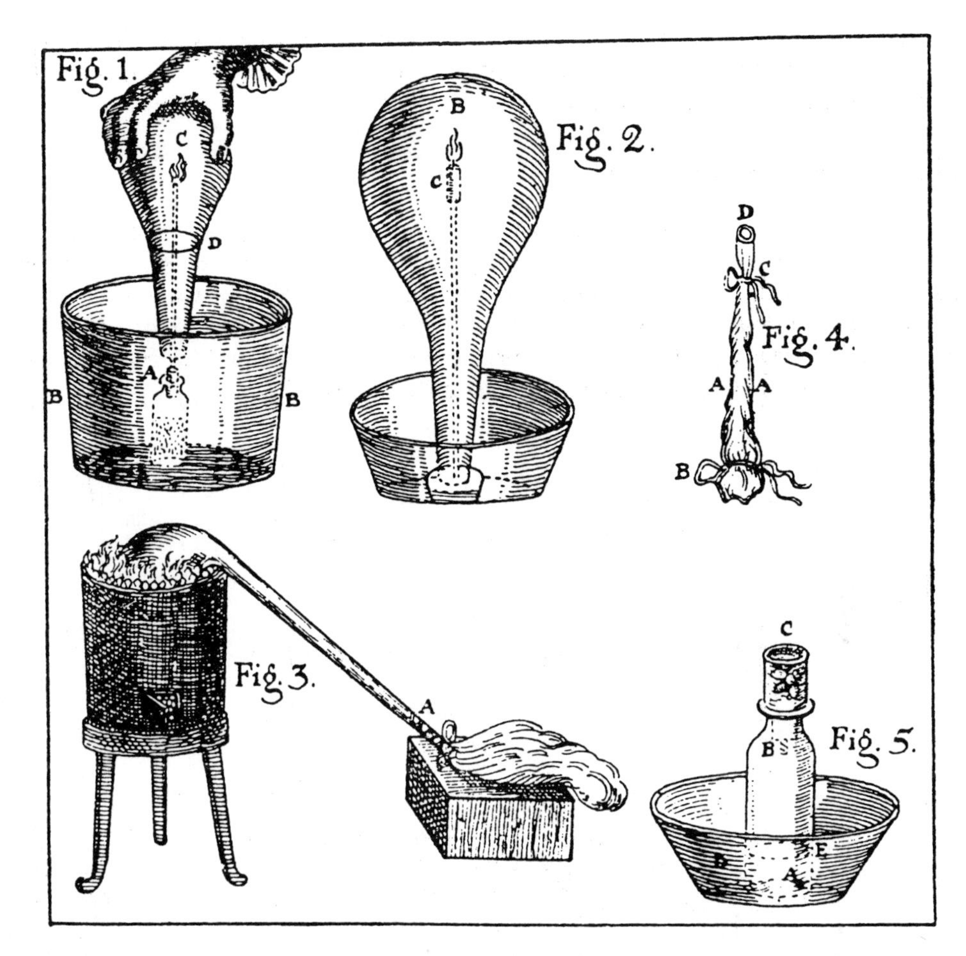

Figure 2.2.3. Scheele's research equipment, from his own sketches. From C. W. Scheele, *Chemische Abhandlung von der Luft und dem Feuer* (1777).

independently described the same year, Scheele prepared a mixture which did not differ at all from ordinary air. He thus realized that atmospheric air is a mixture of gases, which later were to be known as nitrogen and oxygen. This seems to us to be an easy conclusion, but due to the phlogiston theory, Scheele's findings were difficult to explain.

Scheele discovered his "fiery air" knowing nothing about Priestley's experiments, and informed Lavoisier about it in a letter on September 30, 1774. The real understanding of Scheele's experimental results was later made possible by Lavoisier.

Laughing Gas, N_2O

In the early 1770s so-called saltpeter gas was known. It turned out that it can be transformed (by reaction with iron dust) into another variety of air supporting combustion, but not supporting respiration. Thus, Joseph Priestley discovered another nitrogen oxide, N_2O, and named it, according to the logic of the phlogiston theory, "dephlogisticated saltpeter air."

Hydrogen + Oxygen = Water

The greatest of Cavendish's discoveries might be his finding that water is a compound of hydrogen and oxygen, and not a primary element as had been thought for the previous 2000 years. By passing an electric spark through a mixture of air and combustible air (hydrogen), Cavendish in 1781 was able to observe the formation of water, and he later repeated the experiment using pure oxygen and hydrogen in the correct proportions. He published his results only in 1784.

TRAITE

ÉLÉMENTAIRE

DE CHIMIE,

PRÉSENTÉ DANS UN ORDRE NOUVEAU

ET D'APRÈS LES DÉCOUVERTES MODERNES;

AVEC FIGURES:

Par M. LAVOISIER, de l'Académie des Sciences, de la Société de Médecine, des Sociétés d'Agriculture de Paris & d'Orléans, de la Société de Londres, de l'Institut de Bologne, de la Société Helvétique de Basle, de celles de Philadelphie, Harlem, Manchester, Padoue, &c.

SECONDE ÉDITION.

TOME PREMIER.

A PARIS,

Chez CUCHET, Libraire, rue & hôtel Serpente.

M. DCC. XCIII.

Figure 2.2.4. Title page of Lavoisier's "*Traité élémentaire de Chimie*," 2nd edition (1789).

Figure 2.2.5. Giant burning lens developed for the French Academy of Science in 1774. Initiated by Antoine Lavoisier among others and used for his gas experiments. With a similar lens he was able to show that a large amount of fixed air was released when a metal calx (oxide) was roasted with charcoal.

Lavoisier knew about these experiments and, after repeating them in 1783, he concluded that water is not a simple substance, but a mixture of inflammable and dephlogisted air. He showed that the weight of the water produced was equal to the sum of the weights of the two gases. He interpreted the reaction as the simple combination

$$\text{Hydrogen} + \text{Oxygen} = \text{Water}$$

Determination of the composition of water made it possible to get an insight into the nature of hydrogen.

The Birth of Modern Chemistry[4,5]

The greatest contributors in the field of gas analysis all worked at about the same time, around 1770. It often happened that they discovered the same things independently of one another. The discoveries that Cavendish, Priestley, Scheele, and others made about oxygen, nitrogen, hydrogen, and the composition of air and water formed the basis of the new chemical theory. All these scientists were, however, so steeped in the traditions of the phlogiston theory that they could not draw the necessary conclusions from their discoveries. It was left to Lavoisier to incorporate them into his antiphlogiston theory.

Figure 2.2.6. Antoine Lavoisier conducting respiration experiments. Drawing by Lavoisier's wife, Anne Marie, depicting the experiments, with herself taking notes. From *Oeuvres de Lavoisier* (1862–1893).

	Noms nouveaux.	*Noms anciens correſpondans.*
Subſtances ſimples qui appartiennent aux trois règnes, & qu'on peut regarder comme les élémens des corps.	Lumière	Lumière.
	Calorique	Chaleur.
		Principe de la chaleur.
		Fluide igné.
		Feu.
		Matière du feu & de la chaleur.
	Oxygène	Air déphlogiſtiqué.
		Air empiréal.
		Air vital.
		Baſe de l'air vital.
	Azote	Gaz phlogiſtiqué.
		Mofète.
		Baſe de la mofète.
	Hydrogène	Gaz inflammable.
		Baſe du gaz inflammable.
Subſtances ſimples non métalliques oxidables & acidifiables.	Soufre	Soufre.
	Phoſphore	Phoſphore.
	Carbone	Charbon pur.
	Radical muriatique	Inconnu.
	Radical fluorique	Inconnu.
	Radical boracique	Inconnu.
Subſtances ſimples métalliques oxidables & acidifiables.	Antimoine	Antimoine.
	Argent	Argent.
	Arſenic	Arſenic.
	Biſmuth	Biſmuth.
	Cobalt	Cobalt.
	Cuivre	Cuivre.
	Etain	Etain.
	Fer	Fer.
	Manganèſe	Manganèſe.
	Mercure	Mercure.
	Molybdène	Molybdène.
	Nickel	Nickel.
	Or	Or.
	Platine	Platine.
	Plomb	Plomb.
	Tungſtène	Tungſtène.
	Zinc	Zinc.
Subſtances ſimples ſalifiables terreuſes.	Chaux	Terre calcaire, chaux.
	Magnéſie	Magnéſie, baſe du ſel d'epſom.
	Baryte	Barote, terre peſante.
	Alumine	Argile, terre de l'alun, baſe de l'alun.
	Silice	Terre ſiliceuſe, terre vitrifiable.

Figure 2.2.7. Lavoisier's nomenclature of the elements. From A. Lavoisier, *Traité élementaire de Chimie* (1789).

The True Nature of Combustion

The discoveries of gases in the last decades of the 18th century made the understanding of the nature of combustion possible. Now the necessary information was at hand for liberation from the phlogiston theory. Lavoisier pointed out that when a substance burns, the products always weigh more than the original substance, and that the increase in weight is due to the combination of the oxygen in the air with the substance that is burned. He also created a fundamental criterion for deciding whether substances were elements or compounds.

The phlogiston theory quickly lost ground, and by 1800 practically everyone recognized the correctness of Lavoisier's oxygen theory. Lavoisier also became the

CHEMICAL SYMBOLS Elements	Discovered	Bergman 1780	Hassenfratz, Adet 1787	Dalton 1803	Berzelius 1812
Copper	ca 3500 BC	♀	Ⓒ	Ⓚ	Cu
Lead	ca 3400 BC	♄	Ⓟ	Ⓑ	Pb
Iron	ca 3000 BC	♂	Ⓕ	Ⓔ	Fe
Gold	"	☉	☉	Ⓖ	Au
Silver	"	☽	Ⓐ	Ⓢ	Ag
Mercury	"	☿	Ⓗ	Ⓠ	Hg
Tin	"	♃	Ⓢ	Ⓩ	Sn
Sulphur	ca 2000 BC	🜍	◡	⊕	S
Carbon	ca 1425 BC	🜍	(	●	C
Arsenic	ca 1250 AD	o—o	(As)	(Ar)	As
Bismuth	ca 1450 AD	♉	Ⓑ	Ⓦ	Bi
Antimony	"	♁	(Sb)	(An)	Sb
Zinc	ca 1500 AD	Ô	Ⓩ	(ZK)	Zn
Platinum	"	☽☉	Ⓣ	Ⓟ	Pl !
Phosphorus	1669	🜍	∩	☮	P
Cobalt	1735	♌	Ⓚ	(Cob)	Co
Nickel	1751	⚲	Ⓝ	Ⓝ	Ni
Hydrogen	1766	⧗	)	☉	H
Nitrogen	1772		/	⦶	N
Oxygen	1774	⌧	—	○	O

Figure 2.2.8. Nomenclature of the elements according to Bergman (1780), Hassenfratz/Adet (1787), Dalton (1803), and Berzelius (1812).

founder of the modern theory of oxidation and reduction and demonstrated that respiration in animals is a special form of oxidation. He performed the earliest true biological experiments in his studies on respiration (Fig. 2.2.6).

The First Modern Chemical Nomenclature System

The first list of chemical elements that was at all similar to our modern list was published by Lavoisier in 1786 (Fig. 2.2.7). His list included 33 substances, most of which we still recognize as elements. The list includes hydrogen, oxygen, nitrogen, sulfur, phosphorus, and carbon, plus 17 known metals. Also on this list, however, were the "muriatic, fluoric, and boracic radicals" (from hydrochloric, hydrofluoric, and boric acids) and certain oxides (lime, magnesia, baria, silica, and alumina). Most surprising, in view of Lavoisier's interest in analytical weighing, is the fact that he included light and heat (caloric) in the list of elements. His "caloric," the weightless material of fire, was, believe it or not, a kind of modified phlogiston.

References

1. Anderson, R. G. W. (1884). From chaos to gas: Pneumatic chemistry in the 18th century. In *Proceedings of the 3rd BOC Priestley Conference*, pp. 392–407.
2. Ramsay, W. (1905). *The gases of the atmosphere*. London: Macmillan.
3. Weeks, M. E. (ed.). (1935). *The discovery of the elements*. Easton, PA: Mack Printing Co.
4. Kauffman, G. B. (1989). The making of modern chemistry. *Nature* **338**:699–700.
5. Brody, J. (1987). Marie Lavoisier. Behind every great scientist. *New Scientist* **1987**(December 24/31):19–21.

Further Reading

Ihde, A. J. (1964). *The Development of Modern Chemistry.* New York: Harper & Row.
Szabadváry, F. (1966). *History of Analytical Chemistry.* Oxford: Pergamon Press.

2.3. BIOGRAPHIES OF THE EARLY PIONEERS IN THE STUDY OF GASES[1]

Paracelsus

Aureolus Philippus Theophrastus Bombastus Paracelsus von Hohenheim (1493–1541), Swiss physician and alchemist, started iatrochemical science, which was later to become medical chemistry (Fig. 2.3.1).

Up to his time the ancient alchemical concepts of the nature of substances had not changed essentially. Paracelsus used alchemical methods in order to find medicines which would cure disease. He searched for the "elixir of life," a substance which would prolong life and good health indefinitely. During his lifetime Paracelsus acquired quite a reputation as a drug manufacturer and physician throughout much of Europe. He thought of life as fire and believed that air is required for the burning of wood and fuel.

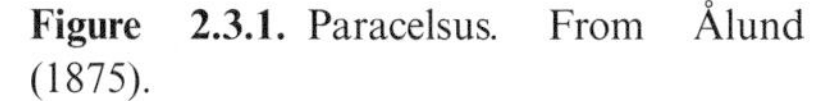

Figure 2.3.1. Paracelsus. From Ålund (1875).

Figure 2.3.2. van Helmont. From J. van Helmont, *Oriatrike or Physik refined* (1662).

In his theory Paracelsus replaced the four elements from Aristotle's time with three principles: sulfur, mercury, and salt. According to him, these three principles compose not only the metals, but all the organic and inorganic substances, which are themselves each composed of varying proportions of the four elements: earth, air, fire, and water. Paracelsus' research started a new period of scientific progress in chemistry, and slowly the fruitless experiments of 1000 years involving meaningless attempts to make gold from other metals ceased.

He was admired as Europe's greatest scientist by some and considered a charlatan by others. His dual character is reflected in his varied life. From a vagabond he became a professor at the University of Basle, but because of his extreme views he lost his occupation and died in a hospital for the poor in Salzburg.

Johann Baptista van Helmont

Johann Baptista van Helmont (1577–1644), born in Brussels, studied medicine at the University of Louvain (Fig. 2.3.2). At first he was a follower of the old, classical school, but was later influenced by the ideas of Paracelsus, which he developed.

He later modified the chemical view of Paracelsus, rejecting the three principles, and created a new theory of the nature of the elements, considering only water as a primary element. The early experiments of van Helmont showed that water was formed when organic substances were subjected to combustion, a fact that he used to support his theory that the ultimate composition of all substances is water.

When burning coal, an "air" (carbon dioxide), which van Helmont called "*spiritus sylvestris*," was created. He introduced the word "gas," derived from the Greek word "*chaos*," and realized that there must be different gases.

He combined mystical ideas with true scientific research, and spent much of his life as a wandering courtier and scholar. A major part of his research was devoted to examining the products of the combustion of solids and liquids. He was offered the post of Imperial Court Physician of Vienna, but refused it, preferring to devote the rest of his life to research at his estate near Brussels. He died in 1644 and most of his works were published after his death by his son in 1648.

Otto von Guericke

Otto von Guericke (1602–1686) of Magdeburg made the first pneumatic machine (vacuum pump) (Fig. 2.3.3). During the mid-17th century, it became generally known that the atmosphere was a fluid possessed of weight and could be made to do mechanical work when a vacuum was formed. Von Guericke set out to prove that Aristotle had been wrong when he had said that air had no weight. He carried out a famous experiment at Magdeburg in 1657 with a metal globe made up of two hemispheres which fitted exactly into each other. These became known as Magdeburg hemispheres. With his air pump he emptied the globe of air and in the presence of a large group of people he invited anyone to pull it apart. Twenty-five horses were necessary to accomplish this. Then Guericke showed that, by turning a little tap at one end of the globe and admitting the air, the two hemispheres could be easily separated by one person.

Figure 2.3.3. von Guericke. From Neudeck (1905).

Robert Hooke

Robert Hooke (1635–1703), English experimenter and natural philosopher, was employed by Robert Boyle in 1655 to construct the Boylean air pump. He discovered the law of elasticity, which states that the stretching of a solid body is proportional to the force applied to it. In *Micrographia* (1665) he presented his theories of combustion. He found among other things that all matter expands when heated and that air is made up of two different types of substances, the particles of which are separated from each other by relatively large distances.

Johann Joachim Becher

Johann Joachim Becher (1635–1682) was born in Speyer, Germany (Fig. 2.3.4). In 1660 the Archbishop of Mainz nominated him for the Professorship of Medicine at the University of Mainz and also appointed him as his personal physician. Becher believed substances to be composed of three "earths" (vitreous, fatty, and fluid earths), and supposed that when a substance burned, a combustible earth was liberated. His experiments on minerals and combustion are described in his *Physica Subterranea* (1669). G. E. Stahl summarized Becher's views on combustion in a book entitled *Specimen Becherianum* (1702).

Becher was very restless and could not stay for very long in one place. From Mainz he went to Munich, where he became the Court Physician to the Prince of

Figure 2.3.4. Becher. From Weeks (1935) with permission from the *Journal of Chemical Education*, copyright © 1935, Division of Chemical Education, Inc.

Bavaria; from there he went to Vienna, to the post of Court Chancellor. A few years later he was in Holland and from there he went to England, where he died.

Georg Ernst Stahl

Georg Ernst Stahl (1659–1734) was born in Anspach, Germany, and studied medicine in Jena (Fig. 2.3.5). In 1693 he became Professor of Medicine at the newly established University of Halle. In 1716 he became the personal physician to Frederick I of Prussia and held that appointment until his death.

Stahl developed the phlogiston theory, which provided an explanation of processes taking place during combustion and the calcination of metals and also related to biological processes such as respiration, fermentation, and decay. This theory dominated chemical knowledge for almost a century until Lavoisier in 1777 began to convince the scientific world of its faultiness.

Robert Boyle

Robert Boyle (1627–1691), Anglo-Irish natural scientist, was born the 14th child of an aristocratic family and was given the best possible education (Fig. 2.3.6). His health was poor throughout his life and he never married. Boyle worked with many problems in physics, chemistry, and the medical sciences, and his theories were always based on experiments. He severely criticized alchemical thought and is considered to be the creator of early modern chemistry and the science of gases. In spite of his disputing of alchemy, he believed that the transmutation of base metals into gold might be possible.

Figure 2.3.5. Stahl. From Weeks (1935) with permission from the *Journal of Chemical Education*, copyright © 1935, Division of Chemical Education, Inc.

Figure 2.3.6. Boyle. From Ramsay (1905).

Boyle was one of the founders of the Royal Society, of which he later became President. In 1662 he discovered the simple relationship between the volume of a gas and its pressure. This relationship is known in much of the world as "Boyle's law," but is also called "Mariotte's law" because the same relationship was discovered by the French physicist Edmé Mariotte in 1676. Boyle also showed that sound cannot be transmitted in a vacuum and that air is necessary for animal life.

His work with different gases and liquids led to the development of many ingenious new experimental techniques, which included measurements of high precision. His most famous work, *The Sceptical Chymist* published in 1661, explains that the proof of the existence of elements, as well as their identification, rests on chemical experimentation.

Stimulated by von Guericke's "way of emptying glass vessels," he planned a similar development, and in 1658 his assistant Robert Hooke built a single-barreled air pump. Boyle attempted to present a mechanical picture of chemical reactions, explaining the behavior of substances by analogy with machines. He considered air to be composed of corpuscles. To explain its compressibility, he likened the particles to small coiled springs.

John Mayow

John Mayow (1640–1679), English chemist and physiologist, explained combustion by saying that air contains a gas that is consumed in respiration and burning, with the result that substances no longer burn in the air that is left (Fig. 2.3.7). Mayow was the first to identify two gas components of air, of which one supported combus-

Figure 2.3.7. Mayow. From Ramsay (1905).

Figure 2.3.8. Hales. From Ramsay (1905).

tion; however, his ideas were later forgotten and had to be rediscovered by others. To investigate respiration, Mayow placed a mouse in a cupping glass sealed with a bladder. He studied the breathing process and found that the diminution in volume of air was due to the removal of nitro-aerial particles caused by respiration. His experiments suggested to him that there were two kinds of particles which made up air. Mayow also adopted Boyle's view that air particles are like little springs, but claimed that they were not all alike.

Stephen Hales

Stephen Hales (1677–1761), English clergyman and botanist, showed that air takes part in chemical reactions (Fig. 2.3.8). He performed many quantitative experiments with gases, burning measured weights of natural substances and recording the volumes of the products generated. He also constructed the earliest form of the gas-collection apparatus, which is still in use today. He devised another apparatus specifically for measuring the volume of gas evolved in distillation or combustion.

In 1727, Hales showed that large volumes of "air" could be released from the pores of solids and liquids. He recorded the results of heating 24 gas-producing substances. He did not, however, distinguish between the gases evolved, but simply recorded the quantity of "air" and discarded it, considering "air" to be the same element in each case and explaining the differences between one gas and another by impurities.

Figure 2.3.9. Black. From Ramsay (1905).

Joseph Black

Joseph Black (1728–1799), Scottish chemist, prepared "fixed air" (carbon dioxide) in 1756 (Fig. 2.3.9). He was born in Bordeaux in 1728, the eighth child of a rich Scottish wine merchant. He went to school in Belfast and studied medicine at the University of Glasgow under William Cullen until 1751, when he moved to Edinburgh to complete his famous doctoral thesis, *Experiments Upon Magnesia Alba, Quicklime, and Some Other Alcaline Substances.* In his investigations of the heating of magnesia alba (magnesium carbonate), Black anticipated Lavoisier and modern chemistry by investigating the existence of a gas, carbon dioxide, distinct from common air. In 1766 Black became Professor in Medicine at Edinburgh University, where he worked until 1797.

Black is also known as the discoverer of the bicarbonates. He introduced the expression "heat of evaporation" in 1764 and discovered the latent heats of fusion and vaporization. He also established the theory of specific heat and invented the ice calorimeter.

Mikhail Vasilievich Lomonosov

Mikhail Vasilievich Lomonosov (1711–1765) was only 54 years of age when he died, but during this relatively short life he accomplished much (Fig. 2.3.10). He was a poet, historian, sociologist, and factory manager, but most of all he was a great natural scientist. Even though he was only an amateur in most of these fields, he made important contributions to all of them. Besides enunciating the principle of the

Figure 2.3.10. Lomosonov. From a painting by L. Miropolsky (1787).

indestructibility of matter, Lomonosov also investigated the phenomenon of combustion, preceding the work of Lavoisier.

Joseph Priestley[2–5]

Joseph Priestley (1733–1804), British theologian and natural scientist, was born at Birstal, Yorkshire, the son of a cloth weaver (Fig. 2.3.11). He entered the Church, which was the only profession that offered him any opportunity for study. At the age of 19 he began his studies at the Daventry Academy, where he read mathematics, physics, and philosophy, and where his dissenting religious views developed. In 1755 he was appointed minister to a congregation at Needham Market in Suffolk.

In 1758 he moved to Nantwich in Cheshire because of problems with what his congregation considered to be his extreme theology. There he also started a school. In 1761 he became a teacher in languages and belles lettres at the dissenting academy at Warrington. He had a gift for languages and learned nine. In 1765 he obtained a Doctor's Degree of Laws from Edinburgh University. His scientific interests grew, and when he met Benjamin Franklin in 1765, who told him about his electricity experiments, Priestley decided to write a book on electricity, which was published in 1767. He was elected a fellow of the Royal Society in 1766.

Figure 2.3.11. Priestley. From Ramsay (1905).

In 1767 Priestley moved to Leeds to be minister of the Mill Hill chapel; here his scientific interests turned to optics and gases, while he continued his theological work. The friendship with Benjamin Franklin led Priestley into scientific experimentation. He first studied the solubility of carbonic acid in water. The success from this produced the further discovery of about 10 gases, including laughing gas (nitrous oxide) in 1772 and oxygen in 1774.

In 1773, Lord Shelburne offered him a generously salaried post as secretary and literary companion in Wiltshire. It was during this period of employment that Priestley published his most important chemical work, but after publishing an essay criticizing the upper classes, in particular the aristocracy, he had to change employment. After a period of great poverty Priestley at last obtained a post as vicar of a poor Free Church in Birmingham, where he moved in 1780.

Here he was able to discuss scholarly matters with a brilliant circle, which included Matthew Boulton, James Watt, Josiah Wedgwood, and Erasmus Darwin. This was the famous Lunar Society, so called because the members met for dinner meetings once a month on the night of the full moon so that members who lived far away could see to drive home.

Priestley's radical views on politics became widely known and during the anti-French "Church and King" riots on Bastille Day, 1791, his house and laboratory were wrecked and Priestley had to flee the city. After a period of teaching at a dissenting college at Hackney he decided, at the age of 61, to emigrate to America. He settled in a remote community at Northumberland, Pennsylvania. Though he refused to take the Chair of Chemistry offered by the University of Pennsylvania, he continued to experiment in chemistry. He died in February 1804.

Priestley was the only major chemist who refused to abandon the phlogiston theory in favor of Lavoisier's new chemistry. In 1796, he published his *Consideration on the Doctrine of Phlogiston* and 4 years later his last scientific work, *The Doctrine of Phlogiston Established.* In a letter of 1800 referring to his belief he wrote, "I feel perfectly confident of the ground I stand upon . . . though nearly alone, I am under no apprehension of defeat."

Carl Wilhelm Scheele[6]

Carl Wilhelm Scheele (1742–1786) Swedish chemist, was born in the Prussian town of Stralsund, the capital of Swedish Pomerania, where his father was a merchant and a burgess (Fig. 2.3.12). He was the seventh of 11 children. After receiving his education, partly in a private school, partly in the public school (gymnasium) at Stralsund, he became apprenticed to the apothecary Bauch in Gothenburg at the age of 15. In those days an apothecary was a manufacturer as well as a retailer of drugs and had to prepare medicines in a pure state from very impure materials. In the evenings young Scheele read all the chemistry books he could obtain, and carried out experiments with the limited number of chemicals available to him.

He kept very detailed records of his experiments, which he took with him when he became an apothecary's assistant in Malmö, 1765–1768, later in Stockholm,

Figure 2.3.12. Scheele. From Ramsay (1905).

1768–1770, and finally in Uppsala, 1770–1776. In Uppsala, Torben Bergman, who was Professor of Chemistry at the University of Uppsala, obtained his chemical supplies from the apothecary where Scheele was employed. Bergman noticed Scheele's great talent and a friendship and collaboration developed between them. Scheele began to publish the results of his research and at the age of 33, still an apothecary's assistant, he was elected a member of the Swedish Academy of Science. When Hindrich Pasher Pohl, an apothecary at Köping, died in 1775 Scheele took over the shop. In 1777 Pohl's widow sold it to him, and it was here that Scheele was to show the world how great his work was in spite of his insignificant position.

Scheele resisted many offers of important and well-paid posts, saying "I cannot eat more than enough, and I earn enough for me to eat here in Köping." Scheele's health finally deteriorated due to the constant unhealthy work with poisonous chemical compounds in draughty and cold premises. He died at only 44 years of age.

Scheele's greatest contributions to science were made in the field of organic chemistry, which in fact he originated. He prepared numerous organic acids and discovered molybdenum, tungsten, manganese, chlorine, and barium. He discovered oxygen, hydrogen, and nitrogen independently of his English colleagues, and he also analyzed air.

Daniel Rutherford

Daniel Rutherford (1749–1819) was a Scottish doctor educated at Edinburgh University as a pupil of Joseph Black, and later was Professor of Botany at Edinburgh University (1786). In his dissertation of 1772, *Dissertatio Inauguralis de aere fixo dicto, aut mephitico*, he studied nitrogen. Together with Scheele, he is cred-

Figure 2.3.13. Rutherford. From Ramsay (1905).

ited with its discovery. Rutherford does not seem to have pursued the study of chemistry further because his duties led him into other fields.

Antoine Laurent Lavoisier[7,8]

Antoine Laurent Lavoisier (1743–1794) was a French physicist and chemist born in Paris (Fig. 2.3.14). He received an excellent education at the Collège Mazarin, including training in the sciences. He completed his law studies in 1764, but his talents drew him to science. He became a member of the Academy of Science and gained wide recognition for his work in different scientific fields. His job of administering the tax-gathering system financed much of his scientific interests, but it was also to be the cause of his death during the terror after the French Revolution; he was guillotined in 1794.

In the early 1770s Lavoisier started to experiment on the role of air in combustion. After repeating and developing the work of others, he published these results in 1774 in his first book, *Opuscules physiques et chimiques.*

Lavoisier carefully designed his experiments so that, unlike most of his predecessors, he could precisely weigh both the combustibles and their products. He was able to show that when a substance corrodes in a sealed container, the resulting gain in weight is compensated by a corresponding loss in weight of the air in the container. Thus, reasoned Lavoisier, when a metal corrodes, something in the air enters, or combines with, the metal. By disproving the phlogiston theory, Lavoisier laid the foundation of modern chemistry.

Figure 2.3.14. Lavoisier. From Ramsay (1905).

Lavoisier did not make as many great experimental discoveries as the earlier phlogiston chemists, although he repeated many of their experiments, sometimes claiming their discoveries as his own. He clarified the combustion process; he established that oxygen is needed for burning, and that when nonmetallic elements are burned an acidic product is formed, and when metallic elements are burned a basic product is formed. He also made the first clear distinction between elements and compounds and explained the process of dissolving metals in acids.

Lavoisier's *Traité élémentaire de Chimie* of 1789 described in detail the experimental basis for the rejection of phlogiston and for the new theory of combustion in which oxygen was the central element. The antiphlogiston theory was gradually accepted by many chemists.

All of Lavoisier's works are based on the principle of the indestructibility of matter, which he first postulated in 1789, and is considered the greatest of his achievements.

Henry Cavendish

Henry Cavendish (1731–1810) was a British physicist and chemist (Fig. 2.3.15). He produced and analyzed hydrogen gas in 1766, and in 1784 showed that water was formed when the gas was burned in oxygen.

Figure 2.3.15. Cavendish. From Ramsay (1905).

He was born into an ancient aristocratic family, and was related to the British royal family. Although he studied at Cambridge University, he did not pass any examinations. He dealt with scientific problems, drawn from almost all branches of natural science, from astronomy to chemistry. Cavendish lived in London, but he was a lonely and solitary man who avoided meeting people whenever possible.

Cavendish was an outstanding investigator in the study of gases. His observational capacity was great and his experiments were carefully designed. The most important of his many published works is *The Experiments on Air* (1784). He is generally regarded as the discoverer of hydrogen, as he was the first to establish that it is a separate gas quite different from air. Cavendish was also the first to measure the density of gases and to use this property to differentiate among them.

Cavendish never hurried to make public his experimental results and sometimes several years passed before his articles appeared. It is therefore difficult to find out when he observed and described the liberation of "inflammable air." This was published in 1766 in *Experiments with Artificial Air*, based on research probably performed under the influence of Joseph Black.

William Ramsay

William Ramsay (1852–1916), a Scottish scientist, around the end of the 19th century found the inert gases helium, neon, argon, krypton, and xenon in extremely small concentration in the atmosphere (Fig. 2.3.16). There were no empty spaces in Mendeleev's table for these new elements, but it was obvious from their atomic weights that they would fit easily into the periodic table as a new vertical column.

John William Strutt Rayleigh (1842–1919) found that different ways of measuring the density of nitrogen led to different results. On Rayleigh's request, Ramsay began a study in 1894 which led to his discovery of the presence of an unknown component in air, argon. Later in these studies Ramsay also isolated helium, which earlier

Figure 2.3.16. Ramsay. From Weeks (1935) with permission from the *Journal of* Chemical Education, copyright © 1935, Division of Chemical Education, Inc.

had been found in the solar spectrum. Together with his assistant, Morris Travers (1872–1961), Ramsay found krypton, neon, and xenon in air in 1898.

References

1. Weeks, M. E. (ed.). (1935). *The discovery of the elements*. Easton, PA: Mack Printing Co.
2. Andersson, R. G. W. and Lawrence, C. (eds.). (1987). *Science, Medicine and Dissent*: Joseph Priestley (1733–1804). London: Wellcome Trust/Science Museum.
3. Wilson, M. (1954). Priestley. *Scientific American* **1954**(10):68–73.
4. Austerfield, P. (1983). Dr Phlogiston, the 'honest heretic'. *New Scientist* **1983**(March 24):812–814.
5. Neville; R. G. (1974) Steps leading to the discovery of oxygen, 1774. *Journal of Chemical Education* **51**(7):428–431.
6. Cassebaum, H., and Schufle, J. A. (1975). Scheele's priority for the discovery of oxygen. *Journal of Chemical Education* **52**(7):442–444.
7. Duveen, D. I. (1956). Lavoisier. *Scientific American* **1956**(5):85–94.
8. Holmes, F. L. (1994). Antoine Lavoisier. The conservation of matter. *Chemical and Engineering News* **1994**(Sept 12):38–45.

3

INDUSTRIAL GASES

Background

3.1. HYDROGEN: THE GAS THAT HAD A FLYING START

On dissolution of some base metal, especially iron, in acidic alcohols, inflammable vapour [hydrogen] shoots out from the opening of the flask . . . —which is phlogiston.

Mikhail V. Lomonosov ("On Metallic Lustre", 1745)

The Dream of Flight

Humanity's dream of flying is probably as old as humanity itself. Despite brave and often foolhardy attempts, it was long before that dream became a reality. In its stead, myths were created in the form of sagas and legends in which flying beings, gods or people, were gifted with wings by supernatural forces. By the 17th century people had begun to understand more of the nature of air pressure and the atmosphere through the work of such individuals as da Vinci and Galileo, and came to appreciate the impossibility of imitating the flight of birds using human muscle power. The principle of "lighter-than-air" flight was discovered, and interest in "air-sailing" blossomed.

The first to openly discuss this principle was most probably the author Cyrano de Bergerac. In his *Voyages to the Moon and Sun*, published around 1650, he has Enoch, the principal character, rise up to the heavens with the aid of a smoke-filled container.

The invention of the air pump in 1654 by Otto von Guericke of Magdeburg led to many new ideas. In 1670, the Jesuit priest Francesco de Lana-Terzi visualized a flying ship borne aloft by four spheres of riveted thin copper sheets. An air pump was to empty these spheres of air and thereby lift the ship. The project was abandoned when de Lana realized that the invention could have military usage. He had not realized, however, that the spheres would have been compressed as a result of the evacuation of the air (Fig. 3.1.1).

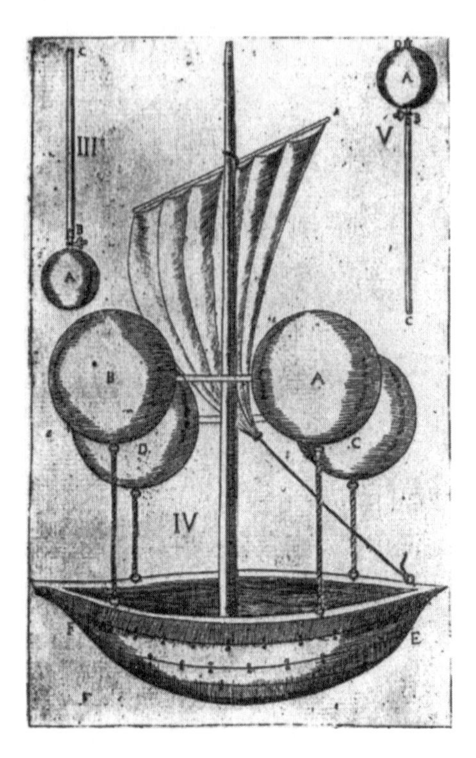

Figure 3.1.1. Possible means of flying according to the Jesuit priest Francesco de Lana-Terzi (1670).

The Discovery of Hydrogen[1–3]

During the 18th century many worked on ideas surrounding the lighter-than-air principle, but suitable means were not available until in 1766 Henry Cavendish succeeded in producing hydrogen gas in pure form (then known as inflammable air), and discovered that it was 14 times lighter than air. The dream of the century, that of "air sailing," could now become a reality.

Hydrogen: "Inflammable Air" as Fuel

Alchemists had observed the gas created when pieces of iron were thrown into oil of vitriol (sulfuric acid). Chronicles from 13th century Paris told of the Rabbi Ezechié, "an avid reader of books on black magic, a man known to the Devil, and an expert sorcerer," who worked with a lamp having a bluish light, burning without a means of lighting and without oil. During the 16th century, Paracelsus established "that the gas burnt with a hot, blue flame." Henry Cavendish described the flammability limits of hydrogen in air in 1766. His results were close to the modern limits of 4–75 percent hydrogen by volume. The very wide flammability limits of hydrogen make it easy to burn over a wide range of conditions, a great asset in a fuel. The same property, however, makes hydrogen hazardous to handle. The property of hydrogen that is of even greater importance than flammability is the large amount of energy released during combustion.

In 1783–1784 Antoine Lavoisier and Pierre Laplace measured the heat of combustion of hydrogen and found that it was more than double that of gasoline.

Lighter-Than-Air Experiments

Cavendish's newly discovered gas evoked the interest of his colleagues. Joseph Black of Edinburgh was inspired by Joseph Priestley's thesis on hydrogen gas written in 1774, and described an attempt with a balloon made of a hydrogen-filled calf's bladder before friends and students. One of the students, Tiberio Cavallo, investigated the theoretical and practical problems surrounding manned balloons in 1781, but was unable to manufacture a balloon casing that was sufficiently gas-tight. Instead, it was development in France that was to attract world attention.

The findings of these pioneers not only opened up a new chapter in the history of gases, but also attracted attention to hydrogen as an alternative to hot air as a bouyant gas. In Paris, Joseph and Étienne Montgolfier, sons of a paper manufacturer, worked systematically on using a balloon for flying. They studied Priestley's report and had intended to use hydrogen gas enclosed in a lightweight casing of paper and woven silk. Unable to successfully find a method of enclosing the gas, however, in July 1782 they turned to hot air as a medium, and on 5 June 1783, in Annonay, they made the first successful public attempt with a balloon.

To Jacques Charles (Fig. 3.1.2), member of the French Academy of Science, the lighter-than-air attempts were nothing new. He had put to the test Cavallo's findings regarding the lifting power of hydrogen, and immediately set out to build a full-scale hydrogen balloon. Thanks to the newly awakened interest in ballooning in Paris, he raised the necessary sum of 10,000 francs in just a few days.

Figure 3.1.2. Jacques Charles, the hydrogen balloon pioneer. From Ålund (1875).

Figure 3.1.3. Balloon filling by using iron and sulfuric acid. From A. Ganot, *Traité de Physique* (1853).

The Robert brothers, Ainé and Cadet, who had recently developed a process for coating cloth with rubber, had the responsibility of producing the balloon casing. One big obstacle remained, however. To that point, hydrogen had only been produced in quantities of less than 1 liter. To fill the 4-meter-diameter balloon, 50 cubic meters of hydrogen gas would have to be produced and stored.

Charles arrived at a simple method for the large-scale production of hydrogen, using the reaction of sulfuric acid and iron. Water and iron filings were poured into a wooden barrel, the lid of which had two holes. One was for a leather pipe that led the gas to the balloon, the other for adding the sulfuric acid (Fig. 3.1.3). Problems were experienced with sulfur dioxide and acid condensation in the balloon as a result of considerable heat from the chemical reaction. The balloon was two-thirds filled in 4 days and the raw material consumed amounted to 500 kilograms of iron and 250 kilograms of sulfuric acid (Fig. 3.1.4).

The balloon was sent up on the evening of 27 August 1783 before about 300,000 Parisians. *The Globe* sailed away and came down in the village of Gonesse, some 22 kilometers from Paris, where local farmers, unaware of its importance, saw the descending object as a dangerous monster and promptly destroyed it.

Competition between the "Montgolfiéres" and the "Charliéres" to have the first manned flight had begun. The first manned flight using a Montgolfiére was achieved by the Marquis d'Arlandes and the physicist Pilatre de Rozier on 21 November 1783. The journey lasted 26 minutes and they landed 8 kilometers from their starting point.

Less than 2 weeks later Charles built a balloon 9 meters in diameter (400 cubic meters) and a gas production plant, with the basic components that are still found in today's gasworks, with facilities for washing, purifying, and heat exchanging. On 1 December the hydrogen balloon *Charlér* with Professor Charles and Ainé Robert aboard took to the air from the Palace of the Tuileries before a gathering of over 400,000 people (Fig. 3.1.5). The descent took place 2 hours later in Nesle, 50 kilometers away.

Figure 3.1.4. Filling of the first hydrogen balloon by the brothers Robert in 1783. From de Saint-Fond (1783).

Figure 3.1.5. Manned flight by a hydrogen balloon from the Tuileries in Paris, December 1783. From Ålund (1875).

Hydrogen's flammability became the cause of the first air tragedy, which happened when two men in 1785 attempted to cross the English Channel in a hydrogen balloon, carrying a small hot-air balloon for altitude control. After 30 minutes the hydrogen balloon ignited and the two men perished. Nevertheless, the attractiveness of hydrogen as a buoyant gas remained for more than 160 years of lighter-than-air flight. The increased demand for hydrogen for balloon flight following Charles's successes also brought an early improvement in the technology of hydrogen production.

Methods of Production[4,5]

The balloon era quickly aroused military and scientific interest, and considerable attempts were made to find alternative methods for producing pure hydrogen.

Figure 3.1.6. Production of hydrogen by sulfuric acid according to Gillard, Paris, 1867. From Ålund (1875).

Despite these, the Charles method using iron and sulfuric acid was to be the one used for the greater part of the 19th century (Fig. 3.1.6).

This method was expensive. The Balloon Commission in Paris, of which Antoine Lavoisier was a member, was given the immediate task of finding other ways of producing hydrogen. Two methods were recommended: reduction of water vapor using iron, and using coal.

The Iron Method (Lavoisier and Meusnier, 1784)

A new idea was put forward in 1784 by Antoine Lavoisier and army officer and inventor Charles Meusnier, whereby water vapor was passed over a red-hot bed of iron at about 600°C, thereby producing pure hydrogen at about one-third of the cost of the acid method.

The Coal Method (Fontana, 1780)

This method involved passing water vapor over a red-hot coal or coke bed at about 810°C to produce water gas containing about 50 percent hydrogen and 40 percent carbon monoxide. First presented by Abbé Felice Fontana in 1780, the method produced an impure gas and was also more expensive than the iron method. The water gas method was, however, industrially developed during the 19th century for applications in which the whole of the energy content of the gas could be utilized.

The LFC Process

When Linde AG began producing liquid air in Germany, Adolf Frank suggested that the company could produce pure hydrogen by removing the carbon dioxide and hydrogen from water gas by condensation. The method was developed in 1909, and became known as the LFC process (Linde–Frank–Caro). It was extensively used from 1913 on in the Haber process for ammonia synthesis.

The Electrolysis Method

This method is believed to have been first used by the Dutchmen Adrian Van Trootswijk and Jan Deiman in 1789. At that time there were only electrostatic machines for producing electricity. One such machine was used to charge a Leyden jar, which was then periodically discharged via gold electrodes in a vessel of water, thereby producing a mixture of oxygen and hydrogen.

In 1800, Alessandro Volta presented his so-called voltaic pile, the forerunner of the electric battery. A few weeks later, William Nicholson and Anthony Carlislc constructed a voltaic battery and manufactured considerable quantities of oxygen and hydrogen. The electrolysis method remained expensive until the Belgian Zénobe Gramme invented the first steam-driven dynamo in 1873. From 1890 on, when large hydroelectric power stations were built, the method was used on a large scale wherever cheap hydroelectric power was found (Fig. 3.1.7).

Large quantities of hydrogen have also been achieved using other processes, such as the chlorine–alkaline process (Griesheim Elektron, from 1899). Since the 1920s, hydrogen gas has been produced mainly from fossil fuels by liquefaction of natural gas, partial oxidation of heavy oil, or carbon gasification.

Figure 3.1.7. Production of hydrogen by electrolysis. Early electrolysis plant, Chalais Meudon, 1890. From Renard (1890).

Nowadays for small-scale generation of hydrogen (up to 200 cubic feet per day, 60 cubic meters per day), the electrolytic method is most convenient, even if the price of electricity is high. In the intermediate range (200–20,000 cubic feet per day), the cracking of ammonia is a very practical method if the presence of nitrogen is not objectionable. In larger capacity applications hydrogen is now most economically made by steam reforming of natural gas or naphtha.

Hydrogen Airships

As early as 1784 thoughts were turning to ways of making a balloon steerable. The French scientist Jean Baptiste Meusnier built a cigar-shaped balloon and demonstrated a method for regulating altitude, but it lacked a means of forward propulsion, which was not to become attainable until two generations later.

The military found uses for balloons and a balloon company was formed during preparations for the revolutionary wars in France. Military observation balloons were to continue in use until the 1930s, when they were replaced by observation aircraft and autogiros. During World War II, large numbers of hydrogen-filled antiaircraft balloons were used.

The first steerable balloon came in 1852 in the shape of Henri Giffard's 44-meter-long, coil-shaped design powered by a 3-horsepower steam engine weighing 159 kilograms. In 1884, Charles Renard and Arthur Kreb produced a design with an electric motor. At the turn of the century, Graf Ferdinand von Zeppelin invented the first dirigible with a fixed framework and equipped it with two of the newly developed Daimler petrol engines. A Zeppelin, as they were to be called, first flew in 1900 (Fig. 3.1.8). Following a decade of development, they were used for passenger transport. From 1910 to 1914 about 35,000 passengers were carried and in 1919 the first Atlantic crossing was achieved. From 1915, Zeppelins were used by the military for long-distance raids, including the bombing of London.

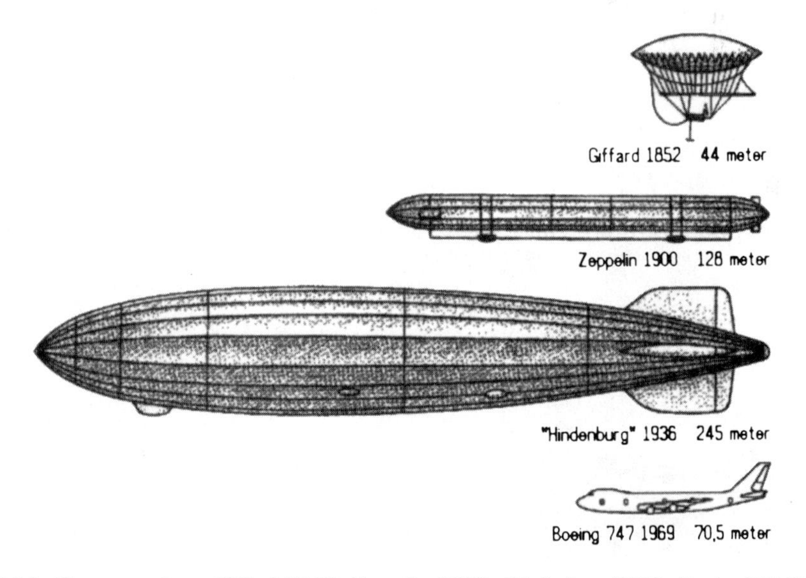

Figure 3.1.8. Size comparison: Giffard (1852), Zeppelin (1900), *Hindenburg* (1936), Boeing 747 (1969).

From the middle of the 1920s, Zeppelins were used in regular intercontinental traffic. The airship era came to an abrupt end with the crash and explosion on landing of the *Hindenburg* at Lakehurst, New Jersey, on 6 May 1937, killing 33 of the 95 on board (Fig. 3.1.9). One of the survivors, saving himself by jumping from a height of 6 meters from the blazing airship, was the Swedish Chamberlain Rolf von Heidenstam, who was to become the president of AGA 7 months later (Fig. 3.1.9).

Figure 3.1.9. The *Hindenburg* crash of 1937 (used with permission from Corbis Images, New York). Chamberlain Rolf von Heidenstam escaped without being hurt and was elected president of AGA 7 months later (used with permission from the AGA Historical Picture Archive).

Twentieth-Century Applications (see Fig. 3.1.10)

Chemical Applications: Hydrogenation

Hydrogen had only minor uses before the 20th century. Present large-scale chemical usage derives from the beginning of the 20th century with hydrogenation processing of mineral oils (P. Sabatier in 1897, the Bergius process in 1913) and of vegetable and animal oils (fat hardening, William Normann in 1903). In Germany in 1913, Badische Anilin- und Soda Fabrik (BASF) began to synthesize ammonia using hydrogen from water gas (about 50% H_2, 40% CO; the Haber–Bosch method), and from 1922 they produced synthetic methyl alcohol (the Patart method). The Fischer–Tropsch process for producing synthetic petroleum from brown coal was developed from 1923 and used on a large scale from 1935. Oil companies began using the hydrogenation method for removing sulfur from oils in 1929.

In P. Sabatier's hydrogenation method (about 1897) a mixture of hydrogen and the vapor of an organic compound was passed over a metallic catalyst, normally finely divided nickel. This process brought about a large variety of reactions, notably the reduction of double bonds between carbon atoms to single bonds in organic substances. In 1902 a German patent was granted for Sabatier's invention.

William Normann, in 1903, patented the simple process of bubbling hydrogen through heated oil in which finely divided nickel was suspended. This process converted vegetable oils, such as cotton-seed oil, very common in the United States, and also whale-oil and fish oil in Europe, into hard, white fats that could be used as lard substitutes for making margarine, which had been invented by Mège Mouriés in the late 1860s. The use of these fats as foodstuffs was at first very limited, but the blockade of 1914–1918 forced their use in Germany. They were found to be excellent foods and the fat hardening process has continued to be important ever since.

During the World War I the hydrogenation process was further developed based on the fact that the reaction of hydrogen with organic compounds is much accelerated by pressure. From about 1913 Frederick Bergius worked out methods of hydrogenating substances by reaction with hydrogen under great pressure and without a metallic catalyst. The effort, especially in Germany, to make motor fuels resulted in processes for hydrogenating coal and tars which proved invaluable as war-time expedients.

In Great Britain, Imperial Chemical Industries (ICI) acquired the Empire rights to the Bergius process and after further experiment in 1935, a large-scale plant for conversion of coal into motor fuel was started up. A number of industrially important processes depend on the hydrogenation of carbon monoxide. The Fischer–Tropsch process for synthetic motor fuels consists in passing purified water gas over certain catalysts. At 180–200°C and atmospheric pressure, about 80% of the carbon monoxide forms hydrocarbons, which are very useful as diesel fuels. The Fischer–Tropsch process was mainly developed in Germany from about 1935, producing enormous quantities of motor fuel from brown coal.

BASF started work on hydrogenation for the synthesis of methanol based on water gas enriched with hydrogen by various means. The work began in 1913, but only after 1922, when the French chemist G. Patart patented a process, did production start on a large scale. Since 1924 this process has been used in most manufacturing

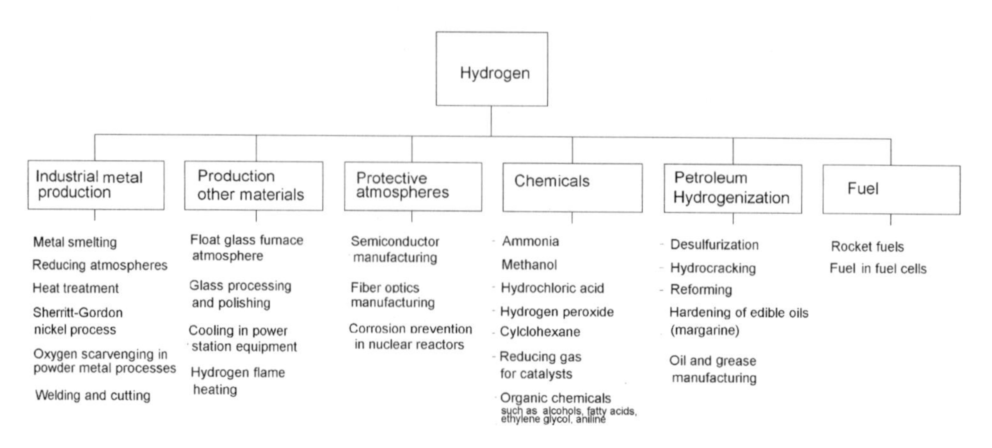

Figure 3.1.10. Hydrogen applications.

countries as a source of methanol for conversion into formaldehyde required for the plastics industry.

About 1929 the oil companies took up the hydrogenation process as a means of removing sulfur and nitrogen from oil and of cracking and reducing heavy hydrocarbons to lighter ones.

Metals Applications

In the metallurgical and metal treatment fields hydrogen was first used for welding and smelting of metal. Lead welding using hydrogen and air was developed by the Frenchman Desbassayns de Richemond in 1839. In 1847, Henry Sainte Claire Deville smelted platinum, iron, and silver using an oxyhydrogen blaster, and the cutting of iron and steel using hydrogen and oxygen was reported by the Englishman Thomas Fletcher in 1890.

Compressed hydrogen gas was marketed in steel cylinders from 1901 by Griesheim-Elektron in Germany. Following the introduction of oxyacetylene gas for welding, the use of hydrogen gradually diminished, and it is now used as a shielding gas. Since World War II, hydrogen has been used in metallurgy as a reducing agent on metallic oxides to produce pure metal.

Production and Uses of Liquid Hydrogen[6]

In 1898 James Dewar in London announced that he had liquefied hydrogen at 20.3 K with a multistage process based on Joule–Thomson expansion and heat exchange (Fig. 3.1.11). Using liquid nitrogen, he precooled gaseous hydrogen under 180 atmospheres, then expanded it through a valve in an insulated vessel also cooled by liquid nitrogen. The expanding hydrogen produced about 20 cubic centimeters of liquid hydrogen, about 20 percent of the hydrogen used.

Dewar measured the density of liquid hydrogen as 0.07 kilogram per liter, the modern value, which is 1/14 the density of water and about 1/12 the density of kerosene or gasoline.

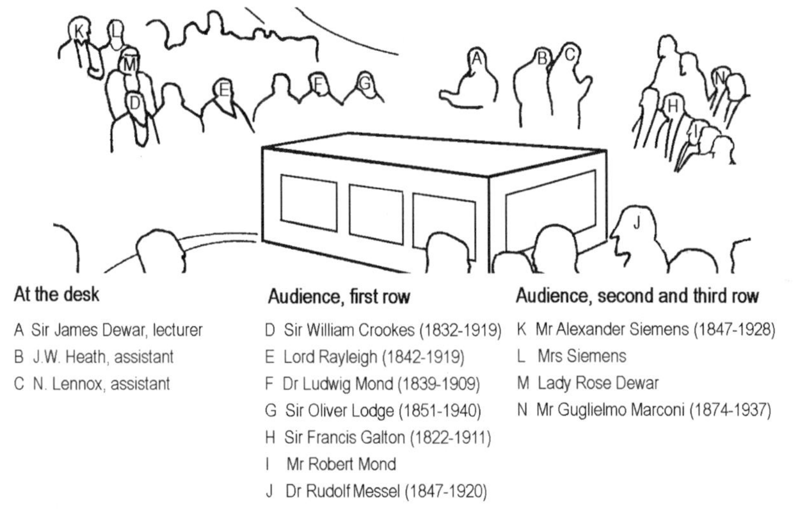

Figure 3.1.11. Sir James Dewar demonstrating liquid hydrogen at the Royal Institution in London. *A Friday Evening Discourse at the Royal Institution; Sir James Dewar on Liquid Hydrogen, 1904* (oil on canvas) by Henry Jamyn Brooks (1865–1925). Courtesy of The Royal Institution, London, UK/Bridgeman Art Library. Reproduced with permission.

By 1900 many of the major properties of gaseous and liquid hydrogen were known. Hydrogen had been liquefied, and Dewar flasks made its storage and transportation feasible. In 1904 Brin's Oxygen Company exhibited a hydrogen liquefier at the Louisiana Purchase Exhibition in St. Louis. The National Bureau of Standards purchased it for $2400 for use in low-temperature thermometry.

Ortho- and Parahydrogen

When liquid hydrogen was used for scientific cooling applications at the beginning of the 20th century, it was found that during its production the hydrogen boiled away faster than was anticipated. The reason for this anomaly was discovered by the German physicist Werner Heisenberg in 1926. Hydrogen is a mixture of two allotropic forms, a parallel form, orthohydrogen, and a nonparallel form, parahydrogen; their proportion changes with temperature. Balance is normally achieved after a few days; by catalytic conversion, it can now be achieved directly. In 1932 Heisenberg was awarded the Nobel Prize in Physics for his part in the creation of quantum mechanics.

Fuel Applications[7]

Hydrogen-enriched gases such as water gas have been used commercially for heating and lighting since about 1820. A hydrogen-driven combustion engine was designed by the Munich Court Clockmaker Christian Reithmann in 1852, and the first hydrogen-driven car was made in France in 1917.

Suggestions for using liquid hydrogen also were not long in coming. Of great interest to history were the ideas of an obscure Russian schoolteacher. In 1903, 5 years after Sir James Dewar successfully produced liquid hydrogen, Konstantin Tsiolkovskiy presented his theories on the use of liquid hydrogen as a rocket fuel for space flights. His ideas became a reality in 1963, when NASA launched the Centaur rocket using liquid hydrogen/oxygen (see Chapter 7.5).

Interest in hydrogen liquefaction began with U.S. research on moderators for atomic weapons, and rapidly accelerated in the 1950s with thermonuclear research. Catalysts were developed to convert ortho- to parahydrogen as part of the liquefaction process, and hydrogen storage and transport became practical. Low-loss air-transportable Dewars and surface-transportable Dewars for liquid hydrogen were developed by the mid-1950s, but total hydrogen liquefaction capacity remained too low for propulsion use.

Industrial production of liquid hydrogen started at a government-run experimental plant in Boulder, Colorado, in 1952. The first commercial plant was built in 1957 by Air Products in Painesville, Ohio, on a U.S. Air Force contract, to be followed by plants in Florida and California. The most important applications are air and rocket fuel trials for the space program.

Concurrent with the building of a series of four hydrogen liquefaction plants, the Air Force developed road trailers with a capacity of 26,500 liters of liquid hydrogen, with a loss rate of about 2 percent per day. The combination of hydrogen liquefaction plants and trailers for transporting liquid hydrogen was a large step forward in both liquid hydrogen availability and handling experience.

The Hydrogen Economy

Following the oil crisis in 1973, the subject of the "hydrogen society" was investigated—a society of the future in which hydrogen is the predominant source of energy. Due to global threats to the environment, these aims have been accentuated. Hydrogen gas is currently too expensive to produce and store. Considerable hope is placed upon future technological breakthroughs.

The Zeppelin Story[8]

Airships would probably never have been anything but a technical curiosity if Count Zeppelin had not devoted his life to developing the technology. Ferdinand Graf von Zeppelin (1838–1917) (Fig. 3.1.12) founded his airship-building company in 1897. He was convinced of the military importance of airships, and the German military authorities purchased some of the very first Zeppelins. Germany was the country that did the most to develop the airship and the only one ever to run regular passenger services.

The Zeppelins were the first rigid airships. After initial difficulties, airships eventually won the confidence of the military chiefs, and by the end of the World War I more than 100 airships had been built by Count Zeppelin's industrial concern. During that period it faced competition from another German airship builder, Luftschiffsbau Shütte-Lanz, who built 20 airships for military purposes.

The very first Zeppelin was built and flown around the turn of the century. It was powered by two engines with an aggregate output of around 60 horsepower and a maximum speed of just under 30 kilometers per hour. Passenger traffic had already begun before the war broke out. Count Zeppelin started the world's first airline, Deutsche Luftschiff AG, in 1909, and by 1914 it had carried 34,000 passengers on 1600 flights. Between the late 1920s until 1937, Zeppelins flew regularly between Germany and North and South America (Fig. 3.1.13). In 1929 an airship, the *Graf Zeppelin*, even flew around the world.

Figure 3.1.12. Ferdinand von Zeppelin. Used with permission from the archive of Luftschiffbau Zeppelin GmbH, Friedrichshaten, Germany.

Continued

Figure 3.1.13. Early advertisement for commercial Zeppelin airship flights. Used with permission of the archive of Luftschiffbau Zeppelin GmbH, Friedrichshafen, Germany.

By the end of 1936 the *Graf Zeppelin* had made 500 flights carrying 20 passengers each time. Zeppelins made 2300 flights and carried more than 50,000 passengers over the three decades for which they were in service. In 1936 Germany launched a new, larger, and faster airship, the *Hindenburg*. It was designed to be inflated with helium, but in 1935 the United States refused to export this valuable and precious raw material, in accord with the *Helium Control Act* of 1927. So the *Hindenburg* was filled with hydrogen for flights to Rio de Janeiro and to the United States (it took 62 hours to fly from Germany to New Jersey). A terrible explosion occurred during the landing operation in New Jersey on 5 May 1937. This was the only accident involving a German passenger Zeppelin where passengers were killed. World War II put a definitive end to commercial hydrogen airship services.

Military Uses

Zeppelins launched the world's first long-distance bombing attacks in 1915. These bombing raids often had disastrous consequences for the airships and their crews. It is said that German aircrews were frequently instructed to bomb London on weekends to avoid civilian casualties.

Naval airships were more successful because they were used almost exclusively for reconaissance purposes, for which they were considered extremely effective. The Zeppelins of the latter war years were large and surprisingly fast. The biggest were more than 200 meters long and 24 meters in diameter, could travel at speeds of more than 100 kilometers per hour, and had remarkable endurance. One airship covered almost 8000 kilometers in one flight lasting 4 days.

Other Airships

Airships were also built in countries other than Germany, particularly in Great Britain and the United States, but also in France, the Soviet Union, and Italy. Of the more than 1100 airships built until the 1980s throughout the world, 75 percent were soft airships, around 15 percent were rigid airships, and around 10 percent were semirigid.

Germany made the largest number of rigid airships, mostly through the Zeppelin companies. Great Britain and the United States built the largest numbers of soft airships, most of the British ones being produced before 1920 and most of the American ones after 1920. More than 100 airships (all soft) have been built since the World War II.

Short Biographies

Jacques Charles (1746–1823) was a French mathematician and physicist, Professor of Physics at the Sorbonne University. He was the first to produce hydrogen on a large scale. He designed hydrogen gas balloons and made the second manned flight in history in 1783. He is also known for Charles's law (1798), in which he formulated the connection between the pressure, the volume, and the temperature of a gas.

Sir James Dewar (1842–1923) was a British physicist and a professor at Cambridge and London. He invented the Dewar flask in 1892, a double-walled thermos vessel with the walls silver-coated. He produced liquid hydrogen in 1898.

Konstantin Tsiolkovskiy (1857–1935) was a Russian teacher at a girls school. From 1885 he made systematic studies of the theoretical and practical problems related to space flight. In 1903 he proposed the use of liquid oxygen and liquid hydrogen as rocket propulsion fuels, a vision that was to be realized 60 years later.

References

1. de Saint-Fond, F. (1783). *Description des experiances de la machine aerostatique de MM. de Montgolfier.* Paris.
2. Jones, G. (1983). The day the French took to balloons. *New Scientist* **25**:534–538.
3. Scott, A. F. (1984). The invention of the balloon and the birth of modern chemistry. *Scientific American* **250**(1):102–111.

4. Pottier, J., and Bailleux, C. (1986). Hydrogen: a gas of the past, present and future "a drama in 3 acts". *Hydrogen Energy Progress VI*. In *Proceedings of the 6th World Hydrogen Energy Conference*, Vol. 1:197–214. Vienna, Austria.
5. Aureille, R. (1984). Deux siecles de production d'hydrogene. *Hydrogen Energy Progress V*. In *Proceedings of the 5th World Hydrogen Energy Conference*, Vol. 1:21–42. Toronto, Canada.
6. Scott, R. B., Denton, W. H., and Nicholls, C. M. (eds.). (1964). *Technology and use of liquid hydrogen*. Oxford: Pergamon Press.
7. Cox, K. E., and Williamson, K. D., Jr. (eds.). (1979). *Hydrogen: Its technology and implications*. Vol. IV *Utilization of hydrogen*. Boca Raton, FL: CRC Press.
8. Pigge, H. (1999). *Ferdinand Graf von Zeppelin*. Ullstein Buchverlage.

Further Reading

Winter, C. J., and Nitsch, J. (1986). *Wasserstoff als Energiträger*. Berlin: Springer-Verlag.
Taylor, F. S. (1957). *A History of Industrial Chemistry*. London: Heinemann.

3.2. OXYGEN: "FIRE AIR"

> *The feeling of it [oxygen] to my lungs was not sensibly different from that of common air; but I fancied that my breast felt peculiarly light and easy for sometime afterwards. Who can tell but that, in time, this pure air may become a fashionable article in luxury.*
>
> Joseph Priestley (*Experiments after Discovery of Oxygen*, 1774)

Background

Pure oxygen is one of our most important basic raw materials, with a global yearly production of about 250 million tons. The greatest increase began after World War II, when new metallurgical processes came into being. It was during the early 20th century that oxygen found a practical use when the combination of oxygen and acetylene came into use for welding and cutting. Prior to this the discovery of oxygen had contributed to several interesting lines of development in technological history.

The Discovery

In Uppsala, Sweden, Carl W. Scheele extracted the "air of fire" (oxygen) in at least six different ways during the years 1771–1773. He studied the qualities of the gas in order to discover the essence of combustion. He perceived that ordinary air was a mixture of the "air of fire" (oxygen) and of "rotten air" (nitrogen). In Leeds, England, Joseph Priestley extracted and described oxygen on August 1, 1774. At first he was not aware that he had found a new element, however, but thought that the new gas was a complicated mixture consisting of saltpeter, earth, and phlogiston.

A New Theory of Combustion

The Theoretical Revolution

In Paris in November 1772, Antoine Lavoisier determined that a substance increased in weight during combustion, in contrast to what the dominating phlogiston theory maintained. He suspected that the remarkable heating substance phlogiston was humbug, but was unable to prove it.

Chance brought Priestley and Lavoisier together in Paris in October 1774. Some weeks previously, Lavoisier had received a letter from Scheele in which he had described the hitherto secret discovery of oxygen. Based on Priestley's similar experiences, Lavoisier was able by his own experiments to determine that oxygen was a basic element and that it united with other elements during combustion.

In this way the part played by oxygen in the processes of combustion and oxidation were explained and the centuries-old phlogiston theory could be overturned. This theoretical revolution was to be a cornerstone of Lavoisier's foundation of modern chemistry, presented in 1789 in *Traité élèmentaire de Chemie* (*Treatise on Chemical Elements*), which met tremendous resistance from the established chemists of the time. Lavoisier, however, had the pleasure of seeing the success of his opinions before he died by the guillotine during the French Revolution in 1794.

Oxygen Gives Oxygen = Untrue

Not all of Lavoisier's theories were correct, however. Since "sour oxides" were formed when, for example, coal, phosphor, and sulfur were burnt in oxygen, Lavoisier concluded that the newly discovered gas was the "oxygen-generating principle," which constituted a component of all acids. Therefore it was given the name "oxygene" (acid-builder). Analogous to this, the gas was later named *Sauerstoff* in German and *syre* in Swedish.

In 1787 Lavoisier's colleague Claude Louise Berthollet found that hydrogen cyanide contained no oxygen. He was unable to convince the newly converted circles surrounding Lavoisier of the errors in the theory. The task of proving this was to come more than 20 years later, and fell to the Englishman Sir Humphry Davy, who in 1810 demonstrated that hydrochloric acid gas contained no oxygen. By this time the teachings of the new chemistry had developed so strongly that Davy's results were looked on as an attack on the whole teaching establishment. The "threat" was contested strongly by, among others, Berzelius, and Davy's analytical capability was questioned.

Production of Oxygen[1]

In the 19th century numerous chemical methods of preparation had been found. Some of them were commercialized in the form of small oxygen generators, but only one had been used to make a gas which could be compressed and sold in cylinders, mainly for sundry medical uses and limelight (see below).

The oxygen that was used during the first years following its discovery was extracted by laboratory-type methods of heating various substances, and the gas was transported in sacks. In 1800, Volta's battery was introduced, which meant that oxygen and hydrogen could be extracted from water by electrolysis. Oxygen produced during the most of the 19th century was extracted in small quantities and had low purity, and it was not until about 1890 that gas bottles made of iron were used for its transport.

In the second half of the 19th century several attempts were made to separate oxygen from air on a commercial basis by means of a chemical process, for example, using a thermal cycling process with alkaline manganates as proposed by Tessié Du Motay and C. R. Marechal in 1866. Companies were formed and works erected at Paris, Lille, Brussels, Vienna, and New York. They apparently were temporarily successful in the United States, but the company there failed in 1871. In a similar process G. Kassner (1890) used a sodium manganate/plumbate mixture, but he and all others working seem to have suffered a similar fate. The Mallet process (1867) used cuprous chloride and moist air to extract the oxygen. Production rates were poor and a temperature swing of 200–300°C was required.

All the above processes used a metal as the oxygen absorption site and worked in the solid state. These systems all required high temperatures (hence high energy) because they broke the oxygen molecules into atoms and then reformed them. Many systems have been proposed using transition metal complexes that would be "tuned" so this would not be necessary.

First Industrial Production of Oxygen from 1887[2]

The first industrial process whereby atmospheric oxygen could be isolated cheaply and in large quantities was a reversible reaction of barium oxide/barium dioxide, whereby barium oxide (BaO) was reacted with air at 540°C to give the barium dioxide BaO_2, which in turn decomposed at 870°C (Fig. 3.2.1). It was used by the French scientist Jean Baptiste Boussingault (Fig. 3.2.2) in 1851. He saw the possibilities behind the process, but because of certain technical problems, combined with the diversion of other research interests, he did not develop the idea. In 1879, he presented it during one of his lectures at the Conservatoire de Agriculture in Paris. Two of the students, the brothers Arthur and Leon Quentin Brin, set about improving the process and started the company Brin Frères at Passy, close to Paris, in 1884.

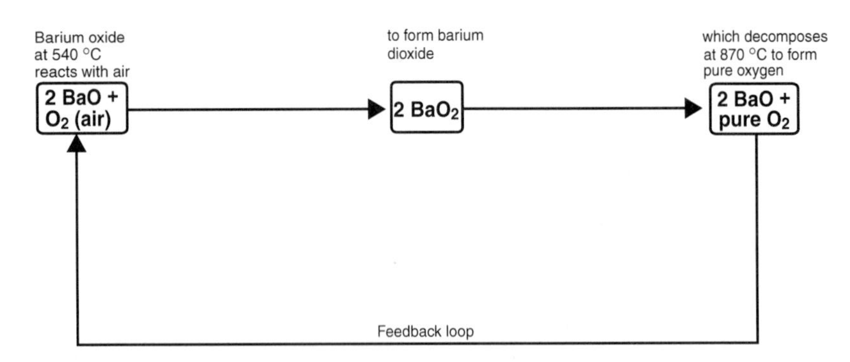

Figure 3.2.1. The Brin barium oxygen process cycle.

Figure 3.2.2. Jean Baptiste Boussignault. Edgar Fahs Smith Collection, University of Pennsylvania Library. Used with permission.

After they showed the process at an exhibition in London in 1885, the Brin brothers were contacted by a sponsor, Henry Sharp, who was willing to invest money in a new company, Brin's Oxygen Co., founded in 1886 in Westminster, England (Fig. 3.2.3). The company, which produced 6 tons (4000 cubic meters) of oxygen during its first year, was developed further and in 1906 became the British Oxygen Company.

A Scottish engineer, Kenneth Sutherland Murray, was hired as a foreman of the Brin works (later to become chairman of the British Oxygen Company). His mechanical ingenuity and skills converted the Brin process into the first successful commercial production method of oxygen. Oxygen purity was about 95 percent and with this improved process the U.K. output in 1889 was about 39 tons (27,000 cubic meters) of oxygen. The main use of oxygen at the time was for producing limelight (see below).

From 1890 the electrolysis of water, which was made conducting by acid or preferably by alkali, became a competitive process; it consumed about 5 kilowatt-hours of electricity for production of 1 cubic meter of hydrogen and ½ cubic meter of oxygen. Modern industrial electrolyzers may use pressures of up to 30 bars (e.g., the Lonza cell) or just above atmosphere pressures (e.g., the Knowles cell). The method is now mostly used for small-scale hydrogen production.

Figure 3.2.3. Plant interior, Brin Co., Westminister, England, 1886. From BOC Group (1986) with permission.

Liquid Air

The Brin method brought the price of oxygen down, to 20 times cheaper to produce than before. As the brothers Brin began their work with the barium oxide method, however, the next stage in development had already begun. In 1877, the Swiss Raoul Pictet and the Frenchman Louis Cailletet reported that they had successfully extracted liquid oxygen, working independently and using quite different adiabatic methods.

Practical industrial methods for producing liquid air were announced around 1895 by three scientists working independently: William Hampson in England, Carl Linde in Germany, and Charles Tripler in the United States. They all used the so-called Joule–Thomson effect of cooling the gas in stages by adiabatic expansion through a throttling valve. Another method, using the piston expansion machine, was developed by the Frenchman Georges Claude in 1902. In his research into combustion using oxygen and acetylene, he had studied Linde's method for producing liquid air. If it were possible to get the gas to work in a machine instead of a throttling process, then it should be possible to achieve liquefaction at a lower working pressure. Finding machine parts that were capable of working at –200°C was, however, not an easy task. Working further with the methods used by Cailletet and Solvay, and by using kangaroo skin for sealing, Claude succeeded in reaching the low temperatures necessary.

The liquid air that Linde obtained in 1898 by utilizing the quicker evaporation of nitrogen could be made to contain more than 50 percent oxygen. This oxygen-enriched liquid was called "Linde air"; when mixed with fine coal dust it was a cheap, safe, and effective explosive. Around 1900 it was used in the excavation of the Simplon Tunnel and later was employed in mines. The mixture was called Oxylikvit.

Figure 3.2.4. The machine hall, Brin Co., Westminister, England, 1886. From BOC Group (1986) with permission.

Air Gas Separation

By using the system of rectification developed by Linde and Claude it was possible to obtain hitherto-unthought-of quantities of nitrogen and oxygen of high purity. During the period 1902–1905 distillation columns were developed capable of producing better than 99 percent pure oxygen or nitrogen (Fig. 3.2.5). In 1910 the double-distillation column was invented by Linde whereby oxygen and nitrogen could be extracted at the same time. The price of oxygen was now down to one-third of what it had been from the Brin method.

Already before World War I the air-gas separation technique had spread throughout the world, and later even noble gases could be extracted from air. The number of air gas customers increased, making the transport of air in liquid form more interesting. The German Paul Heylandt developed a technique for this during the period 1923–1927. Heat exchanging was improved via Mattias Fränkl's regenerator construction, which was replaced in about 1950 by the even more efficient revex heat exchanger. The piston expansion machine was replaced during the 1930s by the expansion turbine, further developed by the Russian Pjotr Leonidovich Kapitza in 1939.

Other Early Oxygen Production Methods

Around 1900 the autocombustion of potassium perchlorate, which formed the substance Oxygenite, came into use for production of oxygen (Figs. 3.2.6 and 3.2.7). Under pressure, 1 pound (about ½ kilogram) of Oxygenite gave ½ liter of oxygen. Another substance, Epurite, consisting of calcium hypochlorite, iron, and copper

Figure 3.2.5. Early liquid oxygen plant (50 cubic meters per hour), Ougrée, Belgium. From Claude (1926).

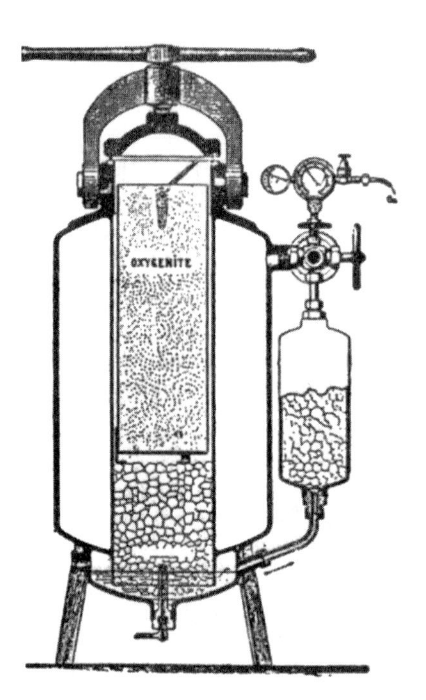

Figure 3.2.6. Apparatus for the production of oxygen by the Oxygenite process. From Granjon and Rosemberg (1912).

Figure 3.2.7. Advertisement for Oxygenite, 1912.

sulfate, gave off oxygen when impregnated with water. A third substance was Oxylithe, which also gave off oxygen in contact with water.

For military purposes hydrogen peroxide decomposition has been used for oxygen production and this method was introduced commercially in Japan to produce respirable oxygen for home care. Oxygen candles are other chemical generators for breathing in emergency situations where the oxygen supply may fail. They are usually based on thermal decomposition of sodium chlorate.

Oxygen Applications: The Use of Oxygen during the Nineteenth Century[3–5]

At first the largest application of oxygen was for limelight in theaters, followed by medicinal uses.

Limelight

Gases achieved considerable importance in lighting. William Murdoch, employed by Watt and Boulton in Birmingham, developed different forms of coal gas for lighting purposes. The first installation was carried out in Murdoch's home in 1792. Street lighting using coal gas came to London in 1808, marking a step toward the modern way of life. New opportunities opened up for activities outdoors and in homes during the nighttime hours. A new type of street lamp was the first commercial application for oxygen. Lieutenant Thomas Drummond (1797–1840) heated a piece of quicklime with hydrogen until it glowed, and then led a flow of oxygen toward the glowing lime. He thereby achieved a spot-shaped source of light, so-called "limelight," that was 37 times brighter than the strongest oil lamp of the time.

The breakthrough for this innovation came in 1825. The English Army had been given the task of mapping Ireland, and needed to make measurements by observing a light source at long distances using a telescope. Attempts using sunlight and mirrors and, at times, oil lamps all failed. Using the Drummond limelight, it was possible to discern the light at a distance of 100 kilometers. Drummond went on to try different configurations of the lime and other fuels such as hydrogen, but all using oxygen. The intensity of limelight rapidly became celebrated and was proposed for applications such as lighthouses. An oxyhydrogen limelight was capable, on a moonless night, of casting the shadow of a hand on a dark wall more than 10 miles (16 kilometers) distant.

In those days, theaters had a great need for strong light sources to achieve new effects. The electric arc lamps (one of Davy's inventions) gave light that was too strong and bluish, and was also noisy. Murdoch's gaslights filled the auditorium with smoke and soot. Limelights came into general use in theaters in the 1850s because they were warmer in tone and cheaper than electric arc lights (Fig. 3.2.8).

Because the gases needed were not available commercially in the early days of limelight, hydrogen was usually made by pouring dilute sulfuric acid on zinc. Oxygen was obtained by heating manganese dioxide, potassium chlorate, or sodium nitrate. Both gases were then filtered and stored in large, bellow-shaped bags of rubberized fabric or leather. Equal weights were applied to the bags to obtain the equivalent pressures required from the two gas streams. In the early days, rubberized bags were filled with gas and sandwiched between weighted wooden boards to create pressure, but the bags were subject to wear and tear, making this a hazardous procedure (Fig. 3.2.9).

W. C. Macready at Covent Garden used limelight starting in 1837 and charged his public 30 shillings per night, a considerable sum at that time. This covered the cost of an operator and of hydrogen and oxygen. At first oxygen and hydrogen were stored in gas bags. The more prosperous theaters later stored the gas in metal tanks at low pressure. In all, it was costly, cumbersome, and dangerous and this delayed the widespread use of limelight for many years.

Magic Lanterns

"Magic lanterns" with limelight for picture projection were used in the second half of the 19th century (Fig. 3.2.10). They became more efficient and more diverse, and lantern slides became more and more sophisticated. In the early days, however,

What is seen by the spectators in a theatre [1883]

The scene as it is actually played [1883]

Figure 3.2.8. Theater performance with limelight, as depicted in *Le Nature*, 1883.

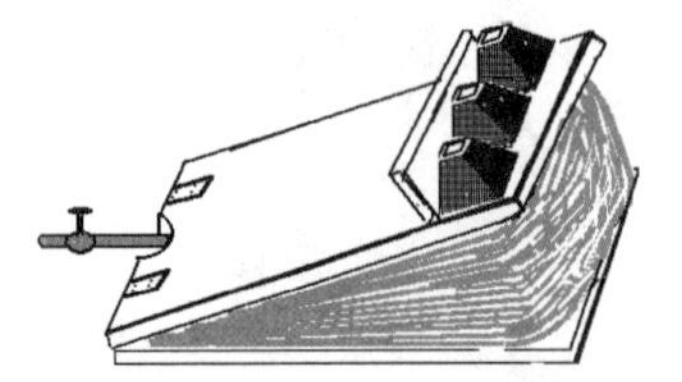

Figure 3.2.9. A gas bag with "pressurized addition," possibly from the 1850s.

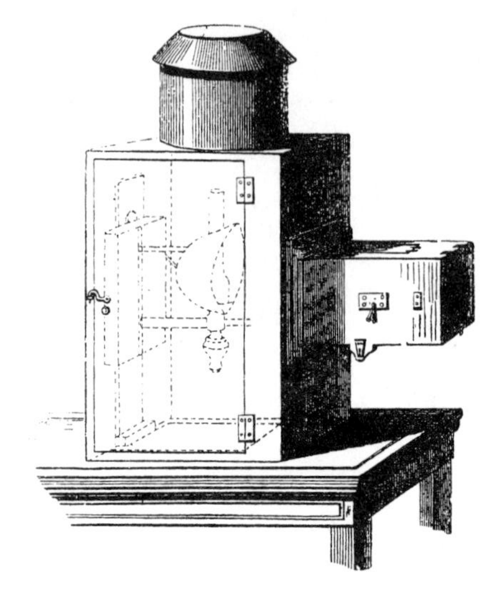

Figure 3.2.10. Laterna Magica. From Ålund (1875).

the lens quality of the lantern did not match the quality of the slides. Toward the end of the century electrical arc lamps began to offer competition, and electric filament bulbs came to replace other forms of lantern illuminants.

Limelight in magic lanterns was still used when the movies began to flourish: For many years the two coexisted. The early cinematographs used parts from magic lanterns (such as light sources and lens systems), but the glass slides were replaced by the film mechanism. In this application the use of limelight caused problems because the lime had to be heated to an extremely high temperature. The first-generation movie projectors were hand-cranked.

Medical Applications[6]

Scheele, Priestley, and Lavoisier all perceived that oxygen was part of the processes of combustion and of life. The belief in the possibility of using oxygen to cure ailments began to blossom during the years following its discovery. In 1779 the Dutchman Jan Ingenhousz discovered that plants give off oxygen when exposed to sunlight. He foresaw that, by the process of photosynthesis, there was a simple and cheap method of producing oxygen in the quantities necessary to cure ailments.

Gas medicine became a popular field of research and even engaged some of the members of the multidisciplinary Lunar Society in Birmingham, such as Joseph Priestley, James Watt, and Erasmus Darwin. One who also took part was a doctor who strongly advocated the use of gases in medicine, the Oxford professor Thomas Beddoes. With technical backing from members of the Lunar Society, Beddoes opened the Pneumatic Institute at Clifton near Bristol in 1799 for research into gas medicine. Based on recommendations from such as Watt, the 20-year-old apothecary Humphry Davy was employed as head of the Institute. He soon learned the experimental methods of Priestley and Watt and studied the effects of gases on people. Laughing gas eased pain, whereas oxygen was toxic, and no gas was able to cure illness. Davy's negative results brought about the closing of the Institute less than 3 years after its inception. Other enthusiasts continued the work, but no results of note were achieved until the middle of the 19th century, when nitrous oxide combined with oxygen began to be used as an anesthetic. The oxygenated water that was thought to cure many ailments was marketed actively by less-scrupulous health prophets. Research into the physiological effects of oxygen were carried out by Louis Pasteur, Claude Bernard, and Paul Bert from 1860 through 1880.

Modern Applications for Oxygen (see Fig. 3.2.11)[7–9]

Chemical and Fuel Synthesis

From the first years of the 20th century oxygen found a rapidly increasing market in oxy-fuel welding and cutting of steel. Prior to World War II, apart from welding/cutting, the main large-scale use of oxygen was for the production of nitric acid from ammonium chloride (from 1914) and the gasification of brown coal (from the 1930s). Germany sought to convert low-grade coal to chemicals and fluid fuels in the 1920s. A number of oxygen plants were built to further the development of synthesis gas (syngas) by the Fischer–Tropsch and similar processes, and also for ammonia.

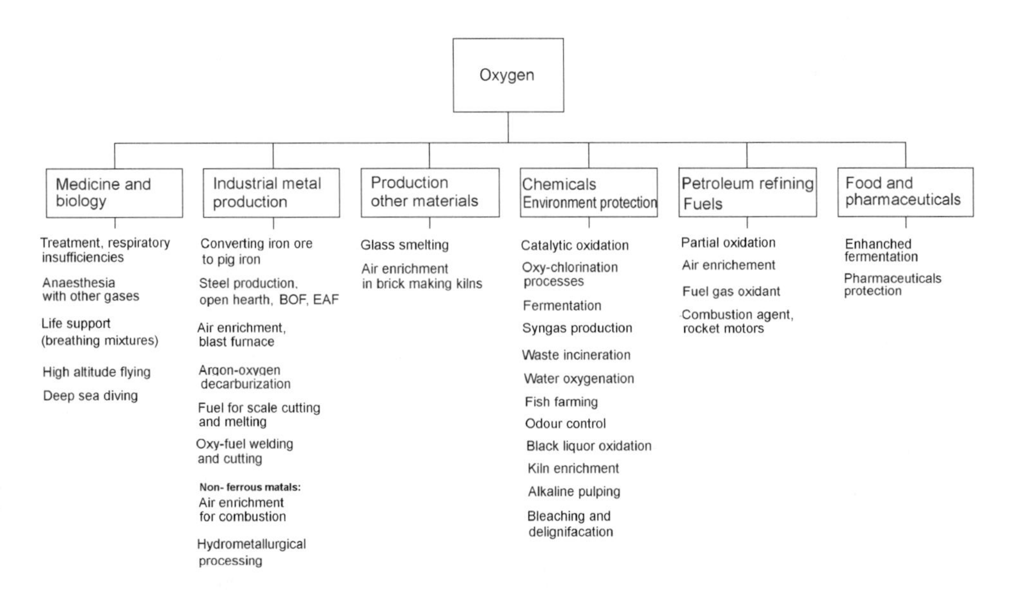

Figure 3.2.11. Applications of oxygen. BOF, Basic Oxygen Furnace; EAF, Electric Arc Furnace.

Synthesis was required for making chemicals and fuels. Syngas processes are in use on a very large scale at the Sasol plant in South Africa, where coal availability and strategic considerations have been major factors, whereas in Europe and the United States natural gas and oil can be used as fuels.

Oxygen in Steel, Glass, and Ceramics Manufacturing

After World War II metallurgical oxygen processes became the main consumer of the world's oxygen production.

The increased availability of relatively low-cost oxygen in the 1930s because of the development of the Heylandt process for liquid oxygen and the Linde–Fränkl process for gaseous oxygen encouraged experiments in its use in electric and open-hearth furnaces. Oxygen was used to remove unwanted silicon, manganese, and carbon from steel by oxidation. In the 1940s the steel produced had a high nitrogen content. Work was started toward the use of pure oxygen as a substitute for air in order to decrease the nitrogen content. In Austria work was conducted using the top blowing of oxygen through a water-cooled lance into liquid iron contained in a vessel similar in shape to the Bessemer converter. This process, known as the Linz–Donawitz (LD) process, now accounts for the main part of the world's steel output.

This process created the first need for on-site tonnage-scale oxygen plants in the early 1950s, which also led to the opportunity of producing large quantities of low-cost liquid oxygen, nitrogen, and argon by adding liquefiers to air separation plants. Use of oxygen has also gradually increased for melting and combustion of glass and for increasing the output from rotary kilns for calcining ceramics and minerals.

Chemical Processes

Many chemical processes use molecular oxygen, traditionally in the form of air as oxidant. There are several fully developed chemical processes which use oxygen in preference to air, with the advantages of faster reaction, consequent smaller reactor, gas-handling, and purge volumes, and improvements in yield and environmental aspects. These include processes for producing acetaldehyde, acetylene from hydrocarbons, titanium dioxide by the chloride route, oxyalcohol, and vinyl chloride monomer.

In 1953 Shell announced the development of a direct ethylene oxide process using pure oxygen, the advantages claimed over the air process being reduced purge loss, lower capital cost, improved yield, and a reduced gaseous effluent.

On a smaller scale oxygen is used as an oxidizing agent for water treatment. It is used as a substitute for air in reducing the basic oxygen demand of waste water streams and for bleaching in the pulp and paper industry.

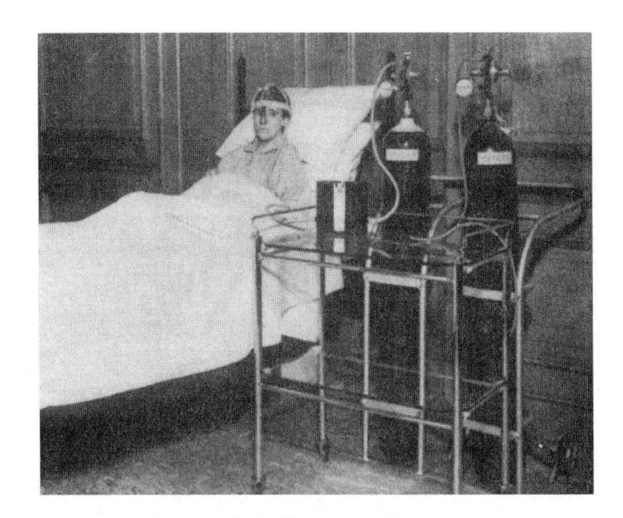

Figure 3.2.12. Oxygen therapy with nasal mask, 1939. From BOC Group (1986) with permission.

Medical Uses

Respiratory uses of oxygen include the following (Figs. 3.2.12 and 3.2.13):

- Inhalation during surgery under anesthesia
- Resuscitation for victims of drowning or accidents
- Life support during malfunction of the lungs or heart, or in conditions such as pneumonia, heart attack, chronic lung obstruction, and respiratory distress in new-born babies

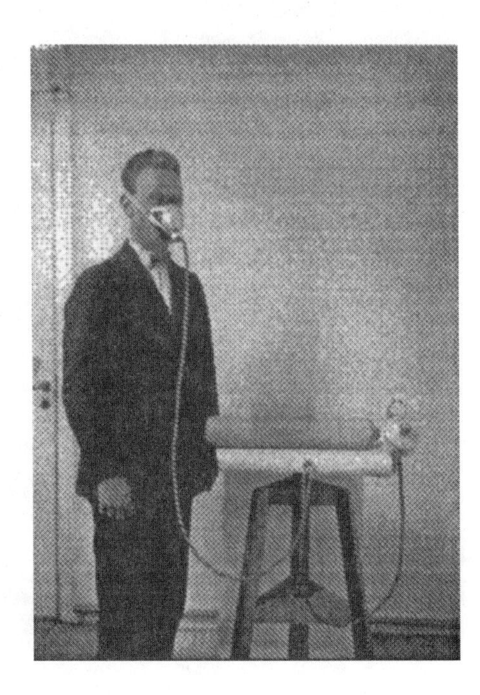

Figure 3.2.13. Breathing equipment, ca. 1925. Courtesy of the AGA Historical Picture Archive.

- Counteracting the effects or carbon monoxide and other poisons
- Breathing gas mixed with helium for deep-sea divers, rather than air
- Breathing at high altitudes, for example, in aircraft or on mountains

Drebbel's Submarine[10–12]

What do we mean by the discovery of a gas? Not the invention or the first time of use. Normally it is a process containing the following:

- Extraction and collection in a pure form
- Identification and separation from similar elements
- Demonstration of basic qualities
- Documentation and reporting to the trade press

That is why the Dutchman Cornelius Jacobzoon van Drebbel (1572–1633) is unable to take credit for the discovery of oxygen. At the beginning of the 17th century he extracted and used the gas in a way that was as remarkable as it was elegant. Van Drebbel is usually credited with building the first true (oar-powered) submarine in 1620 (Fig. 3.2.14).

Drebbel was a typical inventor and is remembered for mercury fulminate, an explosive, and for a temperature-regulated incubator. He only engaged himself in chemistry for a short time during which he extracted oxygen for an underwater boat that he had invented. Oxygen-enriched air provided the ability to remain under water for a long time.

Drebbel extracted oxygen by pyrolysis of saltpeter (Fig. 3.2.15), which he reported in his book *The Nature of Elements* in 1621. Around 1620, in a demonstration before King James I, a crew of 21 rowed Drebbel's underwater boat down the Thames from Westminster to Greenwich in 3 hours at a depth of 4–5 meters. Possibly the King was on board. Drebbel reportedly used a submarine which had "air within being freshened by a subtle spirit [oxygen?] which he had extracted from the atmosphere."

In 1876 an English naval officer, Henry Fleuss, first used oxygen as a breathing gas in a diving rig, thus creating a new application for the gas.

The Lunar Society[13–16]

The Lunar society was a remarkable conjunction of wide-ranging intellectual talent. It started in the 1760s in Birmingham, England, as a club for scientific and cultural thought.

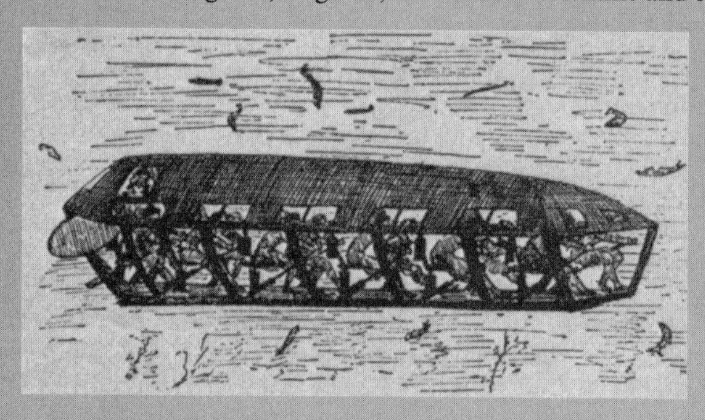

Figure 3.2.14. Drebbel's underwater vessel. From Naber (1934).

Figure 3.2.15. Pyrolysis of saltpeter. From Drebbel (1621).

Its meetings were of a very informal nature. Among the items on its agenda were the production and use of oxygen. The "Lunatics" usually met in industrialist Matthew Boulton's home, the Soho house, or in the home of James Watt in Harper's Hill (Fig. 3.2.16).

The meetings were held each Monday evening at the full moon (Fig. 3.2.17). Members assembled at 2 in the afternoon, dined together, and dispersed at 8. There was nothing really significant or mystical about the choice of date; it was a purely practical consideration, in that the members were easily able to travel home by moonlight after their meeting, and some had many miles to go. The members of the club were all eager to debate such matters as poetry, art, religion, music, and natural science with open minds. From these creative seances were born many of the ideas that contributed to the Industrial Revolution, as well as its financial sponsoring. Among those who attended the meetings during its most active period from the early 1780s were the following:

- Joseph Priestley (1733–1804), minister and scientist (member from 1780 until his departure for the United States in 1791)
- Matthew Boulton (1728–1809), industrialist
- James Watt (1736–1819), engineer and scientist
- Erasmus Darwin (1731–1802), physician, grandfather of Charles Darwin and Francis Galton
- Thomas Day (1748–1789), social reformer
- Richard Lowell Edgworth (1744–1817), educationalist and wealthy land owner
- Samuel Galton (1753–1832), merchant and gunsmith
- James Keir (1735–1829), industrial chemist

Continued

Figure 3.2.16. James Watt's home in Birmingham. The Lunar society met in the homes of Watt and Boulton. From Smiles (1865).

Figure 3.2.17. A Lunar Society meeting in the home of James Watt. From Thurston and Hirsch (1880).

Figure 3.2.18. (a) Wedgewood, (b) Boulton, and (c) Watt, prominent members of the Lunar Society. From Ålund, (1875).

- William Murdoch (1754–1839), engineer
- William Small (1734–1775), mathematician and philosopher
- Jonathan Stokes (1755–1831), botanist
- Josiah Wedgewood (1730–1795), famous potter, also grandfather of Charles Darwin
- John Whitehurst (1713–1788), clock- and instrument maker
- William Whitering (1744–1799), physician who introduced digitalis

Because every member could bring a guest, many distinguished visitors came to the assemblies, among them the builder John Smeaton, the explorer Joseph Banks (President of the Royal Society 1778–1820), the astronomer William Herschel, and many writers, poets, and artists. This contributed to the rise of the scientific culture associated with the early Industrial Revolution. The Lunar Society continued to at least 1807, but was not very active after the turn of the century.

Short Biographies

Jean-Baptiste Boussingault (1802–1887) was a many-sided French statesman and agricultural chemist. He was a professor at Bogotá, Lyon, and Paris, and made fundamental studies of the nutrient intake of plants. He developed the barium oxide method of industrial production of oxygen from air.

Carl Linde (1842–1934) was a German pioneer in the field of refrigeration. He developed the ammonia compressor in 1874 and started manufacture in Wiesbaden in 1879. From 1890 he was a professor at Munich Technical University, where the technology for the industrial extraction of liquid air was developed in 1895. His first air separation column came in 1902, followed by the double column in 1910.

Georges Claude (1870–1960) was a French engineer who invented dissous gas in 1896. He developed the piston expansion machine for condensing air in 1902. Together with Paul Delorme, he founded l'Air Liquide in the same year.

References

1. Anonymous. (1911). The manufacture of oxygen. A visit to British Oxygen. *Acetylene* **1911** (November):252–257.
2. Gardner, J. (1986). One hundred years of commercial oxygen production. *BOC Technology Magazine* **5**:3–15.
3. Hocking, M. B., and Lambert, M. L. (1987) A reacquaintance with the limelight. *Journal of Chemical Education* **1987**(4):306–310.
4. Greenacre, D. (1986) *Magic Lanterns*. Aytesbury, UK: Shire Publications.
5. Hoffmann, D., and Junker, A. (1982). *Laterna Magica. Lichtbilder aus Menscenwelt und Götterweld.* Berlin: Fröhlich & Kaufmann.
6. Gilbert, D. L. (1981). *Oxygen and Living Processes. 1: Perspective on the History of Oxygen and Life.* Berlin: Springer-Verlag.
7. Miller, M. (1986). The growth of applications for industrial gases. *BOC Technology Magazine* **5**:26–37.
8. Longden, I. P. (1980). Today's applications for industrial gases. *Steel Times* **1980**:796–801.
9. Winfield, G. (1977). Oxygen Today And Tomorrow. In *Proceedings of the 1st BOC Priestley Conference*, pp. 15–38.
10. Drebbel, C. J. (1621). *Van de Natuere der Elementen.* Rotterdam.
11. Naber, H. A. (1934). *De ster van 1572 (Cornelius Jacobszoon Drebbel).* Amsterdam.
12. van Spronsen, J. W. (1977). Cornelis Drebbel and Oxygen. *Journal of Chemical Education* 54(3): 157.
13. Smiles, S. (1865). *Lives of Boulton & Watt.* London: John Murray.
14. Thurston, R. H., and Hirsch, J. (1880). *Histoire de la Machine a Vapeur.* Paris: Libraire Germer Balillière et Co.
15. Lord Ritchie-Calder. (1982). The Lunar Society of Birmingham. *Scientific American* **246**(6):108–117.
16. Schofield, R. E. (1963). *The Lunar Society of Birmingham.* Oxford: Oxford University Press.

3.3. NITROGEN: "ROTTEN AIR" AND ITS CHALLENGES

Fixation of the nitrogen in the atmosphere is one of the greatest challenges confronting the ingenuity of the chemist

William Crookes (1898)

Background: Nitrogen and Saltpeter in Agriculture[1,2]

Fertilizers have been used in agriculture since time immemorial. The ancient Egyptians used sediment from the Nile, and in the *Odyssey*, Homer indicates that the Babylonians understood the value of fertilizer. Until the discovery of nitrogen, the use of fertilizer was based on empirical knowledge. Soil fertility was determined by its humus content.

Nitrogen gas was independently discovered by Joseph Black's student David Rutherford in Scotland and by the Swede Carl Scheele. Scheele called it the "rotten air" as opposed to the "fire air" (oxygen). Claude Louis Berthollet demonstrated in 1784–1785 that nitrogen existed in living plants and animals and that ammonium nitrate consisted of nitrogen and hydrogen. The contribution by the enormous nitrogen reservoir in the atmosphere to the nitrogen content of plant life was first discussed by Jean Antoine Chaptal, and in 1790 he introduced the name nitrogen ("the saltpeter builder").

Figure 3.3.1. Sir William Crookes. From Weeks (1935) with permission from the *Journal of Chemical Education*, copyright © 1935, Division of Chemical Education, Inc.

Malthus and Crookes

In a famous speech to the British Association for the Advancement of Science at Bristol in 1898 relating to the "wheat problem," the eminent chemist Sir William Crookes (Fig. 3.3.1) put a challenge before his scientific colleagues. The background was the fact that the natural fertilizers in current use were running out. Within 20–30 years there would be famine throughout the world if the production of nitrogen fertilizing agents could not be increased. Fixation of nitrogen in the air was a matter of life or death for generations to come. Sir William ended by saying, "It is the chemist who must find a solution to this threat to mankind. Fixation of the nitrogen in the atmosphere is one of the greatest challenges confronting the ingenuity of the chemist."

Crookes knew it was possible to synthesize ammonia on a large scale if a cheap enough source of electricity could be found. In 1892 he had followed the example of Henry Cavendish in the 1780s and Humphrey Davy in 1800 by conducting a public experiment on "the flame of burning nitrogen." He showed that, upon passing an electric arc, air caught on fire, producing nitrous and nitric acids.

Crookes's challenge was not the first of its kind. Exactly a century earlier, in 1798, there had been prophecies of overpopulation and famine. That prophecy, which in part became a reality, was made in England by the priest, professor of history, and national economist Thomas Malthus (Fig. 3.3.2) in his book *An Essay on the Principle of Population*. According to Malthus, population increases faster than food production. In this way, living standards fall below subsistence level, whereby further

Figure 3.3.2. Thomas Malthus. From the *Nordisk familjebok* (1912).

growth in population was restricted by famine, disease, and war. His theory, "Malthusianism," aroused considerable attention and contributed to interest being taken in plant enrichment.

During the first few years of the 19th century chemists began to study the factors affecting the nutritive intake of plant life and the fertility of soil. The Frenchman Theodore de Saussure discovered that a large part of plant substance was derived from carbon dioxide and water and a smaller part was derived from the soil.

Although Malthus did not even know that nitrogen was a plant nutrient, Crookes knew it all too well, the more acutely in view of contemporary crop failures. More than 50 years before, John Lawes and Henry Gilbert had clearly demonstrated the relationship between grain yields and fixed nitrogen.

Guano and Saltpeter from Chile

Shortly after Malthus published his prophecies, the German explorer Alexander Humboldt made a journey to South America. In 1802 he reported that in Peru he had discovered large quantities of guano, a high-grade fertilizer, which contained potassium nitrate, and which had been used since the 17th century for making gunpowder. Humboldt's discoveries led to the export of guano to Europe, which began in 1810 and reached its peak in 1856.

In the desert regions of Chile, Thadeus Haenke discovered huge deposits of caliche, which could be used for the extraction of sodium. The final product, called

Chilean saltpeter, contained about 15 percent nitrogen. The Spaniards began mining caliche in 1813, and export to Germany commenced in 1825, when a shipload arrived in Hamburg. Because it was not known what the cargo could be used for, it was thrown overboard. From 1830 until the 1920s, Chilean saltpeter was the predominant source of commercial fertilizer in Europe, together with ammonium salts, which from the 1860s were obtained as a by–product of the coke industry.

Despite a considerable increase in agricultural production per unit of surface in Europe resulting from the import of saltpeter, food supply was still insufficient. The rural population diminished as many left the land to work in industry. Food was imported from, for example, the United States and Ukraine, yet even so, large numbers of young people were forced to emigrate to North America during the latter part of the 19th century.

Modern Western agriculture is, to a great part, the result of scientific discoveries made during the 19th century. De Saussure's work was followed up by Jean Baptiste Boussingault, who was the first to make field experiments relating to the intake of nitrogen by plants. The German biochemist Justus von Liebig showed that green plants were dependent upon certain nutritional elements. The soil content, primarily of nitrogen, potassium, phosphorus, and various mineral elements, is what determines the outcome of a harvest. It was soon discovered that to be usable as a fertilizer, nitrogen had to be in the form of nitrates or ammonium salts in order for plant roots to be able to draw benefit from it. A quantity of nutritional elements disappeared with the plants at harvest and therefore further nutrition had to be applied to the soil.

At the end of the 19th century ammonia was being recovered in coke oven plants and gas works as a by–product of the distillation of coke. The ammonia sulfate produced was used as fertilizer. The transformation from ammonia to nitrate takes place by a microbiological process known as nitrification, which was described at about the same time in the 1880s by the Englishman Robert Warrington and the Russian Sergei Vinogradsky. Certain microorganisms existing in leguminous plants such as peas, soy, and clover, as well as in blue–green algae, can utilize the nitrogen in the atmosphere. John Bennet Lawes and J. G. Gilbert discovered in the middle of the 19th century that these plants can provide the soil with a large infusion of nitrogen. By crop rotation, such as using autumn–sown grain and root crops and spring–sown grain and clover followed by a fallow period, the nitrogen supply to the soil could be balanced without resorting to considerable additional fertilization.

Production of Fertilizers[3–5]

The solution that Crookes sought was in fact already under way. In 1895, Carl Linde, William Hampson, and Charles Tripler, working independently, developed industrial techniques for producing liquid air, although a number of problems remained. It was not yet possible to separate nitrogen and oxygen. Also, nitrogen was inert and would only with difficulty combine with other elements.

During the first years of the 20th century, a number of chemists in Europe and the United States were working on the problem of supplying nitrogen; they arrived at three solutions: the electric arc process, the cyanamide process, and the synthetic ammonia process.

1. *The electric–arc process* was based on the production of nitrogen dioxide from atmospheric nitrogen and oxygen by electrical arc discharges. The reaction, already noted by Cavendish and Priestley in the 1780s, takes place naturally during lightning discharges. In 1895, William Rayleigh in England used it to remove nitrogen and oxygen from air in his experiments to isolate noble gases. Industrial attempts using the method were made by the Americans Bradley and Lovejoy, who started a factory at Niagara Falls, New York, in 1899. This closed down in 1903 following financial failure.

In 1903 in Norway Kristian Birkeland and Samuel Eyde succeeded in using cheap hydroelectric power to create an economically competitive process. Nitrogen dioxide was transformed into a calcium nitrate called Norwegian saltpeter. In 1905 the two men founded Norsk Hydroelektrisk Kvaelstof AS, which is currently one of the largest industrial companies in Norway. A similar process was developed some years later by the German Otto Schönherr at BASF, and this company began manufacture in 1907 at Fiskå, Norway.

2. *The cyanamide process* was developed beginning in 1895 by the Germans Adolf Frank and Nicodemus Caro. They discovered that, when mixed with special substances, certain carbides are capable of fixing nitrogen if steam is led over the mixture. In 1902 they used, for the first time, Linde's newly developed single column in their prototype plant in Piano d'Orta, Italy. This effort was a failure, and not until 1907 when their assistant, Fritz Rothe, optimized the process could they start the first plant on an industrial basis in Odda, Norway. Calcium cyanide became an artificial fertilizer under the name of nitrolime, and was the most important source of nitric acid until 1917, when the method for catalytic oxidization of synthetic ammonia developed by Wilhelm Ostwald and Eberhard Brauer took over.

3. In 1904, Fritz Haber (Fig. 3.3.3), a young professor at the Technical University in Karlsruhe, began developing a method for synthesizing ammonia,

Figure 3.3.3. Fritz Haber, Nobel Laureate in Chemistry, 1918, inventor of the first method for synthesizing ammonia. Courtesy of the Nobel Foundation. © Nobelstiftelsen.

sponsored by Austrian businessmen. Aided by Walter Nernst's theoretical studies and practical research into catalysts, in 1908 Haber arrived at a process having an efficiency of about 8 percent. After several years of collaboration with Carl Bosch at BASF, a factory was started at Oppau in 1913. This new process came at just the right moment for the German preparations for war. Shortly after the outbreak of World War I in 1914, the Allies cut off Chilean nitrate shipments to Germany, believing this would surely decrease their enemy's war-making power. German industry countered by accelerating the construction of Haber process plants, and synthetic ammonia production grew from 9000 tons per year in 1913 to 95,000 tons per year in 1918 at war's end.

This was a determining factor that allowed Germany to continue the war for as long as it did. During the opening stages of the war, the Germans also aided their saltpeter situation by capturing 300,000 tons of sodium nitrate at Antwerp.

The process of producing synthetic ammonium hydrate was to be the definitive solution to the nitrogen problem. In 1927, Georges Claude in France presented a high-pressure modification of the Haber–Bosch process, achieving an efficiency 10 times that of the original process. By the 1960s ammonium nitrate containing 35 percent nitrogen was being used as a fertilizer and as a raw material for explosives on a large scale. Since then, liquid fertilizer containing 82 percent nitrogen has come onto the market.

The development of the Haber–Bosch process for synthetic ammonia remains one of the most important achievements in the history of chemical process development. When the first such plant came on-stream almost all raw material for the world consumption of nitrogen–based fertilizers came either from Chile saltpeter or coke oven gases. The uncoupling of the production of ammonia from the supply available in coke oven gas was a milestone. It started a search for raw materials and chemical processes and showed the need for countries to consider alternative raw materials and develope new process technologies to meet their demands for chemical end products.

The principal hydrogen-manufacturing processes that are now in use for synthetic ammonia production are the hydrocarbon (natural-gas) process and the hydrocarbon partial oxidation process. The M. W. Kellogg Co. developed a "reduced energy ammonia process" by examining every stage in the process chain for possible energy savings. In spite of sophisticated design improvements, making ammonia is still very energy-intensive. Therefore areas without access to supplies of coal, natural gas, or petroleum have severely lagged behind the rest of the world in ammonia production.

Ammonia production will continue to fill a need in agriculture. However, if N_2-fixing bacteria could be manipulated into forming direct relationships with crops other than legumes, this would provide fixed nitrogen at no cost in depletion of fossil fuels.

Other Industrial Applications for Nitrogen (see Fig. 3.3.4)[6]

Just after World War II nitrogen in its free form had no use other than that in gas–filled light bulbs. Some decades later, in addition to chemical and metallurgical uses, a large amount was used for deoxygenating and freezing. Today in some

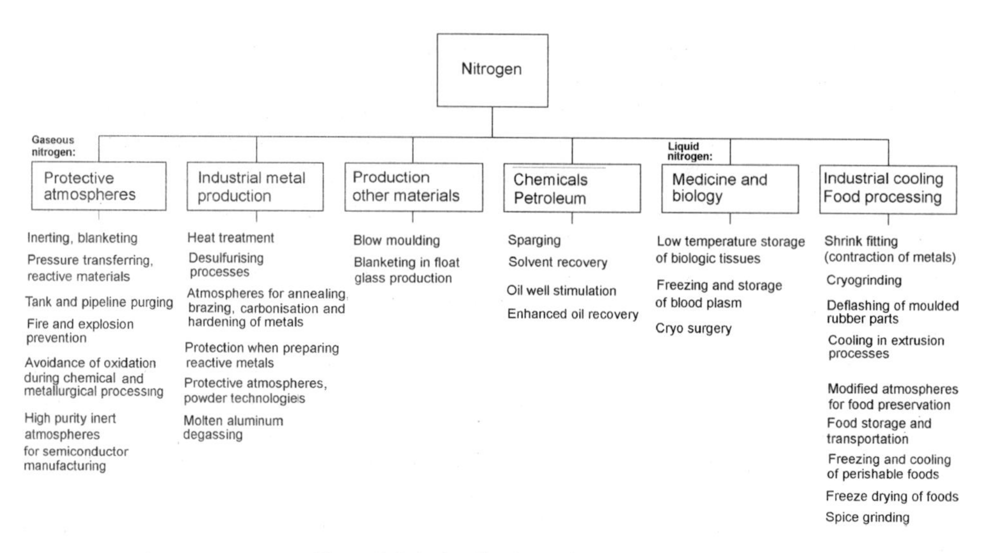

Figure 3.3.4. Applications of nitrogen.

countries industrial gas companies sell more pure nitrogen than oxygen and the yearly world consumption of pure nitrogen is greater than 100 million tons.

The largest uses for nitrogen are in inerting operations in the petrochemical, the oil, and the steel industries. Typical applications in the oil and the petrochemical industries include inerting of storage tanks and reactors and purging of tanks and pipelines. The oil industry also uses nitrogen for enhanced oil recovery.

The iron and steel industries use nitrogen for cooling and purging the charging mechanisms in high–pressure blast furnaces and for purging furnace gas–recovery systems. The metal working industries employ nitrogen in heat treatment processes which require special atmospheres designed to be inert or to react with the surface of the metal in some way. For many years industry has used gases generated by combusting natural gas or liquid petroleum gas with air.

The inertness of nitrogen is also employed in the glass industries. Most sheet glass is produced by the float–glass process, where a wide ribbon of liquid glass floats on a long bath of liquid tin. Oxidation is prevented by inerting the bath with nitrogen mixed with approximately 4 percent hydrogen. Electric lamp bulbs are filled with nitrogen to protect the high–temperature tungsten element from oxidation. Long-life bulbs are filled with mixtures of superinert argon and krypton.

Many food products are spoiled when exposed to air and moisture. Packaging in dry and inert nitrogen prevents spoilage and extends shelf life. In gas packaging, mixtures of gases such as carbon dioxide and oxygen are used to inhibit bacterial growth and maintain the red color of beef. Nitrogen freezing of foodstuffs, introduced at the beginning of the 1960s, meant a considerable improvement in quality and has a rapidly expanding market. Liquid nitrogen is also used as a coolant for industrial applications in which physical changes in a material are required, such as brittleness and shrinkage, or to introduce a large amount of cooling at exactly the moment it is required.

One of the fastest growing applications for nitrogen both now and in the foreseeable future is its use as an inert, high-purity atmosphere in the semiconductor processing industry.

Environmental Problems of Nitrogen[7]

The fact that unlimited quantities of atmospheric nitrogen could be used as a raw material for ammonium-based products brought about the hoped-for growth in the production of grain until the 1950s, but then one began to hear of a dead end. What had happened?

During the 1950s there was, throughout Europe, a trend to depart from rotation of crops and aim at continuous cultivation of one type of crop in a highly mechanized and rationalized one–product factory. The result was increased use of artificial fertilizers as natural nutrients rapidly disappeared from the soil. Too high a concentration of certain nutrients and an impoverishment of others, combined with a high saline content, meant a wiping out of the natural microbes in the soil. The highly developed, quick–growing crops showed a weakening of their cell structures and had to be force–fed with nutrients. The result was lack of resistance to attack, which had to be countered with poisonous spraying.

Nitrogen, converted to nitrates, now overfertilizes the surface water, creating deleterious conditions for marine life and is also transmitted to the subsoil water. Because the automobile engine operates by a form of the Birkeland/Eyde process in miniature, large quantities of nitrous oxides are released in areas of high traffic density. In some parts of the European continent, the releases of nitrogen from vehicles and agriculture alone are sufficient to fertilize the earth, and are a threat to animal life. Too much nitrate in drinking water can cause a blood disorder in babies younger than 3 months, called blue–baby syndrome because it colors the infant's lips and body blue. The cause is the conversion by bacteria of nitrate into nitrite (NO_2^-), which is taken up by the hemoglobin in the baby's blood instead of oxygen.

Nitrates from agriculture and phosphates from industrial and domestic sources also act as fertilizers for aquatic plants. If rain washes nitrates out of soil into streams, rivers, lakes, and then into the sea in large quantities, they can increase the growth of algae and other aquatic plants. This enrichment, which is called eutrophication, may change the balance of aquatic plants and animals so drastically that a particular species may be wiped out. The organisms that survive may grow so well that they clog waterways.

Algae, especially green algae, respond quickest to eutrophic conditions. When they grow rapidly on the surface they prevent light from reaching submerged plants, which may die as a consequence. Bacteria decompose the remains of any plants, algae, and animal that sink to the bottom. The process uses up valuable oxygen and a vicious circle develops, drawing in all forms of aquatic organisms, until rivers, ponds, and lakes become devoid of life.

The Nitrogen Challenge

The world's population is growing at a rate of 2 percent per annum, and grain production is increasing by 3 percent. In the industrially developed countries there is a large grain surplus, which increases the demand of farmers for productivity and price competition. This also reduces the chance of an ecologically balanced agriculture without a nitrogen surplus or spraying.

The new challenge is to produce grain for the world's population while the same time reducing the effect of agriculture on the environment. Once again there is a demand for ingenuity and perspicacity, this time in the search for a grain crop that can take nitrogen directly from the atmosphere. One possibility is that of gene manipulation and the development of new microorganisms. The problem is greatest in industrial countries. Rice, the staple diet of half the population of the world, can be grown without the need for artificial fertilizers because rice fields acquire their nitrogen from the sediment and fertilizing agents in the water. Furthermore, the blue–green algae and the soy that is planted between the rice plants contribute to the fixation of nitrogen in the soil.

History repeats itself. Once again the problem is one where nitrogen plays a major part in destruction or survival. Science and modern agriculture face the challenge of finding a method to feed the 10,000 million people that are expected to inhabit this planet by the end of the 21st century. This cannot be achieved by a thoughtless exploitation of nature that ruins the prospects for coming generations.

Nitrogen as a Raw Material for Gun Powder

Solutions of nitrogen have had technical uses in Europe since the 13th century. Saltpeter (potassium nitrate) was known in China as long ago as 600 BC, but the name of "salpetrae" or "salt of the rock" was first used by the Arabian alchemist Jebir (Geber) around 850. Writings of Arabic origin describing the preparation of aqua regia, or royal water, and aqua fortis, or strong water, were published about 1310 by a Spaniard who also called himself Geber, after his predecessor.

The invention of gunpowder, consisting of saltpeter, sulfur, and coal, came from China toward the end of the Tang Dynasty (about 800 AD), and was written about in Europe by Albertus Magnus around 1250. The development of weaponry during the 12th century increased the need for saltpeter. Based on the need for protection, different areas found it necessary to create their own "saltpeter plantations," centers of decay that resembled today's compost piles in which fertilizer and household waste are mixed together with dry and rich earth, ashes, quicklime, and unslaked lime. After about 2–3 years saltpeter could be extracted from such a "plantation" (Figs. 3.3.5 and 3.3.6).

During the 18th and 19th centuries, peasant farmers in Scandinavia were forced by the crown to create saltpeter beds. Commercial fertilizer is sometimes the raw material in the terrorist bombs of today.

Figure 3.3.5. Saltpeter plantation (Lazarus Ercker, ca. 1550).

Figure 3.3.6. Nitre beds/nitric acid preparation (Lazarus Ercker, ca. 1550).

Short Biographies

Thomas Malthus (1766–1834) in 1798 published his *Essay on the Principle of Population*, which claimed that population growth would be limited by an inadequate food supply. In fact, the global food supply will not meet the requirements of the world's population in the present century if the area of arable land remains constant. Unfortunately, in the present world economy, increasing the arable land area has harmful consequences on the environment. Thus Malthus's theories are still valid.

Sir William Crookes (1832–1919) was an English physicist and chemist who discovered thallium in 1861 and invented the radiometer and spinthariscope in 1874. He founded *The Chemical News* in 1859, which he edited until his death. He lost scientific respect as a result of his beliefs in spiritism during the 1870s, but was knighted in 1897.

Justus von Liebig (1805–1873) was a German professor of agricultural chemistry at Giessen. He carried out pioneering research into the way plants absorb nutrients, and helped provide the impulse to the production of artificial fertilizer.

Fritz Haber (1868–1934) was a German chemist who won the Nobel prize in 1918 for his method of producing synthesized ammonium nitrate from nitrogen and hydrogen. The award was strongly criticized due to his contribution to the development of chlorine gas, which was used during World War I.

In the early 1920s Haber resumed his chemical research by assembling a group of scientists at the Kaiser Wilhelm Institute in Dahlen, where he taught. After Hitler came to power in 1933, Haber's Jewish background and failing health led him to leave Germany. In England he met the Jewish chemist Chaim Weitzmann, who offered him a position with a newly established research institute in Palestine, but Haber never went there and died shortly thereafter. He donated his library to the Kaiser Wilhelm Institute.

Carl Bosch (1874–1940) was a chemist at BASF who developed the high-pressure method for synthetic ammonia production based on Fritz Haber's theories.

References

1. Smil, V. (1997). Global population and the nitrogen cycle. *Scientific American* **1997**(7):58–63.
2. Windridge, K. (1998). Getting out of a fix. *Chemistry and Industry* **1998**(19 October):849.
3. Stock, J. T. (1988). Fritz Haber (1868–1934) and the electroreduction of nitrobenzene. *Journal of Chemical Education* **65**(4):337–338.
4. Craig, P. (1984). Mankind in peace, the fatherland in war. *New Scientist* **1984**(February 2): 15–17.
5. Saull, M. (1990). Nitrates in soil and water. *New Scientist* **127**:U1–U4.
6. Miller, M. (1986). The growth of applications for industrial gases. *BOC Technology Magazine* **5**:26–37.
7. Arena, B. J. (1986). Ammonia: Confronting a primal trend. *Journal of Chemical Education* **63**(12):1040–1043.

3.4. CARBON DIOXIDE: SPIRITUS SYLVESTRE

In consequence of burning coal 'spiritus sylvestris' [carbon dioxide] comes into being. This spiritus, which was formerly unknown and cannot be kept in vessels, and cannot be converted into a visible form, I call by the new name 'gas'.

Johann Baptista van Helmont (*Ortus med.*, 1656)

Background: Spiritus Sylvestre

During ancient times, certain places attracted attention due to the bubbling and "living" water that came out of the ground. The bubbles were of carbon dioxide, which, in water, creates carbonic acid. This acid dissolves certain mineral salts from the bedrock and creates mineral water. Myths were created relating to the "healing powers" of these mineral waters, and the salty, sour taste was appreciated for its thirst-quenching abilities. At many of the springs, temples were built to Asclepius, the god of healing. The sick were healed at these spas by cures which in many ways resemble today's treatment of certain digestive disturbances: light diet, taking the waters, baths, and exercise.

Belief in the healing powers of such waters was gradually abandoned after the fall of the Roman Empire. In the 14th century, the spas of Central Europe came into use as medical baths. With the outbreak of the Thirty Years War, journeys to spas became hazardous and few could afford them. The number of visitors decreased, yet the demand for mineral water remained, now for internal usage. The water was shipped in glazed earthenware jars and sold throughout Europe. In time, health resorts achieved huge popularity and spas came into being in many places.

The alchemists of the Middle Ages prescribed the waters from the health springs as part of their cures. More and more scientists and natural philosophers attempted to analyze the composition of these mineral waters and the sparkling foam.

Johann Baptista van Helmont (1577–1644) investigated the gas he had discovered when charcoal was burnt and lime was slaked with an acid. He found the same gas in air and in the mineral water, and called it "*spiritus sylvestre*," the spirit of the forest. Robert Boyle's gas law was formulated as a result of his experiments with carbonated water. The German Friedrich Hoffmann discovered the oxygen qualities of mineral water in 1702, and was the first to make a systematic breakdown of the natural waters.

Carbon Dioxide: Its Discovery and First Uses

The first to begin the systematic investigation of carbon dioxide was Joseph Black. About 1757 he arrived at the conclusion that this material existed in a bound or fixed form in carbonates and weak alkalis. He therefore called the gas "fixed air" and discovered the lethal effect it had on animal life. In 1782 Antoine Lavoisier identified the "fixed air" as being a combination of oxygen and carbon. He called it "*gaz acid carbonique*" or carbonic acid gas.

The commercial exploitation of carbon dioxide began with attempts to produce artificial mineral waters (solutions of CO_2 in water), which were thought to have medical properties. About 1770 in Uppsala, Sweden, the chemistry professor Torben Bergman began taking an interest in the composition of water from certain springs based upon his own new and precise methods of analysis. The 35-year-old Bergman was seeking a cure for his illness, which included spitting of blood and trembling. He found that foreign mineral water gave relief, but the cure was too expensive. Therefore he began to examine these waters in order to be able to produce them artificially. Bergman's findings were first published in *The Chemical Investigation of Spring Water in and around Uppsala*. Complete descriptions of the method of manufacture appeared in 1773 and 1776.

Joseph Priestley, also experimented with the solution of carbon dioxide in water at the same time as Bergman. His results were published a year before Bergman's. Due to Bergman's theoretical knowledge, however, he had proceeded further and was therefore able to develop the technology to produce a copy of the natural mineral water.

Mineral Drinks

Mineral drinks with carbonic acid produced from lime and sulfuric acid were developed by the Frenchman Gabriel Venel in 1750 and the Englishman Richard Bewley in 1768. Carbonated water was introduced in the British Navy in 1764 as a supposed means of preventing scurvy. The brother of Admiral John MacBride had devised this unfortunate notion and it remained in service until 1795, during which time it was to see the death of many seamen. It was not until then that results of experiments done in 1748 by the Scotsman James Lind were put to the test. He showed that the citrus fruits lemon and lime contained an element that could counteract scurvy. This was later to be known as vitamin C. Soon carbonated lime juice was served regularly in the Navy, and ever since then British seamen have been known as "limeys."

Bergman's work was to result in the start of Swedish mineral water production. The first factory, Nya Mineralvattenfabriken, was started on a small scale in Falun in 1776 by one of his pupils, Johan Gottlieb Gahn. The Falun factory was soon followed by others. In England in 1775 John Mervin Nooth developed the technique for duplicating mineral water, and in 1781 Thomas Henry began manufacture in Manchester. Jacob Schweppe started a factory in Geneva in 1783 and opened a branch in London in 1793. Johan Meyer supplied artificial seltzer water (named after the Nieder Selters region in Germany) in Stettin from 1787 on, and in Paris Jacob Schweppe's partner Nicolas Paul manufactured mineral water from 1799. In the United States from 1807 the mineral water industry was developed by the Yale professor Benjamin Silliman in New Haven and Joseph Hawkes in Philadelphia.

Carbonated Water: Coca-Cola, Schweppes, Ginger Ale, etc.[1]

Up to the beginning of the 19th century, mineral water had been synonymous with health water. Over the years there was a change in the direction of consumption. Belief in the old springs declined, and health spas and "porridge trots" became

Figure 3.4.1. The mineral drink Sprudelsalt from Carlsbad (19th century).

Figure 3.4.2. The inventor of Coca Cola, John S. Pemberton (d. 1888). Courtesy of the Coca Cola Company.

old-fashioned. Health waters were replaced by pharmaceutical medicaments. At the same time, the old, well-known lemonades were being replaced by ready-made mixtures of carbonated water and aromatic substances such as herbal extracts, juices, and wines. Ginger Ale (1852) from Ireland and Coca-Cola (1886) and Pepsi Cola (1893) from the United States are examples of products that quickly became popular (Figs. 3.4.1–3.4.3).

The Industrial Revolution brought about a change in lifestyle. Now that more people were living in towns and cities, there was a need for considerable long-term storage of agricultural products. Cooling was a known and appreciated alternative to the current preserving methods, and there was a large market for the sweet, chilled drinks. The new prosperity created the need for other methods of cooling and producing carbonated water. Trade in ice, mainly from Sweden and Norway, to England and the United States was considerable during the latter part of the 19th century. At the same time cooling machines, under development since the end of the 18th century, were coming onto the market.

Liquid Carbon Dioxide[2,3]

At the beginning of the 19th century there was considerable interest in experimentation with the gases that had been discovered a few decades earlier. Liquefaction of gases by cooling and pressurizing was a challenge. Such experiments were

Figure 3.4.3. Coca Cola was introduced in 1886 and the first bottle prototype (shown here), similar to the present form, was designed in 1913 by a Swedish immigrant, Alexander Samuelsson. Courtesy of the Coca Cola Company.

attempted by Humphry Davy at the Royal Institute in London, who, in 1812, employed Michael Faraday, the son of a poor blacksmith whose genius was soon to exceed that of his master.

The industrial development of the gas really began with Faraday's experiments on the liquefaction of gases (Fig. 3.4.4). In 1826 he succeeded in liquefying CO_2 in a bent glass tube. Charles Thilorier and his colleagues experimented based on Faraday's results and produced larger amounts of liquid (Fig. 3.4.5).Thilorier made extensive experiments to find the properties of carbon dioxide and became the first to produce solid carbon dioxide, which on the evaporation of liquid carbon dioxide appeared as a white "snow." He observed that it disappeared by slow evaporation without first melting to form a liquid (i.e., it sublimed). In 1845, Faraday made larger quantities of liquid and solid carbon dioxide by means of a hydraulic pump. He also used solid carbon dioxide, in a mixture with ether, as a refrigerant for use in further gas liquefaction studies.

Liquefaction took place through the combined action of compression and cooling, and Adolf Martin Pleischl in Vienna (1844) used an air compressor in the

Figure 3.4.4. Faraday's equipment for the liquefaction of gases.

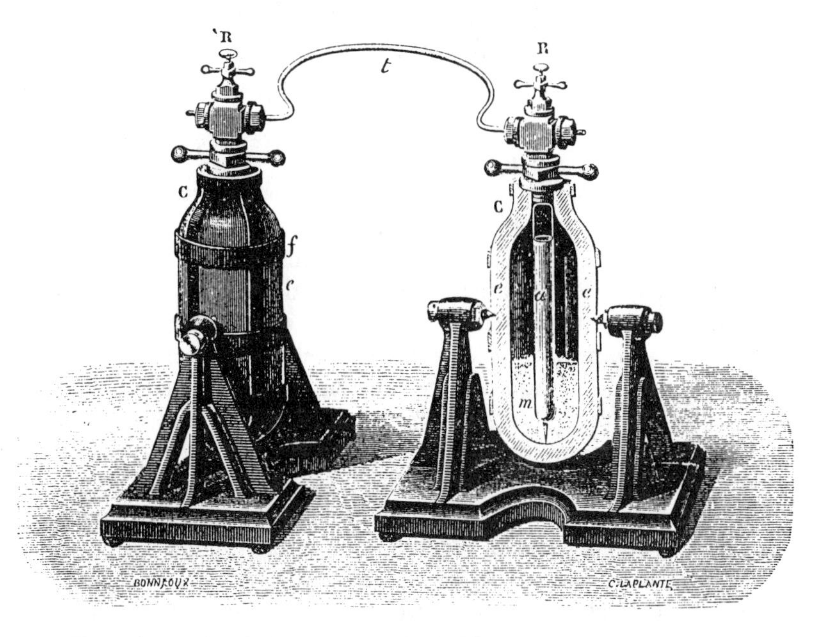

Figure 3.4.5. Thilorier's carbon dioxide liquefier for large-scale production.

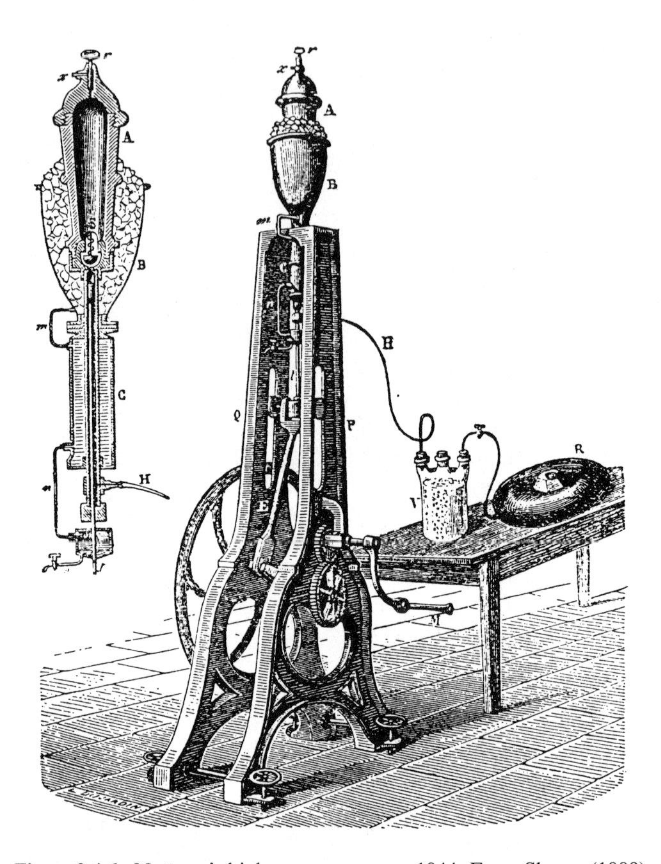

Figure 3.4.6. Natterer's high-pressure pump, 1844. From Sloane (1889).

liquefaction process. The first single-stage compressor for this purpose was constructed by his pupil, Johann Natterer, who succeeded in producing liquid carbonic acid in significant quantities for the first time (Fig. 3.4.6). Soon after, Natterer developed a mechanical compressor (a forerunner of the modern multistage type), which was used to make liquid carbon dioxide. The important commercial value was first realized by the far-seeing physicist and chemist Karl Friedrich Mohr, who about 1870 recognized the significant importance of liquid carbonic acid in the commercial and industrial world. However, for a considerable length of time the development and production of liquid carbon dioxide made little progress. In 1872 Natterer's compressor was further developed by Wilhelm Raydt in Hannover (Fig. 3.4.7).

An unexpected use for liquid carbon dioxide arose when a German warship sank in Kiel harbor. Different methods were investigated for lifting the vessel. Raydt offered to raise the very heavy anchor to the surface using a carbon dioxide-filled balloon. Further efficiency and capacity of the Natterer compressor was now required, and in the summer of 1879 he was able to produce 50 kilograms of liquid carbon dioxide. A balloon 3 meters in diameter was fastened to the anchor and was filled with the liquid carbon dioxide. Eight minutes later the anchor was lifted the 12 meters to the surface.

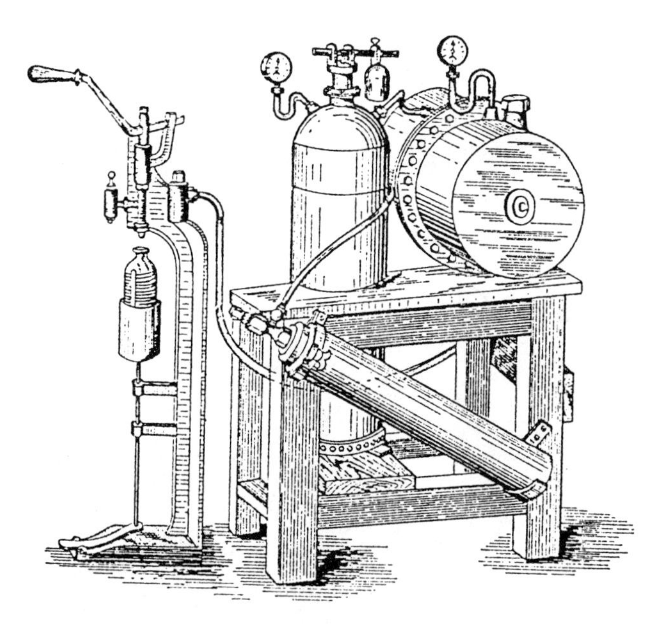

Figure 3.4.7. Carbonator developed by Wilhelm Raydt about 1885. From Goosmann (1906).

The success was much discussed and speeded up the development of applications for liquid gases.

The German Carbon Dioxide Industries

In 1878 Raydt introduced the industrial manufacturing of CO_2. Several production procedures were developed which were used by manufacturers specialized in manufacturing and selling the necessary technological equipment. The best known were the German companies Schütz and Maschinenfabrik Sürth GmbH.

Naturally appearing carbon dioxide had been used for a long time. Modern CO_2 production was based on the extraction of CO_2 from natural mineral water springs. A system of degassing equipment for waters with higher CO_2 content was patented by C. Rommenhöller and E. Leehrmann in 1887. The Sürth company took over the system, improved its design, and launched the manufacturing and delivery of various CO_2 production systems. For soda water manufacturing the carbon dioxide was at first produced from carbonates by decomposition of magnesite or limestone by action of dilute sulfuric or hydrochloric acid. The gas produced by decomposition was blown in distilled water to produce soda water, which could then be flavored to taste by using fruit juices.

The first factory manufacturing CO_2 from generator gas from lime works using an absorption method was built in Berlin by Kunheim and Co. in 1889. As early as 1875 the Schütz company manufactured equipment for obtaining carbon dioxide by fermentation for breweries; it was possible to trap 500 grams (300 liters) of concentrated CO_2 from 25 liters of wort.

Figure 3.4.8. Beer dispensing from the cellar with a carbon dioxide cylinder, circa 1890. From Ålund (1875).

The Extension of the Liquid CO_2 Business

In 1880 Frederich Alfred Krupp, head of the famous Krupp company in Essen, proposed to Raydt that liquid carbonic acid could be a means for compressing fluid steel. Raydt constructed a machine for him, and in 1881, the Chem. Fabriek Kunheim began using Raydt's method for producing liquid carbon dioxide, followed 2 years later by the Wolff & Camberg mineral water factory in Berlin. Raydt commenced his own commercial production in 1884.

The first quantities of liquid carbonic acid commercially used in the United States were imported from Germany in 1884. In the United States Jacob Baur developed technology for the large-scale production, storage, and distribution of carbonic acid. The company, Liquid Carbonic Acid Manufacturing Company, was founded in Terra Haute, Indiana, in November 1888 and is still the world's leading producer, now owned and run by Praxair. In 1889 Liquid Carbonic became the first in the United States to distribute compressed gas in steel cylinders.

As a cooling medium, carbon dioxide was first used in a cooling compressor by an American, Thaddeus Loewe, in 1866. The first commercial carbon dioxide compressors were developed by the German Franz Windhausen. From 1890, cooling machines of the Windhausen type were used on ships transporting foodstuffs, and were common after the turn of the century. Ammonia compressors eventually took over the market. These were cheaper to run, but more dangerous to the environment.

In the beginning, the soft drinks industry was the major customer for carbonic acid, but other applications were soon to appear. The production of soda using the Solvay method required large quantities of carbon dioxide.

Modern CO_2 Applications (see Fig. 3.4.9)

Solid CO_2

Ninety years after Thilorier's first production in Paris of solid CO_2 a factory was opened in Montreal, Canada, in 1924. The CO_2 was extracted from industrial gases (Gaudet's process) or obtained from lime kilns (Hayne's process). The Dry Ice Co. was founded in the United States in 1925, producing 150 tons per year. By 1937 the production was 11,000 tons per day! At first solid CO_2 was mainly used for the distribution of ice cream, and later was also used for transport of quick-frozen products.

Already in 1925 the Swiss had a process, Carba, by which the plastic snow obtained by expansion of liquid CO_2 at the triple point was compressed by kinetic energy, with no press being used. Carbon dioxide ice was introduced in Germany in 1927 by Eifel Trockeneis. In the United Kingdom dry ice manufacture began in 1930, and the Imperial Chemical Industries (ICI) became a major producer. Several freezers using solid CO_2 were put into use in Germany during 1935–1938.

Controlled Atmospheres

Before 1939, "gas storage" was practiced commercially only in the United Kingdom. Controlled atmosphere storage began later in the United States, with its commercial debut in 1942. In the first English "gas stores," fruit packages were made gas-tight and it was left to the fruit itself to create an atmosphere enriched in CO_2 for long storage. It was soon found that the results were better if the oxygen content of the atmosphere was reduced as the CO_2 content increased.

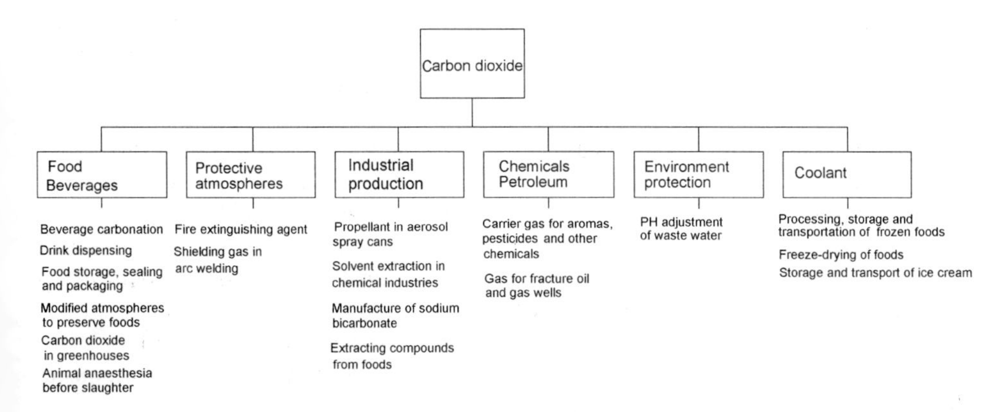

Figure 3.4.9. Applications of carbon dioxide.

Figure 3.4.10. Wrapping of dry ice blocks (Liquid Carbonic, 1958). From Stiegerwald (1988).

Carbon dioxide fertilizer was tried in greenhouses and is now in general use for increasing plant production.

Fire-Fighting Equipment[4]

From 1870, carbon dioxide was used in what were known as “fire grenades” and in patent fire extinguishers. In these, carbonic acid was created by crushing glass ampules containing sulfuric acid onto sodium bicarbonate (Fig. 3.4.11). In 1875 Dick’s Patent Fire Exterminator was designed having a sealed glass phial of sulfuric acid suspended above a solution of sodium bicarbonate. Upon pushing in a metal pin the acid bottle was broken and the acid mixed with the sodium bicarbonate, which expelled the solution. This was the basis of the modern soda–acid extinguisher.

Other extinguishers contained a mixture of acid and alkaline chemicals which reacted with each other to produce carbon dioxide gas. The gas formed kept the contents under pressure, so they were expelled when a tap was opened. This type of extinguisher, however, could leak gas and so might fail to work when required. A further development in the 1930s of the chemical reaction principle led to the foam extinguisher, in which sodium bicarbonate, with a stabilizer to strengthen the bubbles, and a solution of aluminum sulfate are mixed when required to form fill the bubbles and force the foam out of the extinguisher. The foam blankets the flames and extinguishes them by cutting off their contact with air.

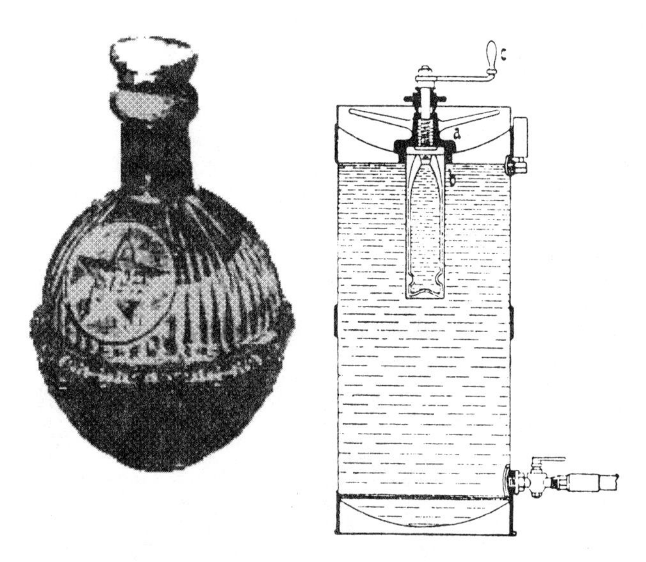

Figure 3.4.11. Carbon dioxide grenade and fire extinguisher, early fire-fighting equipment from the 1800s.

Extraction with Supercritical or Liquid CO_2[5]

Carbon dioxide can be used as a solvent in two different physical states: as a liquified gas and under supercritical conditions. Some of the recent developments use both states.

The first observations of the solvent power of liquid carbon dioxide and of substances dissolving in supercritical gases were made over a century ago. Since then various research ideas have led to practical processes and to the first commercial plants to exploit the solvent power of liquid and supercritical carbon dioxide. In 1861, G. Gore published his observations that camphor and naphthalene were soluble in liquid carbon dioxide. H. Buchner in 1906 published a major study of phase systems of carbon dioxide with many organic liquids. He found that acetone, ether, carbon disulfide, benzene, etc., were miscible with liquid carbon dioxide and that many other compounds showed some solubility. Salts were not soluble in liquid carbon dioxide.

Although much research work was done on supercritical systems, it was not until the 1930s that the first patents began to appear. A U.S. patent was issued to S. Pilat and H. Godlewicz for the use of supercritical carbon dioxide, and F. J. Horvath patented the use of liquid carbon dioxide to concentrate coffee, fruit juices, and essential oils.

In the late 1960s a semitechnical extraction plant was in use at Krasnodar in southern Russia. Their work using liquid carbon dioxide showed that many essential oils could be extracted in greater quantity and with superior quality than with steam distillation. Other experimental work showed that the aroma in tobacco could be boosted by an extract from tobacco waste. At about the same time the staff of the U.S. Department of Agriculture, Western Research Laboratory, in California were working on the extraction of flavors from many fruit and vegetable sources.

In the late 1960s work was done in the Nestlé Company's laboratories which led to the granting of a patent for the carbon dioxide extraction of freeze-dried extract

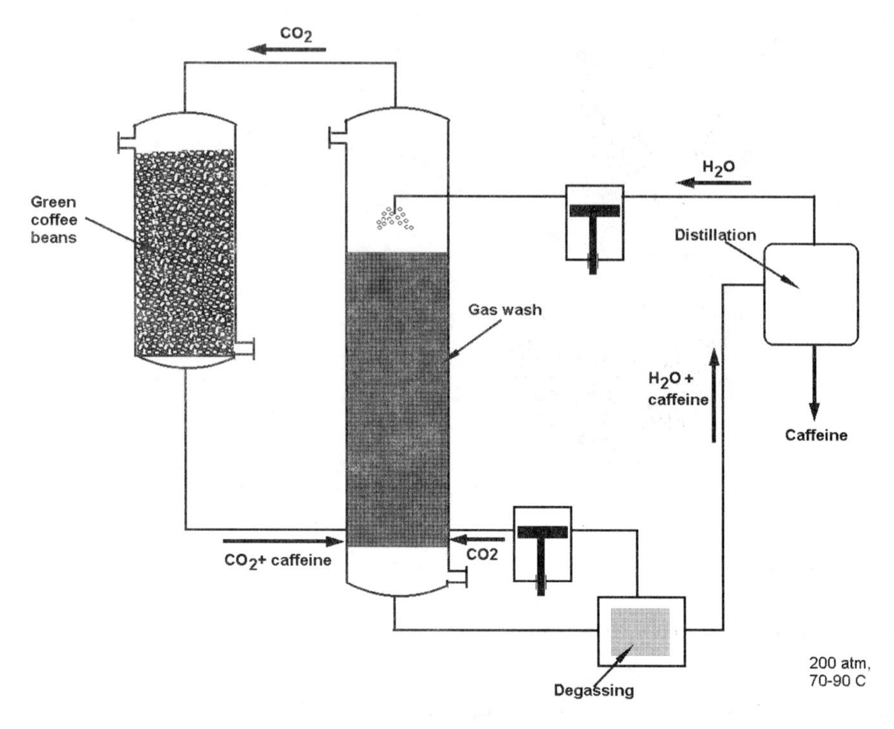

Figure 3.4.12. Sketch of extraction process for the decaffeination of coffee.

from coffee. Instant coffee powder was then given a richer aroma by the addition of this aroma extract. Since the 1970s supercritical fluid extraction has also been used for decaffeination of coffee (e.g., Kaffee Hag in Germany) and tea (Fig. 3.4.12). In the same way nicotine is removed from tobacco for low-nicotine cigarettes. Some successful attempts have also been made at extracting flavors used by the food and beverage industries. Brewers have long used this method for to extract the flavor of hops. Many new applications have appeared to extract flavor, color, and fragrance compounds from spices, herbs, fruits, and other materials. Cholesterol removal is another new market.

From the 1990s supercritical carbon dioxide has been used by Airco/BOC and Lummus Crest as a solvent to treat waste water.

Carbon Dioxide Production

Natural sources of carbon dioxide are found in many places where there are cracks in the earth's crust due to volcanic activity. One such region is in Germany east of the Rhine. It was here that Carl Gustaf Rommenhöller bought a spring called the Victoria-Mineralbrunnen in 1881. Sales grew rapidly and concessions were taken out on several sources in competition with other newly formed companies. As early as 1889, due to the high cost of distribution, Rommenhöller began to manufacture carbon dioxide in Berlin, from coke combustion. Nowadays carbon dioxide is produced from mineral oils by fermentation or combustion.

Baths and Cures

The medical baths of the Romans were of Greek origin. Aristotle's basic terrestrial elements—earth, air, fire, and water—had been united in the hot saline and gaseous waters. Health spas had grown up around such springs, to which the sick made pilgrimages to regain health. The origin of the theory concerning the healing effect of the waters, primarily of cold baths, can mainly be traced back to three Greek natural healers: (1) Asclepiades (124–40 BC) of Prusa, who first advocated cold baths; he founded the "Methodical school," and opposed the gluttonous and emasculating life of the Roman baths; (2) Antopius Musa (active 27 BC–14 AD), a freed slave who became renowned as personal physician to Caesar Augustus, whose ailments were cured by cold baths and raw fruit and vegetables among other things; and (3) Galen (129–199 AD), who investigated the physiological effects of health baths.

Following the Roman period, there was a lapse until the 16th century before the medical properties of the health waters again came under analysis. Such waters were considered as containing a secret life-giving acid, called by Paracelsus "*acidum occultum.*" The acid was actually carbonic acid, and the more of this the waters contained, the healthier and more revitalizing they were considered to be.

The importance of carbon dioxide in this connection is still the subject of discussion today. During a 20-minute carbon dioxide bath the body absorbs about 6 liters (12 grams) of gas. Because the carbon dioxide content regulates the rate of breathing, this can be stimulated. Carbonic acid also creates a considerable irritation of the skin causing the blood vessels to expand. The gas is deposited as small blisters under the skin, which then prevent loss of body heat. The mineral salts released by the carbonic acid probably have a greater importance. Depending on the concentration, certain mineral waters can regulate the osmotic pressure and in this way affect the workings of some of the organs.

The mineral elements of the waters can also be considered to have their greatest effect when one "takes the waters." In the 18th and 19th centuries many people suffered from iron

Figure 3.4.13. Spa in Bath, England. Courtesy of Eric Snook (Bath) Limited, Bath England.

Continued

deficiency due to changes in diet. Swedish springs often contain iron, and people suffering from lassitude were often prescribed to "take the waters." The feeling of renewed health following a sojourn at a spa perhaps depended not just on the mystical healing powers of the waters, but the other beneficial effects, such as environmental and climatic change, a new way of life, exercise, fresh air, a more realistic diet, and a friendly reception, probably meant equally as much.

Greenhouse Effect

In the 1890s the Swedish chemist Svante Arrhenius (Fig. 3.4.14) and the American P. C. Chamberlain came up with the so-called greenhouse hypothesis, whereby an increase in the combustion of coal would cause an increase in the content of carbon dioxide in the atmosphere, which would in turn cause an increase in the average temperature of the earth's surface. A twofold increase in the carbon dioxide content would mean a temperature increase of 4–6°C. The background to this theory is that certain gases in the atmosphere, the so-called greenhouse gases, function in the same way as the glass of a greenhouse. They allow the sun's rays to pass through, but the long-wave rays, the heat radiation, are reflected from the earth and are absorbed by the gases on their way out toward space. Water vapor makes a constant contribution to this effect, whereas the human contribution by way of gas releases has increased rapidly. The increase in temperature is due in part (about 50 percent) to the increase in the carbon dioxide content, whereas methane, Freons, and nitrous oxide make a smaller contribution. The devastation of forests makes a considerable contribution to this increase. Natural phenomena such as volcanic eruptions and high solar activity cause temporary changes (Fig. 3.4.15).

Figure 3.4.14. Svante Arrhenius, the first Swedish Nobel Laureate in Chemistry 1903. Courtesy of the Nobel Foundation. © Nobelstiftelsen.

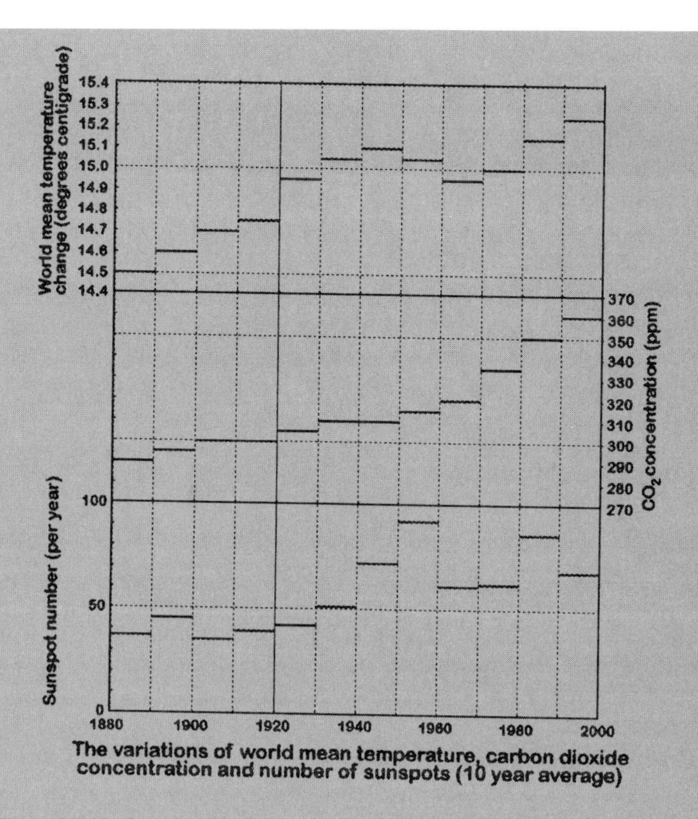

Figure 3.4.15. Some parameters which may contribute to the greenhouse effect.

Should the increase in temperature become too great, then the climate and vegetation zones will be repositioned and the sea level will rise. Over the past 120 years, the carbon dioxide content of the atmosphere has increased by 20 percent, which has brought about an average temperature increase of the earth's surface of half a degree. The sea level has risen by 15 centimeters, mainly as a result of volumetric expansion. The doubling of the carbon dioxide content which can be expected in about another 100 years would raise the temperature at the Equator by about 2°C and at the Polar Circle by about 8°C. The sea level would then be increased by about 50 centimeters and the hot and dry climate zones on the Equator would be displaced toward the poles.

Measurements of the carbon dioxide content in the atmosphere have been carried out in Hawaii and in Antarctica since 1958 and indicate a clear tendency. The limits for reasonable changes are indefinite, but lie above a doubling of the present carbon dioxide level. This means that the western Antarctic glaciers are being displaced and slowly melting and, as a result, the sea level will gradually rise by some meters. These predictions are very unsure, however, because considerable natural variations take place. As late as the 17th to the 19th centuries the planet experienced "a small Ice Age" with a lowering of the average temperature by 2°C as a result of considerable volcanic activity and low solar activity.

Short Biographies

Michael Faraday (1791–1867) was an English scientist best known for his discovery of electromagnetic induction. He paved the way for modern cooling and air-gas technology through his work on the condensation of gases.

Jacob Baur (1856–1917) was an American apothecary who founded the Liquid Carbonic company in 1888 and developed the technology for the production and distribution of liquid carbonic acid.

Carl G. Rommenhöller (1853–1913) was a German merchant who in 1881 purchased the carbonic acid spring Victoria-Mineralbrunnen near Ems, east of the Rhine. He built up the company of Kohlensäurewerke Rommenhöller, which was acquired by AGA in 1987.

References

1. Riley, J. (1958). *A history of American soft drink industry*. Washington DC: American Bottlers of Carbonated Beverages.
2. Thévenot, R. (1979). *History of refrigeration througout the world*. Paris: Institut International du Froid.
3. Wright, B. (1989). *Firefighting equipment*. Aytesburg, UK: Shire Publications.
4. Goosmann, J. C. (1906). *The carbonic acid industry*. Chicago: Nicherson & Collins.
5. Grimmett, C. (1981). The use of liquid carbon dioxide for extracting natural products. *Chemistry and Industry* **1981**(May 16):359–362.

3.5. ACETYLENE: THE GAS THAT TRIGGERED THE INDUSTRIAL GAS BUSINESS

From the brilliancy with which the new gas [acetylene] burns in contact with the atmosphere, it is, in the opinion of the author, admirably adapted for the purposes of artificial light if it can be procured at a cheap rate.

Edmund Davy (1836, in a report to the British Association)

Background

The first method of industrially producing acetylene was discovered quite by chance in 1892. The gas had been discovered about 50 years previously, but had only been regarded as a laboratory curiosity by scientists because the technical conditions for large-scale production did not exist. Following a very slow commercial start, a market was created very quickly during three periods (Fig. 3.5.1). The first was for lighting purposes from about 1897, when the gas competed with coal gas and oil gas. This lasted up until the 1910s, when it lost ground to the electric light bulb. From 1901 welding techniques were developed using oxyacetylene, a method that revolutionized engineering technology until the practical arc-welding technique was introduced after World War I.

Figure 3.5.1. Acetylene cylinders, 1907. Courtesy of the AGA Historical Picture Archive.

The period between the wars saw the chemical use of acetylene as hydrocarbon-based artificial materials and chemicals were derived using acetylene as a raw material. After World War II, however, ethylene took over the position as the hydrocarbon base. The market for acetylene as a welding and cutting gas is still rather large, but has remained stagnant since the 1970s. Few industrial products have experienced such sudden changes between times of prosperity and crisis. Exaggerated expectations were created, only to be followed by setbacks due to poor profitability, competition, and safety and toxicity problems. Acetylene has attracted much interest because of its properties; for example, its flame has a high heat content and gives a greater light intensity than any other gas. The gas also has a strong tendency to decomposition, and is suitable as a base for production of other chemicals.

The Discovery of the Acetylene Process

Acetylene (C_2H_2) was first produced by placing calcium carbide (CaC_2) in contact with water. Calcium carbide is prepared by heating coke and lime in a furnace to a temperature of 2000°C. One kilogram of carbide results in about 300 liters of acetylene. Today, the cracking method, or pyrolysis, of petroleum products is a cheaper method of producing acetylene than the calcium carbide method.

Working independently of one another, the Davy cousins Humphry and Edmund, professors of chemistry in London and Dublin, respectively, played a major role in the prelude to the acetylene industry. Humphry Davy discovered the electric arc in the early 1800s and used this method to separate different elements, including calcium and aluminum. Edmund Davy produced acetylene from calcium carbide in 1836 by heating winestone and charcoal. In a report he foresaw its use as a gas for illumination. Several other scientists produced calcium carbide without being aware of its composition.

Calcium carbide was first identified by the German Friedrich Wöhler in 1862. The French chemist Marcelin Berthelot gave acetylene its name in 1860 and examined its chemical qualities.

The Birth of Industrial Acetylene[1,2]

The first industrial acetylene production process was developed as part of attempts to produce pure aluminum in a Cowles-type furnace. The industrial magnate James Turner Morehead had a generator built to utilize the water power from an inherited water right in Spray, North Carolina. He built two cotton factories, but these did not utilize all the power generated. Then he built a factory with a furnace in order to test the newly invented method for aluminum manufacture. When this failed, Morehead employed a young Canadian, Thomas L. Willson (Fig. 3.5.2), in 1891. Willson had his own ideas about aluminum production and, together with Morehead's son, John Motley, who had studied chemistry at university, formed the Willson Aluminum Company at Spray, Michigan. An electric furnace of brick which had a coal floor was built. Instead of producing aluminum, they wound up with aluminum bronze.

On 2 May 1892 Willson attempted to produce metallic calcium from coal tar and burnt chalk, with the intention of then using the calcium to reduce the aluminum from its oxide. What he achieved was a hard, crystalline solid, which was cooled with water. A large amount of gas was created, which was assumed to be hydrogen gas. When another piece of the material was thrown into the water and the gas given off

Figure 3.5.2. Thomas Willson. From Anonymous (1898).

was lit, they found that it burned with a strong, bright flame. The material was tested by a Dr. Venable, John Motley's lecturer at the University of North Carolina. He declared the material to be pure calcium carbide. By pure chance, the furnace had contained calcium and coal in proportions that fell within the close limits whereby calcium carbide could be created. Willson then patented the method of manufacture and set up five plants for the manufacture of calcium carbide.

It took some time to find markets for their products. They found little or no demand for the aluminum bronze. They also had difficulties in producing high-quality calcium carbide, and there was no market for the acetylene. So the staff dispersed and some months later the works burnt down. Despite economic ruin, they resumed production of carbide. In 1894 the Electro Gas Company was formed, with financiers from New York. During the first phase, acetylene was only used by a coal gas works in Chicago for the enrichment of water gas. After the illuminating ability of the gas was demonstrated, others became interested and in 1896 the rights to use calcium carbide for lighting, heating, and energy were sold. Willson moved back to Canada and began manufacturing carbide there. The Electro Gas Company grew quickly. The establishment at Niagara Falls, the Acetylene Light, Heat and Power Co., with John Motley Morehead at the helm, was reformed on April 1, 1898, and became the Union Carbide Company. Marketing of carbide lighting for cycles had begun in 1897, and the first acetylene town lighting in the United States was installed in New Milford, Connecticut, in 1898. The use of acetylene for lighting spread rapidly throughout the world, helped by another technical novelty, hydroelectric power, which made industrial calcium carbide manufacturing possible.

The Introduction of Acetylene in Europe

At the end of 1894, Armin Tenner took the Willson calcium carbide process to Germany and used burners designed by the Englishman Vivian Lewes. At the same time, Lewes demonstrated acetylene lighting in England.

In France Henri Moissan (Fig. 3.5.3) of the Sorbonne University built a Heroult-type electric arc furnace to produce silicon carbide (carborundum). In December 1892 he discovered the manufacturing process for calcium carbide entirely independently of Willson's work, and Moissan's assistant, L. M. Bullier, succeeded in patenting the method despite both Frederich Wöhler and Willson having published similar findings. The French development was used by some of the first European carbide plants: Neuhausen, Switzerland (1895), three small French plants, and Werk Bitterfeld, Germany (1896).

From Moissan's laboratory, ideas for the use of acetylene spread throughout Paris. Acetylene lighting was shown for the first time at the Industrial Exhibition in August 1895. The main hall was illuminated by over 200 acetylene lamps. As a result of this introduction, many were to become interested in the uses for the gas. The company Societé du Gaz Acétylène was formed and pioneered the installation of lighting in many countries. In 1895, the Frenchman Henri Le Chatelier discovered that a flame from the combination of acetylene and oxygen produced what was the highest temperature then achievable, 3200°C. This encouraged Charles Picard to develop a welding torch using this gas combination in 1901.

Figure 3.5.3. Henri Moissan. From Weeks (1935) with permission from the *Journal of* Chemical Education, copyright © 1935, Division of Chemical Education, Inc.

Dissous Gas

Acetylene gas was easy to manufacture, but its transportation was a problem. The early producers naturally thought of storing and transporting the liquid form of acetylene in steel cylinders, but disastrous explosions occurred. As early as 1877, Louis Cailletet had produced acetylene in liquid form at a pressure of 40 bars. Marcelin Berthelot showed in 1883 that acetylene was explosive at pressures above 2 bars. Nonetheless Raoul Pictet in Berlin attempted to increase the quantity of acetylene in a given container volume by compressing the gas to liquid form and transporting it in steel bottles. In 1896 he started the commercial transport of liquid acetylene, with catastrophic results.

The first attempts to solve the problem of transporting acetylene in compressed form were by Georges Claude and Albert Hess. Claude got his inspiration while studying a siphon bottle with seltzer water (carbonic acid dissolved in water). He looked for liquids in which acetylene was highly soluble, and finally found acetone, with a solubility figure of 25. The resulting gas was called acetylene dissous (dissous gas). Berthelot and Vieille found that these solutions were much less explosive than the liquid, and indeed they considered them safe up to 10 atmospheres of pressure. There was still a high risk of explosion, however, because free gas could arise in a gas bottle during transport.

To prevent this, a porous mass used was used, from an idea by Henri Le Chatelier. The mass created a fine capillary system into which the acetone was absorbed. Since an explosion cannot spread in a space having a diameter of a fraction of a millimeter, in theory, the risk of explosion was eliminated. Claude and Hess launched their invention in 1896, and in the following year the Compagnie Français de l'Acétylène Dissous (CFAD) was founded. Dissous gas, compressed to 10 bars, went onto the market. However, the porous mass used proved to have severe shortcomings. Shaking and vibration gave rise to defects in the porosity. As a result of many serious explosions, acetylene acquired a bad reputation that was to last for a long time.

Early Applications[3–6]

Lighting

The white light produced by acetylene led to the use of acetylene for house and street lighting and in 1897 gave the initial commercial impetus to the acetylene industry. For the first 10 years that carbide was produced, the only large commercial use for acetylene was for lighting (Figs. 3.5.4–3.5.6). The rapid development of safe and efficient generators, together with burners that could utilize the high candle power and intense white light of acetylene, caused the industry to grow rapidly. But it came on the scene just too late to displace coal gas and the new petroleum gases (Pintsch gas, Blau gas, etc.), which used the very efficient Welsbach gas mantle, invented in 1886.

Even though acetylene gave 15 times more light than coal gas and 5 times more light than petroleum gas for an equal quantity of material, competition with the other gases and electricity around the turn of the century was difficut. At that time calcium carbide was being manufactured in the United States, Canada, and 10 countries in Europe.

Figure 3.5.4. Car headlights with dissous gas. From Anonymous (1906).

Figure 3.5.5. Acetylene hand lamp. From Schmidt (1913).

Although acetylene began to compete with other systems of town lighting, the rapid expansion was based on enthusiastic prognostications rather than on any rapid growth of commercial demand. Overproduction of acetylene led to very low prices, followed by the building of cartels in 1900 to control the industry and maintain high prices. Many acetylene companies went bankrupt. The use of acetylene for lighting and signaling was mainly pioneered by Gustaf Dalén of Svenska AB Gasaccumulator, later known as the AGA company.

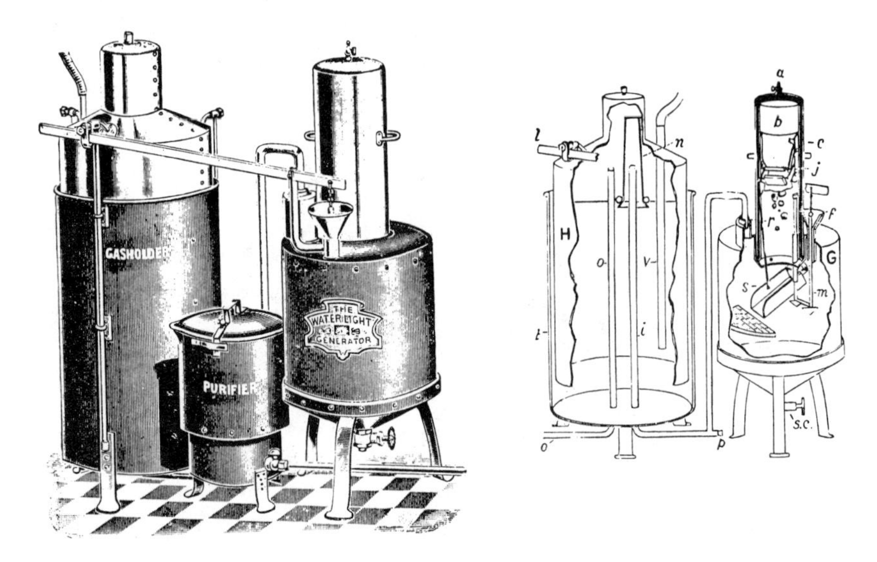

Figure 3.5.6. Acetylene generator. From Anonymous (1906).

Figure 3.5.7. Acetylene heat source, 1910.

Changes in the Acetylene Market

Europe

In Europe a demand crisis came in 1905, just before the start of the cyanamide industry and the beginning of oxyacetylene welding. These new applications had little effect on the demand for calcium carbide until 1909, when prices dropped. A new international syndicate managed to stabilize prices until World War I, when

Figure 3.5.8. Acetylene stove, La Belle, advertisement 1906.

Figure 3.5.9. Picnic with acetylene heater, the early 1900s. Courtesy of the AGA Historical Picture Archive.

war needs kept up the price. By 1911, nearly 1000 towns (one-fourth of them in France) had adopted acetylene for town lighting, but thereafter the number declined rapidly.

By 1921 a price drop occurred because there were over 100 carbide factories producing three times as much carbide as before the war. To deal with this situation again an international syndicate stepped in, in 1924, which kept prices fairly stable until 1929–1930. By about 1925 the use of acetylene for different lighting applications had reached its peak and acetylene lights were being used in homes, mines, buoys, and lighthouses and on automobiles, bicycles, locomotives, and traffic signals.

From 1930 acetylene demand was dominated by developments in polymer chemistry and technology, and acetylene became a raw material in the synthetic rubber and plastics industries.

North America

The calcium carbide plant at Niagara Falls went into operation in 1896. In Canada acetylene lighting became popular for institutions, homes, hotels, and factories. In Ottawa, an extensive plant was erected by the Marine and Signal Company for the manufacture of marine beacons and buoys.

The roots of the acetylene industry in the United States go back to 1904, when P. C. Avery interested the investors James Allison and Carl Fisher in the production and marketing of a container for acetylene. In the United States acetylene had been used for lighting purposes prior to 1904, but up to that time it had to be piped from the generator to the point of consumption. Allison and Fisher recognized the potential of the portable container, and started the Concentrated Acetylene Company.

The growth was fast, and in 1906, when Avery withdrew, the Prest-O-Lite Company was formed at Indianapolis, Indiana, and began the process of compressing acetylene in cylinders filled with porous material containing acetone. This solved the problem of transporting acetylene for use at points distant from its place of manufacture. From 1906 on, growth and development was very fast, partly based on the use of acetylene for welding and cutting. In 1912 a new, large factory was built outside Indianapolis. In addition to the use of oxyacetylene, this company together with other, new acetylene companies played an important part in the development of acetylene chemistry. They sponsored work at the Mellon Institute to develope new sources and uses of acetylene, as in the petrochemical industry. The Prest-O-Lite Company later became part of the Linde Air Products Company.

Frank–Caro Process

In 1895–1899 a new application for calcium carbide was developed. Adolf Frank and Nikodemus Caro developed a process for binding nitrogen in order to produce a nitrogen-enriched fertilizer from carbides for agricultural use. Thus after 1905 the manufacture of calcium carbide was no longer completely dependent on the market for acetylene. The method was developed by the company Firma Cyanid Gesellschaft. Adolph Frank's son found that the cyanamide could be hydrolyzed to ammonia by superheated steam. From this he correctly deduced that cyanamide would act as a fertilizer. A plant was erected at Piano d'Orta, Italy, in 1904, but it was a failure. Manufacture on a large scale began at Odda in Norway and until the 1950s, the lime nitrogen industry was the largest user of calcium carbide.

Acetylene Chemistry[7,8]

The chemical use of acetylene culminated during the interwar years and acetylene was important for the manufacture of polymer products until the 1970s. The development saved the Germans from an early capitulation, because the Allies had managed to cut off supplies of oil and natural rubber to Germany.

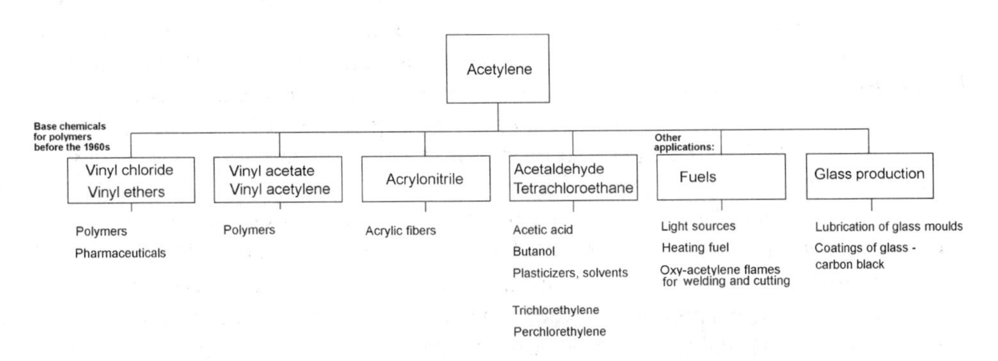

Figure 3.5.10. Acetylene applications.

Chlorinated solvents were the first chemicals produced commercially from acetylene. In 1912–1913, Fritz Klatte at Chemische Werke Hüls discovered that vinyl acetate and vinyl chloride could be synthesized from acetylene. This method was used primarily in Germany and Canada between the wars, and the manufacture of acetaldehyde was begun there in 1916 for the production of acetic acid and acetone, used in rayon and cellulose paints.

About 1914 the acetylene chemical industry was established in North America by the Canadian Electro Products Company, which later was named Shawinigan Chemicals. Like Prest-O-Lite, they took a part in the important work at Mellon Institute, where probably the first thorough economic evaluation of chemicals from ethylene versus acetylene was made. As a result of this study the olefin chemicals industry was first developed. In 1924 the first plant to manufacture chemicals from acetylene in the United States was built by the Carbide and Carbon Chemicals Company at Niagara Falls.

At the Elberfelder Farbenfabrik, Fritz Hoffman realized in 1907 that supplies of natural rubber would eventually vanish. During World War I the company manufactured isoprene from acetylene, and produced "*Ersatzgummi*" (artificial rubber). Since it quickly broke down from contact with the oxygen in air, it was not a success. Usable artificial rubber from acetylene first came in 1932, when Du Pont developed neoprene.

The I. G. Farben cartel was established in 1925 through the combination of Germany's largest chemical firms, including Badische Anilin and Soda Fabrik (BASF), Farbwerke Hoechst, and Farbenfabriken Bayer, which were connected to a number of other firms, including Chemische Werke Hüls. Bunawerke Hüls began producing Buna-S synthetic rubber in 1933. It was based on styrene and butadiene. Buna-S had a better resistance to wear, and had the strength of natural rubber. This plant became the second largest German installation for the manufacture of tire rubber for Hitler's Wehrmacht. The Buna-S technology was known to a number of U.S. scientists because Standard Oil of New Jersey (now Exxon) had, in the late 1920s, set up a technology exchange agreement with I. G. Farben. Project proposals for manufacturing Buna-S-type rubber by Dow Chemicals were rejected by the U.S. Government just before World War II.

German industrial chemists felt that an electric arc process could become an important contributor to their country's acetylene supplies based on the use of coke oven gases and later methane as a feedstock. However, acetylene yields remained too low (5–6 percent) to be commercially viable. Baumann, initially at BASF and later transferred to Hüls, was responsible for developing a commercially successful arc process in the early 1930s. His work led to a doubling of yields. In the mid-1930s, the cooperation between I. G. Farben and Standard Oil Company was extended to this field and led to a commercial process.

Though production facilities were bombed periodically, Germany manufactured much acetylene during the war. By 1942, calcium carbide plants at Hüls, Schkopau, and Knapsack were supplemented by a number of arc process installations. By-products from the arc process included ethylene, hydrogen, and carbon black. Thus the Germans were independent of foreign raw materials for their production of synthetic rubber and aviation gasoline before the start of the Blitzkrieg in 1939.

Between 1929 and 1945, working at BASF, Walther Reppe developed a number of new chemical reactions using acetylene at high pressures and high temperatures. In essence, Reppe's technology was based on the use of calcium carbide-based acetylene and its reactions with compounds such as alcohols, acids, carbon monoxide, and carbonyl compounds. These risky processes enabled polymer materials to be produced without petroleum during the war.

The shortage of key war materials such as rubber and oil was the motivating force for the innovative and hazardous approaches undertaken by Reppe and his coworkers. American and British military–civilian teams brought back details of the new acetylene-based technology based on interviews with Reppe after the German defeat in World War II.

The beginning of the decline of acetylene as a raw material came during the 1930s, when the British company ICI produced polyethylene from petroleum by chance. In the early 1960s the use of acetylene for the production of commodity chemicals (acrylonitrile, acrylates, chlorinated solvents, chloroprene, vinyl acetate, vinyl chloride) reached a peak. Since then, olefins such as ethylene have mainly replaced acetylene for these products. The decline in the use of acetylene from carbide became significant in 1972 when large production facilities were shut down.

Acetylene can be produced by either the calcium carbide (coke plus limestone) technologies or by petrochemicals, such as ethylene or coal (coal plus hydrogen plasma). Several new acetylene production processes were patented during the second half of the 20th century. The principal acetylene processes using hydrocarbon feedstock are:

1. Electric arc (Badische Aniline Soda Fabrik, BASF)
2. Regenerative furnace pyrolysis (Wulff)
3. One-stage combustion (BASF, Sachsse)
4. Two-stage combustion (Union Carbide, Montecatini)
5. Plasma arc or jet (Chemische Werke Huels AG)

The use of acetylene in engines for power, tested before and during World War II, was not successful because of the low ignition temperature of the gas, which by preignition prevented the high compression in the engine cylinders necessary for reasonable efficiency.

Aluminum Manufacture and Its Impact on Acetylene Production

The breakthrough of acetylene in the 1890s was in fact a spin-off from the bull market for aluminum in the 19th century. In 1847, Frederich Wöhler was the first to produce the metal by reducing aluminum chloride with potassium at a very high temperature.

Napoleon III provided the French chemist Henri Saint Claire Deville with considerable funds for research into industrial manufacturing methods for aluminum. The first ingots of aluminum that Deville produced in 1855 cost $600 per kilogram. It is said that the emperor had aluminum plates made for himself, his wife, and his other aristocratic guests, while the other guests had to eat from "ordinary gold plates."

Continued

When Deville later found that the electrolysis of an aluminum oxide solution in molten cryolite was a cheaper way to make aluminum, the price of aluminum went down to about $30 per kilogram. Although two factories were built, the product achieved was impure and interest in industrial production soon faded. Metal smelting technology became available after Werner Siemens invented the dynamo in 1867, and his brother Wilhelm developed an industrially practicable electric arc furnace in 1878. In 1855 the American brothers Alfred and Eugène Cowles built a reduction furnace with fireproof linings. At the same time Paul Heroult in France and Charles Hall in the United States simultaneously optimized the electrolytic furnace for producing aluminum. Large-scale aluminum production was started in the United States, Switzerland, and other countries, and the price was soon down to $0.2 per kilogram.

The way to produce acetylene on a large scale was found by chance when Willson tried to produce pure aluminum in a Cowles-type furnace.

The First Industrial Production of Calcium Carbide and Dissolved Acetylene

Before 1900 calcium carbide was being manufactured in 12 countries, and several forms of furnace construction were in use (Figs. 3.5.11 and 3.5.12):

Switzerland

Neuhausen 1895 (Société pour l'Industrie de l'Aluminum)
Vallorbe 1895 (Société d'Electrochimie)
Vernier 1897 (Cie. Genevoise d'Electricité)
Luterbach 1897 (Schweizer Calciumcarbidfabrik)

Figure 3.5.11. Acetylene gas station for town lighting in Hungary, 1898. From De Szpcynski (1898).

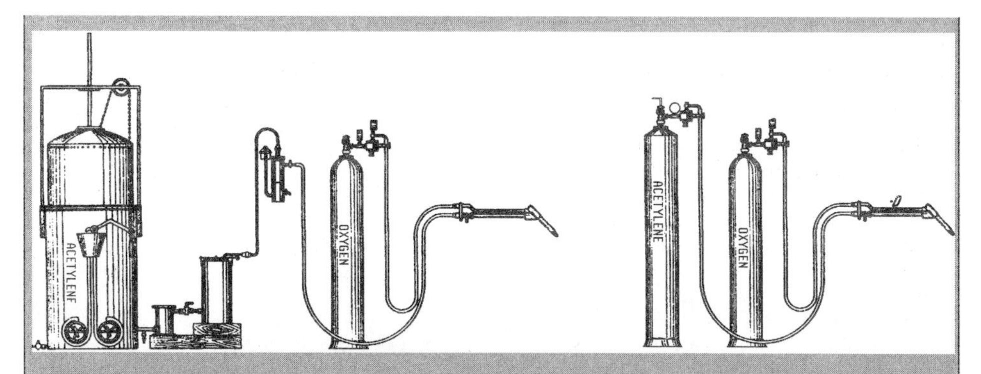

Figure 3.5.12. The two different forms of using acetylene for welding in the early 1900s. Left: Low-pressure, locally produced acetylene. Right: High-pressure dissolved acetylene in cylinder.

Lonza 1898 (Société pour l'industrie à Gampel)
Langenthal 1899 (Elektriziteswerk Wynau)

France
Froges, La Praz 1896 (Société Electrometallurgique Française)
Notre Dame 1897 (Société des Carbures Metalliques)
Arlod 1898 (Ch. Bertolus)

Germany
Bitterfeld 1896 (Griesheim Elektron)
Rheinfelden 1897 (Société pour l'Industrie de l'Aluminum)
Lechbrück 1898 (A. G. Lechbrück)
Lauffen 1900 (Portlandzementfabrik Lauffen)

Austria
Jajce 1897 (now Bosnia) (Bosnische Electrizitets AG)
Meran 1898 (Acetylen Gas AG)
Gastein 1897 (Société pour l'Industrie de l' Aluminum)
Sweden
Trollhättan 1897 (Trollhättan Elektriska AB)
Alby 1899 (Alby Carbidfabrik AB)
Månsbo 1900 (Stockholms Superfosfatfabriks AB)

Norway
Sarpsborg 1898 (A/S Karbidindustri Sarpsborg)
Hafslund 1898 (A/S Hafslund-Karbidfabrik)
Meraker 1898 (A/S Meraker Smeltewerk)
Notodden 1900 (Notodden CalciumCarbidfabrik A/S)

Finland
Siitola Oy, Itala

Spain
Berga 1899 (Sociedad Espaniola de Carburos Metalicos)
Esparraguera (Manufacturas Sedo SA)

Continued

Italy
Terni (Società Italiano per la fabricazione del Carburo di Calcio)

England
Leeds 1899 (Experimental plant from 1895)
Foyer Falls, Scotland, 1899 (Acetylene Illuminating Co.)

Canada
St. Catherine, Shawinigan Falls, 1898 (Willson Carbide Works)

United States
Spray, Michigan 1894 (Electro Gas Company)
Niagara Falls 1896 (Acetylene, Light, Heat and Power Co., ALHPC)
Niagara Falls 1898 (Union Carbide, formerly ALHPC)

Overproduction of calcium carbide occurred soon after 1900. At first the only application was lighting. From about 1905 it was also used as a raw material for manufacturing of cyanamide, and its production increased. After the introduction of dissolved acetylene, manufacture began in France, Sweden, England, and Russia. There were new markets for oxyacetylene welding and cutting. In 1910, 43 known plants existed in France (Paris, Marseille, Bordeaux, Lille, Le Havre), Sweden (Stockholm, Göteborg, Västerås, Karlskrona, Malmö, Hultsfred, Mörby, Halmstad), England (London, Newcastle), Russia (St. Petersburg, 2), Germany (Hamburg, Düsseldorf, Friedrichshafen), United States (New York, Indianapolis), Austria/Hungary (Vienna, Budapest), Italy (Rome, Milan), Norway (Oslo), Holland (Amsterdam), Denmark (Odense), Romania (Bucharest), Spain (Barcelona), Belgium (Cappelin, Liège), Greece (Athens), Algeria (2), Argentina, Brazil, Chile, Japan, Indo-China, and Australia.

Short Biographies

Friedrich Wöhler (1800–1882) was a German professor educated by Berzelius. Wöhler was the first to isolate an organic substance (1826) and to produce and identify pure aluminum (1827) and calcium carbide (1862).

Thomas Willson (1860–1915) was a Canadian engineer who started the acetylene era by producing calcium carbide in 1892. He introduced acetylene for lighting and later became a pioneer of lighhouse technology.

Walter Reppe (1892–1969) was head of research at BASF. He worked with acetylene in its explosive pressure and temperature regions in order to develop polymer materials that were essential for Germany during World War II.

References

1. Thomas Leopold Willson 1860–1915. Acetylene Lighting and Welding Journal, Febr 1916:22–23
2. Scales S. James Turner Morehead 1840–1908. Proc Int. Acetylene Association Conv. 1923:311–325
3. Davies, A. The history and present status of the oxy-acetylene process in America. Proc Int. Acetylene Association Conv. 1911:109–117
4. Morehead, J. M. A century of progress in acetylene. Proc Int. Acetylene Association Conv. 1933:15–24
5. Pitts, C. the history and scope of the acetylene industry. Proc Int. Acetylene Conf. 1954:179–185

6. Byus, D. The history of the acetylene industry and and significance of Prest-O-Lite. Proc Int. Acetylene Conf. 1954: 186–189
7. Spitz, P. H. Chemicals from coal. Chemtech 1989;(2):92–100
8. Yahraes, H. The arrival of acetylene. Scientific American 1949;(1):17–21
9. Miller, S. A. Acetylene—its properties, manufacture and uses. Ernst Benn Ltd, London 1965
10. Tedeschi, R. J. Acetylene and Commodity Chemicals: Acetylene-based chemicals from coal and other natural resources. Marcel Dekker 1981.

3.6. THE NOBLE GASES: AN INACTIVE FAMILY OF GREAT IMPORTANCE[1]

The blaze of crimson light from the vacuum tube told its own story, and it was a sight to dwell upon and never forget. It was worth the struggle of the previous two years; and all the difficulties yet to be overcome before the research was finished. For the moment the actual spectrum of the gas did not matter in the least—for nothing in the world gave a glow such as we had seen.

William Ramsay's assistant, Morris Travers
(after spectroscopic discovery of
krypton, May 1898)

Background

The Noble Gases and Their Delayed Discovery

Up until the end of the 19th century chemists believed that the chemistry of the atmosphere had been totally explored, and few if any thought for a moment about looking there for new elements.

A century earlier, however, the work of Sir Henry Cavendish had hinted at the possibility of an unknown gas in the atmosphere; in 1785 he passed electric sparks through a mixture of oxygen and ordinary air in the presence of an alkali, and discovered that a part of the air had neither oxidized nor been absorbed. He determined the remainder to be less than 1/120 of the total quantity of air. This important experiment had been forgotten when the Cambridge professor John William Strutt (3rd Baron Rayleigh) began measuring the density of atmospheric gases in 1882.

Discovery[2]

Argon

When 10 years later Lord Rayleigh compared the density of nitrogen obtained when the other components of air had been removed with the density of nitrogen prepared from ammonium nitrate he found that the former was greater. He published his findings in *Nature* in 1892 and asked other chemists for their suggestions as to the reason. Sir William Ramsay (see Fig. 2.3.16), Professor of Chemistry at University College in London, responded. He was also asked to assist with further studies (Fig. 3.6.1).

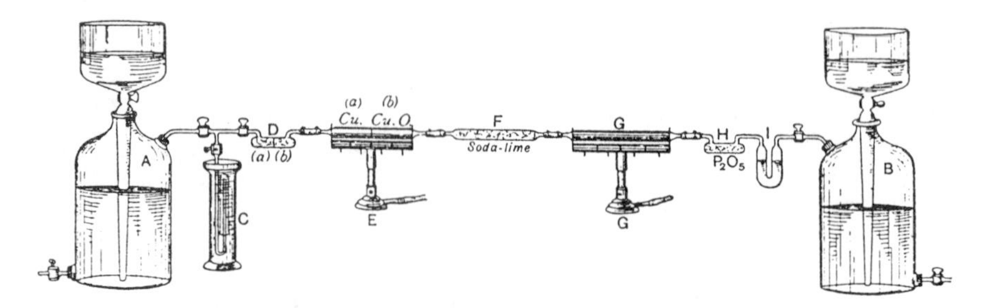

Figure 3.6.1. Ramsay's apparatus for the isolation of argon. A sample of atmospheric nitrogen was passed backward and forward through red-hot magnesium in tube G. The gas was also passed through red-hot copper oxide to oxidize any suspended carbonaceous matter, such as dust, to carbon dioxide. This was then absorbed by the soda-lime. From Ramsay (1905).

In May 1894 Ramsay stated that there must be a further, hitherto unknown, component in air. He was able to isolate it and found its density to be 20 times that of hydrogen. At the same time, Rayleigh (Fig. 3.6.2) showed that the new substance would not react with any other. It was therefore called argon ("the lazy one"). The discovery was the beginning of a long chain of events in which Sir William Ramsay and his assistant, Morris Travers (Fig. 3.6.3), were to play the main part. At first it was thought that the new gas was an allotropic form of nitrogen (much as ozone is a form of oxygen).

Krypton, Neon, and Xenon

Ramsay and Travers were given 1 liter of liquid air by William Hampson, the English pioneer of liquid air technology. They used it to develop new methods in their

Figure 3.6.2. John William Strutt (3rd Lord Rayleigh), Nobel Laureate in Physics, 1904. From Weeks (1935) with permission from the *Journal of Chemical Education*, copyright © 1935, Division of Chemical Education, Inc.

Figure 3.6.3. Morris Travers. From Weeks (1935) with permission from the *Journal of Chemical Education*, copyright © 1935, Division of Chemical Education, Inc.

search for new elements. When most of the air had boiled away, they separated the oxygen and nitrogen using red annealed copper and magnesium. The remainder was collected and examined to see if it contained any unknown gas. The sample mostly contained argon, but they also found unknown lines in the spectrum. The gas from which these originated was given the name krypton ("hidden").

With more liquid air they condensed 3 of the 15 liters of argon that had been previously fractionated. The amount that was first boiled away was collected and under spectral analysis gave a complicated spectrum of red, orange, and yellow lines. Besides the yellow lines of helium there were clearly lines that belonged to a new basic element. Ramsay's 13-year old son Willie suggested the name neon ("new").

Xenon ("strange") was then discovered by further fractionation of liquid air. Thus, within a very short space of time, Ramsay and Travers had discovered three new gaseous elements: krypton on 30 May, neon on 7 June, and xenon on 12 July 1898. Bearing in mind the minute quantities in which these gases appear in nature, this is one of the really great scientific achievements. They went on to initiate a study of the chemical and physical properties of these noble gases which was to continue past the turn of the century.

Radon

Following a visit to Henri Becquerel in Paris in 1896, Ramsay became interested in the disintegration of radium, and especially in the reason helium was found just in radioactive minerals. In 1903, Ramsay and his assistant, Frederick Soddy, discovered that alpha particles were composed of helium nuclei. In 1900, while working with Ernest Rutherford, Soddy had found a gaslike disintegration product which had been assumed to be a noble gas. This was called niton, but the name was later changed to radon. In 1910, Ramsay determined its density, atomic weight, and chemical properties, and was able to place it in the final position of noble gases in the periodic table, thus fulfilling his final objective.

Helium

Helium has its own special story. After hydrogen, it is the next most abundant element in the universe. Yet by volume, helium is barely 1 part per million of the atmosphere. Because helium is lighter than air and does not combine with other elements, one may wonder how it remains here on Earth. Because helium is formed during certain radioactive processes in Earth's crust, there are places where there is a high content of helium in natural gas and in minerals.

In 1868 William Lockyer noticed a bright yellow band in the solar spectrum, indicating an unknown, mysterious element in the chromosphere of the Sun. He believed that it was produced by an element not present on Earth and named it helium from the Greek word "helios" for sun.

About 20 years later, W. F. Hillebrand of the U.S. Department of the Interior obtained an unidentified gas from the mineral uraninite. In 1895 the Swedish professor Per Teodor Cleve and his assistant discovered helium in minerals. The same year William Ramsay isolated helium while searching for a better source of argon. He heated the uranium-containing mineral clevite and purified the resulting gas. This residue, helium, proved after 27 years of uncertainty that the solar element did indeed exist on earth.

The work of the above three pioneers remained relatively unheralded until a method was developed to obtain and use this substance in volume. The key to this development came in 1905, when natural gas discovered in Kansas was found to contain nearly 2 percent helium. In 1918, 50 years after its discovery, helium was first extracted in large quantities from natural gas. From then until today, the use and importance of helium has expanded at an ever-increasing rate.

Applications for the Noble Gases (see Fig. 3.6.4)

Argon, Neon, Krypton, Xenon

Until 1910, the rare gases of the atmosphere were considered a scientific curiosity. The discovery of the noble gases came at the right time, when electric lighting was making its triumphal progress. The first five noble gases discovered all found uses in lighting. By the early 1920s, however, new and important uses began to emerge for these rare gases. Argon was used to provide an inert atmosphere for electric lights, and was later to find extensive use as a blanketing atmosphere for welding and in modern lasers. Neon began to be used for red glowing discharge tubes or "neon lights" and was also later mixed with krypton to produce a brilliant white light for photography.

Krypton and xenon, the rarest of the rare gases, were used in illuminated signs. The separation of neon from the atmosphere on a large scale was patented by Georges Claude of Air Liquide in 1908 followed by experiments with electric discharge tubes containing neon, argon, helium, krypton, and xenon.

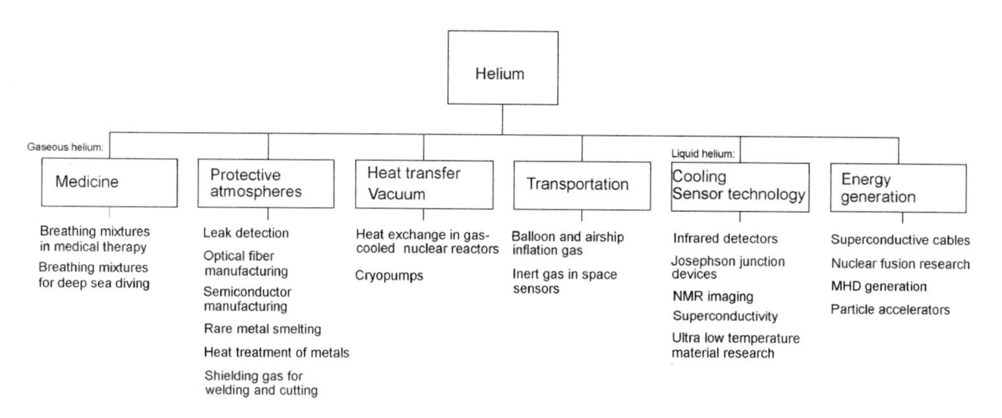

Figure 3.6.4. Helium applications.

Neon manufacture was started by Griesheim Elektron in 1913 and within a decade the neon sign industry had developed. The novel effects obtainable with neon tubes gave them wide use for advertising purposes, their extreme adaptability, high luminosity, and brilliant colors making them particularly suitable in this field. Efficiency of light production by use of neon tubes was too low for general illumination, and none of the colors was satisfactory for interior lighting. The neon lamp began to be used for special purposes, for example, signaling and stroboscopic work about 1918.

Where inertness alone is required, the commonest of these gases, argon, is chosen. It has been used since 1919 in considerable quantity for the filling of electric filament lamps and as a shielding gas in welding of easily oxidizable materials such as aluminum.

Krypton and xenon have a higher boiling point than oxygen and therefore could not be produced in air-gas plants until the 1930s, when Matthias Fränkl's principle of regeneration was introduced. Argon, helium, and neon, alone and with the vapor of mercury, are used for the filling of discharge lamps and fluorescent tubes. Krypton and xenon, despite their rarity, have also occasionally been employed.

Helium

Helium has nearly the same lifting power as hydrogen and because it is nonflammable, it found considerable use in the filling of balloons and dirigibles from about 1919 until these were replaced by heavier-than-air craft. Today it has special importance in welding, leak detection, cryogenic research, and medicine, including respiratory apparatus and helium–oxygen breathing mixtures. Magnetic resonance imaging (MRI), which provides physicians with a look inside the human body without surgery, is made possible by superconducting magnets cooled first with liquid nitrogen and then with liquid helium, which lowers their temperature to near absolute zero. Helium is also mixed with oxygen to an artificial air for divers. Helium is less soluble in the blood than is nitrogen, and is therefore less likely to cause fatal accidents through the formation of bubbles in the blood when the pressure is released.

Helium Production

In 1903 oil workers in Dexter, Kansas, hit a gas well at 400 feet. The escaping gas did not burn. Two years later H. P. Cady of the University of Kansas examined a sample from the Dexter gas well and found that the gas consisted mostly of nitrogen and hydrocarbons, but did contain 1.84 percent helium. This was the first time helium had been discovered to be associated with natural gas. Cady's student Clifford Seibel analyzed samples from other natural gas wells for their helium content in a study for his master's thesis. When he presented the results at an American Chemical Society meeting in 1917 he remarked, "unfortunately this study had no practical application." But Richard Moore, Superintendent of the U.S. Bureau of Mines Station in Golden, Colorado, and a former student of Sir William Ramsay, told the assembly that Ramsay was eagerly searching for a source of helium in England.

In 1915, Ramsay sent a letter to Richard Moore, the head chemist for the U.S. Bureau of Mines, in which he pointed out the possibility of extracting helium from natural gas. At the same time, experimental extraction was going on in Canada. The Germans had used hydrogen-filled dirigibles flying at 5000 meters to bomb London. Because these flew higher than airplanes could reach, the British used rocket-like projectiles to set them on fire. When the British learned that the Germans were seeking to switch to a nonflammable lifting gas, they began a search for a helium source of their own. Moore foresaw that despite its current high cost, helium could serve on a large scale to fill airships. The first plant to isolate helium was built in 1917 by the British in Hamilton, Ontario, Canada. Within four months, in July 1917, the U.S. Government started a helium development project for a demonstration plant. The helium industry was born as an effort to keep ahead of Germany. By the end of the war in November 1918, the United States had taken the first steps for production of helium on a large scale.

The Bureau of Mines chose a gas well near Fort Worth, Texas, which contained 0.84 percent helium. Three small plants for helium extraction were built in Texas, by Linde Air Products and Airco at Fort Worth and by Jefferies-Norton at Petrolia. The helium produced was about 90 percent pure and was first used as a lifting gas for military dirigibles. By the mid-1920s, the available gas source at Fort Worth was empty, and a new plant was built at the nearby Cliffside field in Amarillo, Texas, where the natural gas contained 1.8 percent helium (Fig. 3.6.5).

By the Treaty of Versailles it was stated that the German Zeppelin works should be allowed to make only smaller airships. The Allies divided the nine remaining larger airships (150–250 meters in length) among them. Under the Treaty, the Germans were also to build a new large passenger airship for the United States as partial payment for war reparations. In 1923, an agreement of cooperation was signed between the German Zeppelin works and American Goodyear under which part of the Zeppelin technical staff and patent rights were taken over by Goodyear.

When the airship *Los Angeles* arrived in America in October 1924 its 14 gas compartments containing some 70 million liters of hydrogen were immediately emptied so that they could be filled with helium. But nowhere in the world there was so much helium available, so another large airship, the *Shenandoah*, built in the United States in 1923, was emptied of its helium, and the gas was then used alternately by

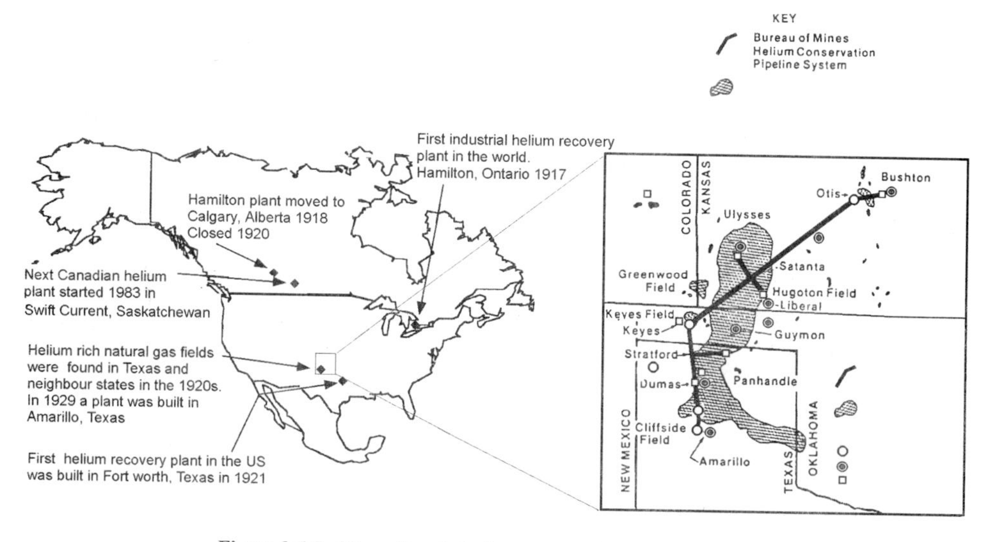

Figure 3.6.5. Map of early helium sources in North America.

both airships. In September 1925 some of the *Shenandoah*'s gas compartments exploded when the crew attempted to go above a storm and was wrecked. This grounded the *Los Angeles* for a long time, until sufficient helium could be produced to fill it. From 1925 on, Goodyear built a series of smaller, 30- to 50-meter-long airships, the Pilgrim series (Fig. 3.6.6).

The military maintained sufficient interest in helium's usefulness as a lifting gas to sustain production until the advent of World War II, when demand skyrocketed. The *Hindenburg* disaster in 1937 forever demonstrated the dangers of hydrogen-filled airships and the Navy used helium in their blimps, which were extensively used on antisubmarine duty in World War II. Toward the end of the war the material used in the development of the atomic bomb also required large amounts of helium.

Since World War II the use of helium for nonmilitary applications has continued to grow. In the mid-1950s, as natural gas demand increased, some helium-rich

Figure 3.6.6. Goodyear began building a series small, commercial airships in 1925.

sources began to dwindle. This led to helium conservation in accord with the 1960 *Helium Act* and the development of a private helium industry in the United States. Attempts to use a miniature sealed coal mine for helium storage were unsuccessful, as the gas leaked through the smallest of cracks. Finally, use of a partly depleted Cliffside, Texas, deep gas well was found satisfactory, and enough helium was stored there to meet U.S. Government needs for up to 100 years.

Industrial Production[3–5]

The quantities of helium produced before World War II were relatively small. After the war a sharp upward trend began as new uses developed in rapidly expanding U.S. Government and civilian activities. Virtually all helium production in the period 1918–1961 was by the U.S. Department of the Interior's Bureau of Mines to meet the needs of federal agencies. All helium production was by extraction from helium-bearing natural gas sources in Texas, Oklahoma, New Mexico, and Kansas. For short periods during 1928–1937 only one private company operated two small plants, one in Kansas and one in Colorado. It was not until the early 1960s that private industry entered the helium production field.

A large helium plant in operation in Saskatchewan, Canada, was started in 1963. In Europe small quantities have been produced as a by-product of air separation. The production of helium from natural gas began in Europe during the 1960s from wells in the North Sea off the coasts of England and the Netherlands and in Poland. Helium may also be available in the uncondensed gas from a natural gas liquefaction plant supplying liquefied natural gas for shipment from Algeria to France and England. In Russia helium is produced in Orenburg and Dubna, from which export to the West began in the early 1990s (Fig. 3.6.7).

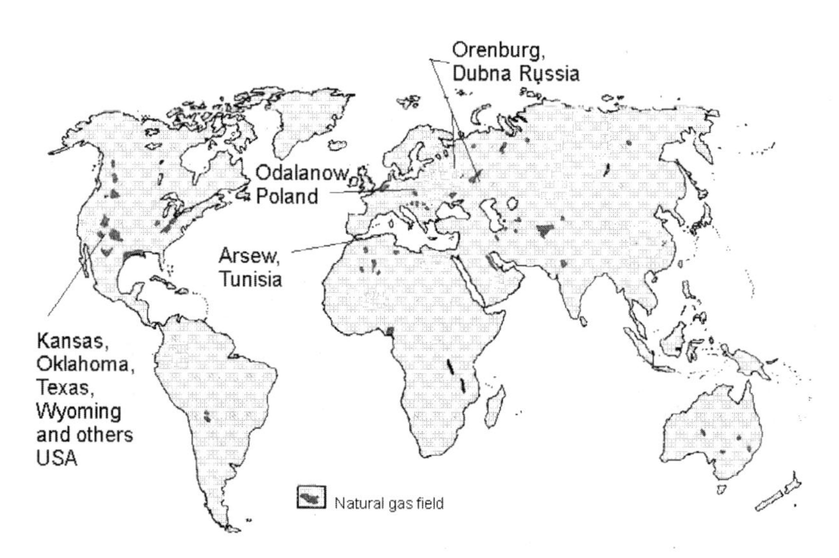

Figure 3.6.7. Main helium sources in the world.

Other Helium Uses and Markets[6]

In 1925 another important use for helium was discovered. After a decade of development it began to replace nitrogen in the gas used for deep-sea diving, in order to avoid decompression sickness. Less soluble in the blood when under pressure than nitrogen, helium does not form bubbles that can block the smaller blood vessels. Following authorization of sales for non-Government purposes in 1937, small quantities were used in synthetic breathing mixtures for temporary relief of respiratory disorders and for reduction of explosive hazards in anesthesia. Shortly thereafter, the use of helium in shielded-arc welding began. Helium also found an important use in early nuclear energy investigations and developments. Nevertheless, about 99 percent of the annual helium production before 1946 was used in blimps, dirigibles, and balloons, and non-Government sales were small.

Inerting and Shielding

In the late 1940s the purity of produced helium was raised, first to about 99.5 percent and shortly thereafter to nearly 100 percent. Larger quantities were used for shielded arc welding and nuclear energy development, and new uses developed and grew rapidly.

In the 1950s, large volumes of helium were used in the development and operation of ballistic missiles and in the initial program for space exploration. At the same time, helium was finding a variety of uses in rapidly expanding research programs. The usefulness of helium in these applications is due to its unique physical properties, especially its inertness, low density, and very low liquefaction temperature.

Liquid Helium[7–9]

One of the most interesting uses of helium is in the field of cryogenics. Because of its very low boiling point, 4.2 K, helium is used in liquid form to provide the refrigeration for various types of cryoelectronic and other devices, especially those involving the phenomenon of superconductivity. Helium itself is scientifically interesting because of its transformation to helium II at 2.18 K and the absence of a triple point. The unusual characteristics of helium II, such as superfluid creep, abnormal viscosity and sound propagation, have been studied extensively by low-temperature physicists.

Several attempts were made prior to 1908 to liquefy helium, notably by Karol Olszewski and Sir James Dewar, but these were unsuccessful. Kamerlingh Onnes made several attempts to liquefy helium by cooling compressed helium gas with liquid hydrogen and then making a rapid expansion. He finally achieved success with a circulation technique in which compressed helium gas is cooled to near the triple-point temperature of hydrogen and then piped through a regenerative heat exchanger prior to expansion (Fig. 3.6.8).

It was not until the early and middle 1920s that other laboratories (first in 1923 by John McLennan at the University of Toronto, and then at cryogenic laboratories

Figure 3.6.8. Heike Kamerlingh Onnes (left) with his assistant (right) in his laboratory at the University of Leiden. Courtesy of the University of Leiden.

in Oxford, Cambridge, Munich, Breslau, and Moscow) reported success in liquefying helium. F. Simon and P. Kapitza found new methods to liquefy helium in the early 1930s, and after World War II Samuel C. Collins of the Massachusetts Institute of Technology developed a highly successful helium liquefier which made it possible for low-temperature research and development labs to liquefy helium on their own. Up to 1968 over 350 Collins liquefiers had been sold commercially (Fig. 3.6.9).

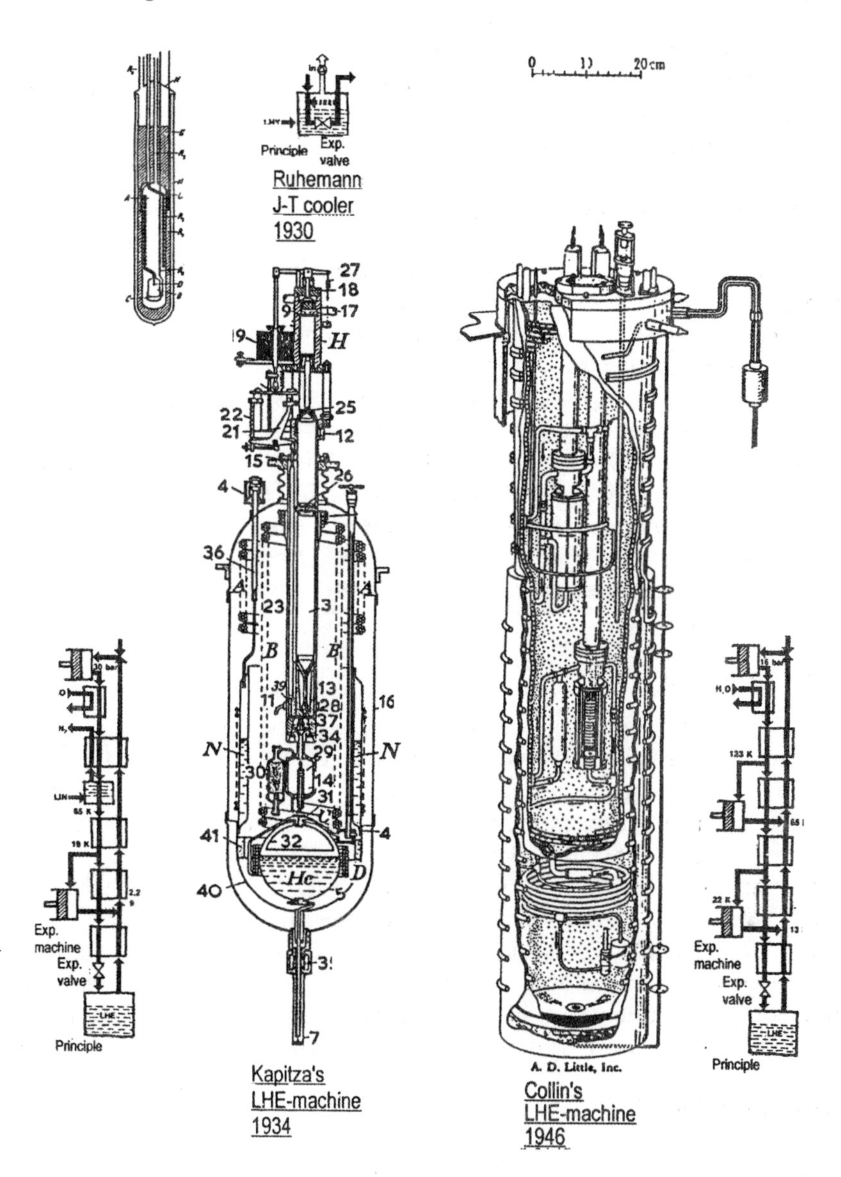

Figure 3.6.9. Collins, Kapitza, and Ruhemann helium liquefiers.

"High-Tech" Research and Development

Liquid helium was used in the world's first superconducting particle accelerator, at Fermi Labs, for a magnetohydrodynamic (MHD) water propulsion system without moving parts at Argonne National Labs. The use of liquid helium for magnetic resonance imaging continues to increase as new uses for this equipment are developed in medicine.

Helium has also been used in several Strategic Defense Initiative (SDI) applications such as antisatellite (ASAT) rockets, where liquid-helium-cooled infrared sensors are used for target location and guidance and for cooling tracking telescopes. Gaseous helium is used in lasing gas mixtures of chemical lasers. Superconducting magnetic energy storage (SMES) has also been investigated as a means to provide power for defense laser systems.

Helium Conservation in the United States

Helium production rose to 600,000 equivalent liquid liters per year by 1932, but by 1937, the production of helium was down to about 200,000 liters per year, and in that year, the *Helium Act* allowed Government plants to sell helium to non-Government users for medical, scientific, and commercial use. Large-scale helium production began in World War II with the need for helium-filled blimps to protect convoys, and consumption later grew with the use of liquid-fueled Atlas and Titan 1 missiles deployed in the Intercontinental Ballistic Missile Force. The 1960 amendment to the *Helium Act* authorized the Secretary of the Interior to continue the operation of Government helium production plants and to enter into 25-ycar contracts for the purchase of crude helium for conservation. At that time, helium prouction was about 20 million liters per year, of which 70–80 percent was consumed by Government projects.

Helium Distribution

Traditionally, helium has been shipped in quantities of about 1000 cubic meters compressed to approximately 160 bars in high-pressure tubes mounted on truck trailers. Newer designs using longer trailers and tubes made of higher strength steels have increased the carrying capacity to nearly 3000 cubic meters. Rail cars have for many years carried about 9000 cubic meters of helium at 270 bars.

Commercial shipment of liquid helium was started in the late 1950s with shipments to various consumers for research purposes. Technological advances in cryogenic insulation and helium liquefaction made the transport of liquid helium practical even in cases when the end use is gaseous helium. In May 1966, Airco started transporting liquid helium in semitrailers from a new helium separation and liquefaction facility to four distribution centers within the United States.

Short Biographies

Sir William Ramsay (1852–1916) was a Scottish chemist and scientist. As a gifted 18-year-old he studied in Germany under Bunsen and others. He discovered five of the six noble gases and determined the atomic weight of the sixth.

John William Strutt Rayleigh (3rd Baron Rayleigh) (1842–1919) was Professor of Physics at Cambridge. His measurements of the density of air led to the discovery of argon.

Morris Travers (1872–1961) was an English physical scientist who worked with Ramsay on the discovery of krypton, neon, and xenon. His experimental expertise contributed to a high degree to the research into the noble gases.

Per Teodor Cleve (1840–1905) was Professor of Chemistry at Uppsala, Sweden. Together with his assistant Langlet he discovered helium in minerals in 1895.

Frederick Soddy (1877–1956) was Professor of Chemistry at Glasgow University. Together with Rutherford he did research into radioactive disintegration and on radon gas. In 1903, together with Ramsay, he found that alpha particles are made up of helium nuclei.

References

1. Ramsay, W. (1905). *The gases of the atmosphere. The history of their discovery.* London: Macmillan.
2. Cook, G. A. (ed.). (1961). *Argon, Helium and the Rare Gases* Vol. II. New York: Interscience.
3. Hammel, E. F. (1980). Helium: Its past, present and future. Advances in Cryogenic Engineering, Vol. 25:810–821.
4. Hammel, E. F., Krupka, M. C., and Williamson, K. D., Jr. (1984). The continuing helium:saga. *Science* **223**:789–792.
5. U.S. Bureau of Mines. (1984). *Helium. Mineral Yearbook 1984.* Washington DC:
6. Hopps, H. B. (1993). From the sun to the moon and beyond. *Chemtech* **1993**(2):447–451.
7. Kropschot, R. H., Birmingham, B.W., and Mann, D. B. (eds.). (1968). *Technology of Liquid Helium.* National Bureau of Standards Monograph 111.
8. Simon, F. (1926). Eine neues einfaches Verfahren zur Erzegung sehr tiefer Temperaturen. *Physik Zeitschrift* **27**:790–792.
9. Kapitza, P. (1934). *Royal Society 147A*:189–190.
10. Collins, S. C. (1947). *Physical Review* **70**:98.

4

THE DEVELOPMENT OF INDUSTRIAL GAS TECHNOLOGY

4.1. LIQUEFACTION OF GASES[1,2]

Early Liquefaction

The liquefaction of gases was an issue of great interest to physicists during the latter part of the 18th century. In about 1790 the Dutchman Martinius van Marum liquefied ammonia by compressing it to 6 bars. At the same time sulfur dioxide was also liquefied, by Louis Clouet and Gaspard Monge, by compression and cooling with ice/salt mixtures. In 1823 Jacob Perkins, one of the pioneers of cooling technology in England, tried to liquefy air with a pressure of 1000–1200 bars, but failed. At the same time his countryman Michael Faraday began systematical work with gases that had not previously been liquefied.

Because Charles Thilorier succeeded in 1834 in producing solid carbon dioxide and mixing it with ether to reach –110°C, Faraday used the "Thilorier mixture" until 1845 to liquefy all the then known gases (the rare gases were discovered later) except six: hydrogen, oxygen, nitrogen, nitric oxide, carbon monoxide, and methane. Faraday called these six gases "permanent gases."

In 1854 the Austrian Johannes Natterer attacked the air liquefaction problem by constructing a compressor for 3000 bars (300 megapascals), but failed. Air really appeared to be a permanent gas. The reason for these liquefaction failures was that the concept of critical temperature was not known: Every gas has a critical temperature above which it cannot be liquefied. This critical state had been predicted by Charles Cagniard de la Tour in 1822, but was not explained until 1860, by Dimitrij Mendelejev's theory of the absolute boiling point of liquids and Thomas Andrews's repeated experiments with CO_2 gas in 1862. When the theory of the critical state of gases was proposed in 1869, the last obstacle on the road to liquefaction of the six "permanent" gases had been eliminated.

Liquid Oxygen Observed

On Christmas Eve 1877 it was announced to the French Academy of Science in Paris that some drops of oxygen had been liquefied for the first time. This was a major breakthrough, with consequences which could not even be imagined at the time. The announcement meant that the prior boundary to achievable temperatures, which was later defined as the cryorange (<120 K), could be passed. This was a starting point for significant developments in many scientific and technical areas, all the time since then. A scientific competition to reach absolute zero also started, which led to a deeper understanding of the innermost characteristics of materials.

At that Academy meeting on Christmas Eve 1877, the metallurgist and Academy member Louis Cailletet announced that he had observed liquefied oxygen. The method was developed as a result of an accident: Cailletet was trying to liquefy acetylene with the (at that time) common method of compressing to high pressure. A leak developed in the apparatus, and the compressed gas flowed out. Cailletet observed that a cloud of acetylene developed and then rapidly disappeared, and drew the conclusion that the pressure drop had resulted in a cooling of the gas, leading to liquefaction (Figs. 4.1.1 and 4.1.2).

As a result of his observations, Cailletet constructed an apparatus which provided the necessary combination for liquefying oxygen: cooling (SO_2 vapor, –29°C) and compression (300 bars, or 30 megapascals) followed by expansion. Cailletet soon liquefied nitrogen as well, and his apparatus was very important for the development of cryotechnology during the next few years, and thus to the initiation of research at various institutes, including those in London and Krakow.

At the same Academy meeting a telegram was read from Raoul Pictet in Geneva stating that, some days later than and quite independently of Cailletet, he had

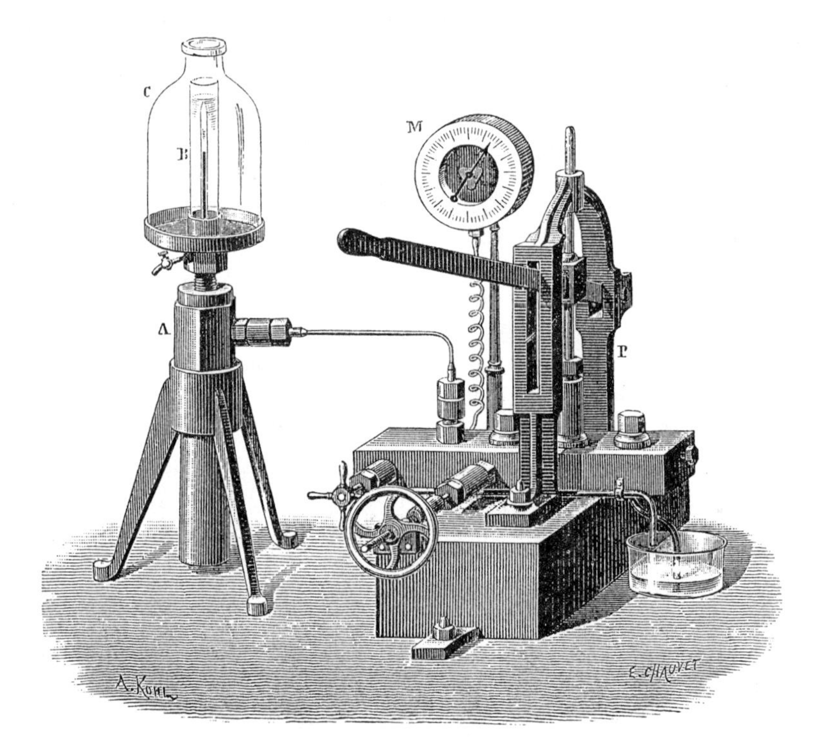

Figure 4.1.1. Cailletet's liquefaction machine. From Guillemin (1884).

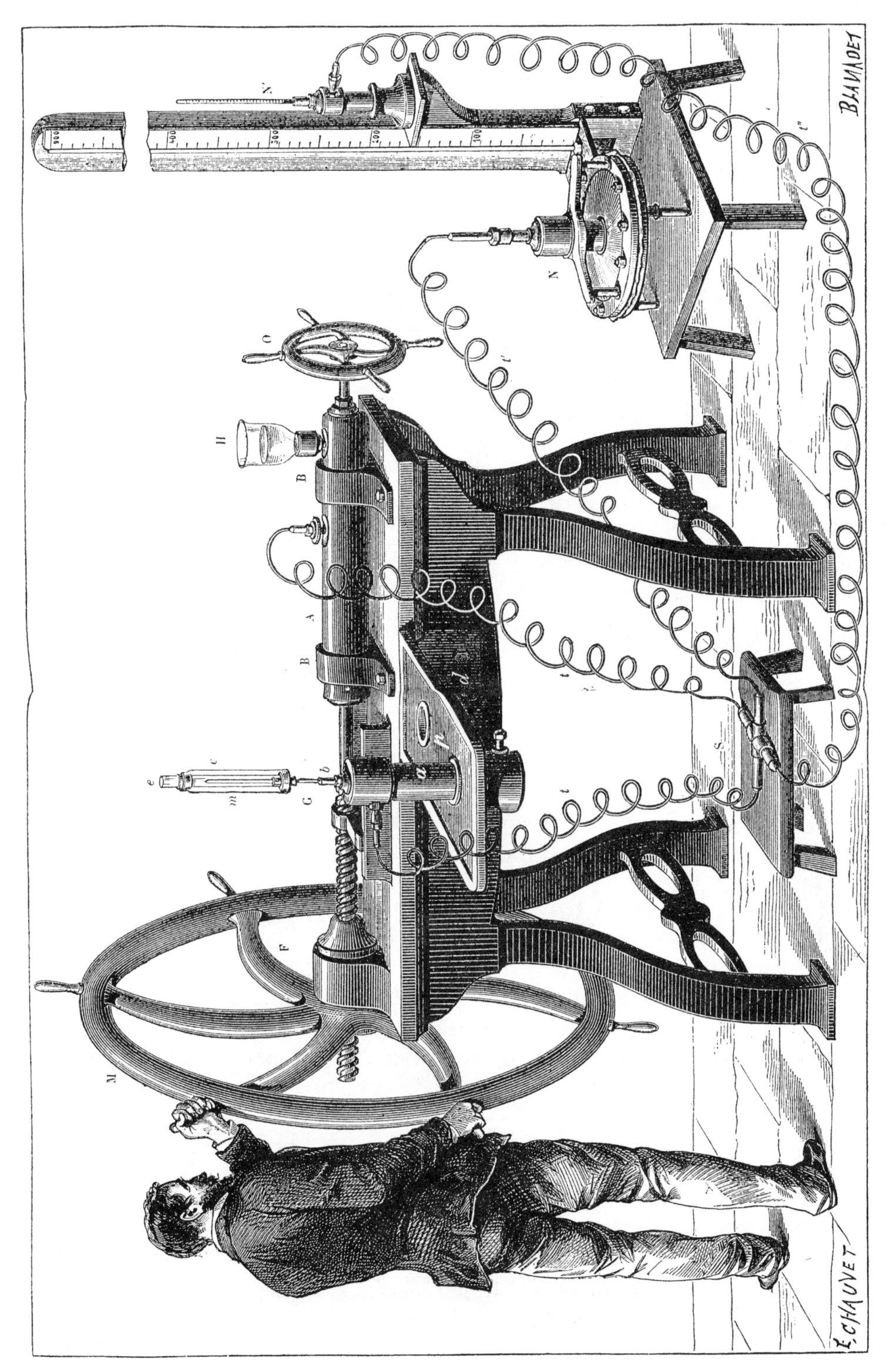

Figure 4.1.2. Liquefaction by the Cailletet method. From Girardin (1880).

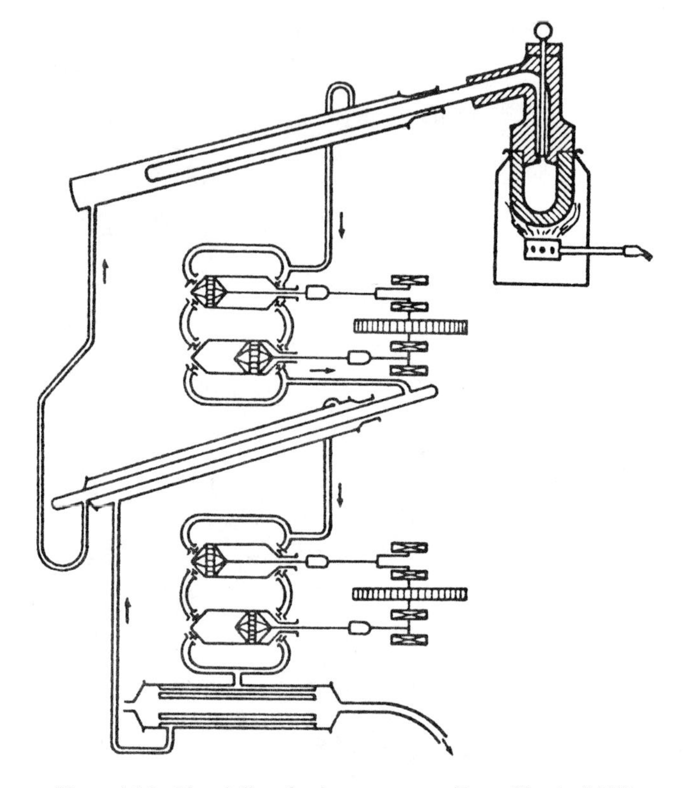

Figure 4.1.3. Pictet's liquefaction apparatus. From Claude (1926).

succeeded in liquefying oxygen. Pictet used another method, and proceeded rather more systematically. He performed the cooling in steps, by compressing and liquefying different gases (SO_2 and CO_2) with higher boiling points than oxygen, and succeeded in liquefying oxygen at 320 bars (32 megapascals) and –140°C. This multistage method, the "cascade method," was later used for liquefying hydrogen and helium (Figs. 4.1.3 and 4.1.4).

Many discoveries during the first years of cryotechnology were made at the same time and independently by several different scientists. The background to the broad interest in this area of research is to be found in the early history of industrialization, especially in the spheres of gas physics, gas theory, thermodynamics, and cooling technology.

Development of Cryotechnology up to 1910

Cailletet's and Pictet's epoch-making cryogenic experiments of 1877 were soon followed by others. In 1883 in Krakow the Poles Szygmunt Wroblewski and Karol Olszewski, using a machine purchased from Cailletet, succeeded in liquefying oxygen and nitrogen in larger quantities (Fig. 4.1.5). Wroblewski attempted to liquefy hydro-

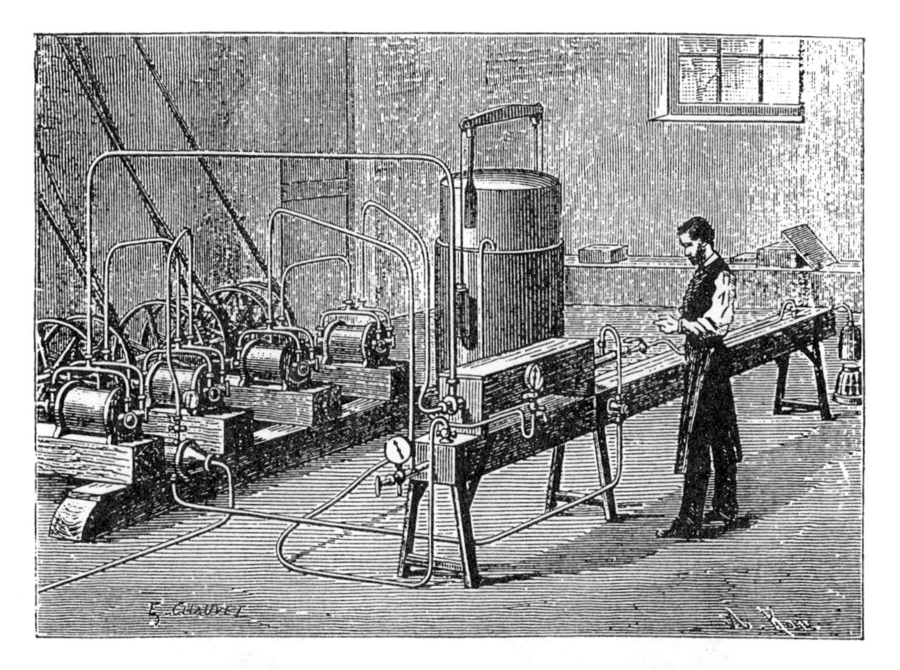

Figure 4.1.4. Pictet liquefaction laboratory. From Guillemin (1884).

gen by cooling compressed hydrogen to 73 K, but this only resulted in a transient vapor. However, in 1885 he published some remarkably accurate physical data. He gave hydrogen's critical temperature as 33 K (modern value, 33.3 K), its critical pressure as 13.3 atmospheres (modern value, 12.8 atmospheres), and its boiling point as 23 K (modern value, 20.3 K).

Gas liquefaction techniques up to 1895 involved:

- Compressing the gas to a high pressure, usually 50–200 atmospheres
- Chilling the compressed gas to as low a temperature as possible
- Expanding the chilled, compressed gas from a high to a lower pressure by means of an expansion valve

The cooling methods included, for example, evaporating an ether–solid carbon dioxide mixture or evaporating liquid ethylene, which Wroblewski used in liquefying oxygen. The third step made use of the Joule–Thomson effect for gases, based on experiments by J. P. Joule in 1845 and later refined by J. J. Thomson. They found that a gas, upon slowly expanding from a higher to a lower pressure, undergoes a change in temperature.

The Start of Industrial Air Gas Liquefaction[3]

In 1895, a breakthrough occurred in gas liquefaction techniques, employing regenerative cooling in the liquefaction process. Regenerative cooling means using a

Figure 4.1.5. The air gas liquefaction pioneers Raoul Pictet, Louis Cailletet, Szygmund Wroblewski, and Karol Olszewski. From Sloane (1899).

fluid as the coolant in a process in which the fluid is itself involved. In the liquefaction of gases, it means that the gas that is cooled by the Joule–Thomson expansion process is later used to cool the incoming compressed gas before expansion. The regenerative cooling concept was an old idea, first introduced by William Siemens in 1857 and used by Kirk, Coleman, Ernest Solvay, and others in refrigeration apparatus. In 1895, William Hampson in England and Carl Linde in Germany almost simultaneously obtained patents for equipment for liquefying air using the Joule–Thomson expansion process and regenerative cooling. Linde described his apparatus to physicists and chemists at Munich in 1895 (Fig. 4.1.6).

Linde's first idea was to use Siemens' regenerative procedure combined with the method that Solvay had tried in 1886, to use the cold air machine in steps toward lower temperatures. He only reached –95°C. Linde recognized the problems of lubricating and tightening machine parts at very low temperatures, so instead he used the method of using compression to 200 bars and throttling (Joule–Thomson effects in steps) to obtain sufficient cooling.

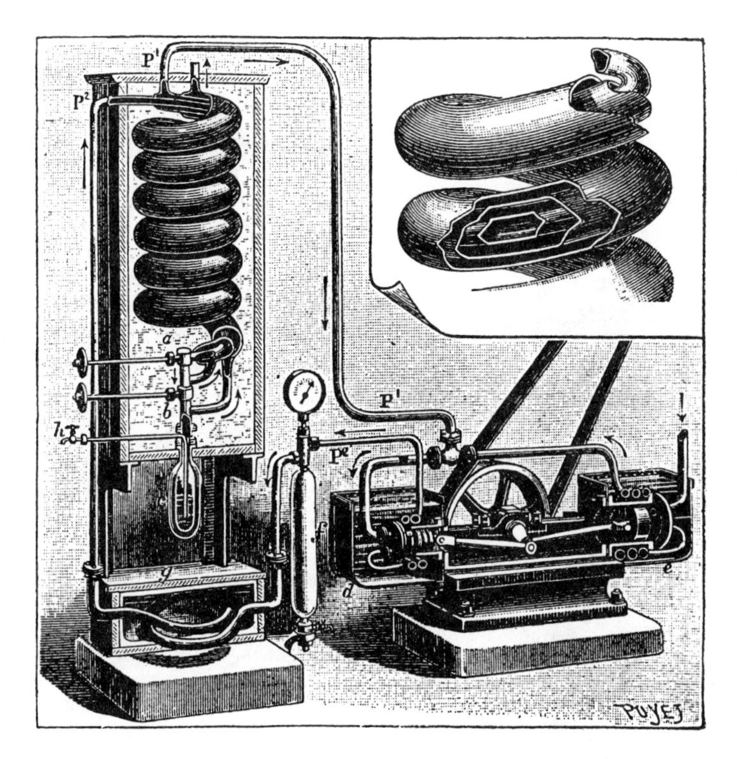

Figure 4.1.6. Linde liquefier. From Sloane (1899).

Hampson's process for liquefying air was simple. He compressed air to 200 bars, expanded it to 1 bar, and passed the expanded, cooled air through a baffled heat exchanger to cool the incoming compressed air. His method was not very efficient, using power at a rate equivalent to 3.7 kilowatts to produce 1 liter per hour of liquid air. Hampson had his apparatus working at Brin's Oxygen Works by April 1896 (Fig. 4.1.7).

Linde's approach was more complex than Hampson's, but it was also more efficient and suitable for large-scale production of liquid air. He used two stages of gas compression, precooled the air with a separate ammonia refrigeration system, and employed a coiled-tube heat exchanger having three concentric tubes for regenerative cooling (Fig. 4.1.8). The heat exchanger was insulated by a wood case filled with wool.

Regenerative cooling was also used to liquefy hydrogen for the first time. In 1892 James Dewar in London invented the vacuum vessel and started work on the liquefaction of hydrogen. On 10 May 1898, he succeeded in statically liquefying hydrogen. He first precooled gaseous hydrogen using liquid nitrogen, under 180 bars of pressure, then expanded it through a valve in an insulated vessel, also cooled by liquid nitrogen. The expanding hydrogen produced about 20 cubic centimeters of liquid hydrogen, about 1 percent of the hydrogen used.

In 1882 the Dutchman Heike Kamerlingh Onnes founded a cryogenic laboratory in Leiden with the aim of achieving ever lower temperatures. When Dewar

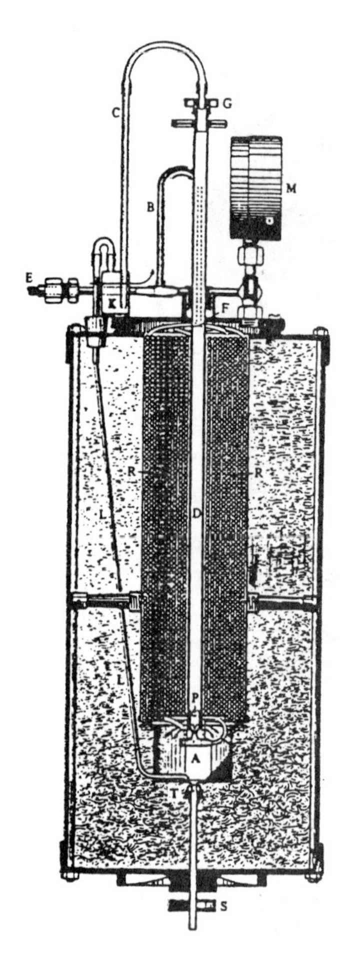

Figure 4.1.7. Hampson liquefier. From Sloane (1899).

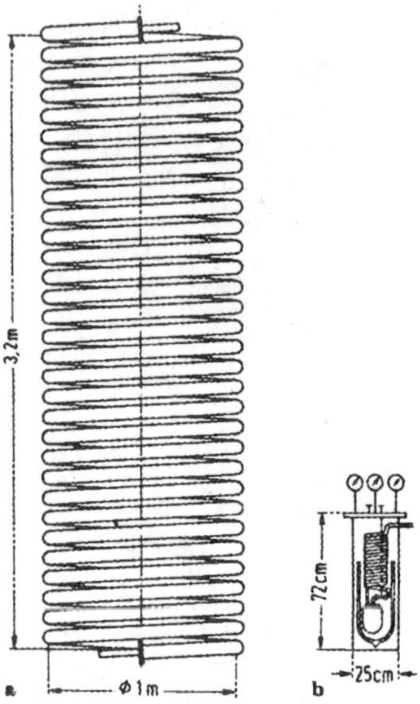

Figure 4.1.8. The size of heat exchanger used in Linde's first air liqufier compared to the size needed to obtain liquefaction today. Courtesy of Linde AG, Wiesbaden, Germany.

announced that he had liquefied hydrogen with a multistage method based on Joule–Thomson expansion and heat exchange, Kamerlingh Onnes started using that method in his attempts to liquefy helium. At last he succeeded in 1908 by using liquid hydrogen at reduced pressure (about 90 millibars, 14 K) in the last cascade step (Figs. 4.1.9 and 4.1.10). Thus all known gases had been liquefied, and the race toward absolute zero could begin.

In Paris in 1899 Georges Claude further developed Cailletet's and Solvay's methods of using piston expansion engines for liquefying air, and by improving insulation and lubrication, succeeded in 1902 through expansion from 40 bars and −100°C. Claude developed an efficient and reliable liquefier with a claimed performance of 0.75 liter of liquid air per kilowatt-hour compared with 0.70 liter for the Linde liquefier.

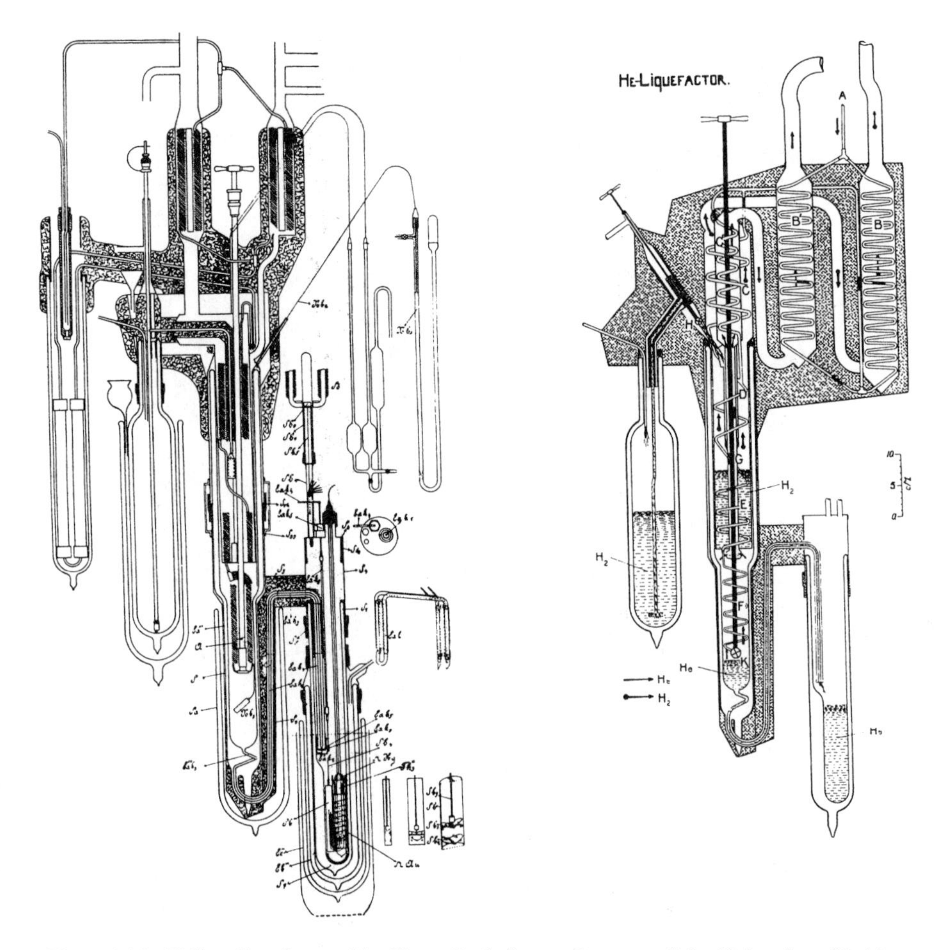

Figure 4.1.9. Helium liquefier used by Kamerlingh Onnes. Courtesy of the University of Leiden.

Production and Use of Liquefied Gases[4,5]

The use of industrial air gases, hydrogen, helium, and natural gas increased sharply during the years after the World War II. Due to advantages deriving from weight considerations, these gases are largely handled in the liquified state. The increased availability of liquified gases at cryotemperatures (<120 K) led to the development of industrial handling techniques.

Liquid Air

The work of Cailletet, Pictet, and Dewar showed that it was possible to convert the so-called permanent gases, including air, into the liquid state at low temperatures.

Figure 4.1.10. Kamerlingh Onnes and Johannes Diderik van der Waals in Onnes' laboratory. Courtesy of the University of Leiden.

For some years liquid air was produced in laboratories in very small amounts, but William Hampson in England, Carl Linde in Munich, Germany, and Charles E. Tripler in the United States succeeded independently in 1895–1897 in developing techniques for producing significant amounts of liquid air. Tripler in fact produced it in considerable quantities and shipped it by rail in insulated containers on occasion. Hampson's initial patent covered only air liquefaction, but Linde went a step further and described a process of fractional vaporization which produced gas enriched with oxygen, so-called "Linde air." The Linde process did not go unchallenged. Georges Claude in Paris soon became a competitor, improving the efficiency of his liquefaction process. After some 6 years of experimenting he succeeded in developing a satisfactory expansion engine. Early units were lubricated with petroleum ether, but finally Claude found that the use of degreased leather washers as piston rings

Figure 4.1.11. Industrial liquefaction pioneers Carl Linde, Georges Claude, James Dewar, and Heike Kamerlingh Onnes.

provided an operating solution. His liquefaction method was further developed from 1906 onward by the German Paul Heylandt.

Liquid Hydrogen (LHY)

Hydrogen is normally produced by electrolysis of water or from petroleum processes. Production of hydrogen through fractional distillation of flue gas at low temperature was carried out on a small scale by Linde in 1909.

Liquid hydrogen on a large scale was first needed in the United States for thermonuclear research, and in 1952 the U.S. Atomic Energy Commission constructed a plant in Boulder, Colorado, with a capacity of 300 liters per hour. The space program in 1959 increased the need, as liquid hydrogen and liquid oxygen came into use in large quantities as rocket fuel. In 1963 Philips' Stirling engine was developed for liquefaction of hydrogen on a laboratory scale. Since the "energy crisis" in 1973 great development effort has gone into methods of using hydrogen in different ways as a power fuel and source of energy.

Liquid Natural Gas (LNG)

Natural gas was first liquefied on a pilot scale by Godfrey Cabot in 1915. Separation of helium from liquefied natural gas was started 1917 and was first mainly used for filling airships until they were replaced some decades later by the development of heavier-than-air craft.

A new application of LNG as stored energy in peak-shaving plants was temporarily stopped by the disastrous accident at a Cleveland plant in 1944. Liquefaction plants were again constructed around 1965, first according to the Linde principle and then with the cascade principle, in which three fractions—propane, ethylene, and methane—were used in separate cycles. In 1959 the Russian A. P. Kleemenko developed a cooling cycle (the mixed refrigeration cycle [MRC] process) where the three fractions are not separated. In 1959 LNG was transported for the first time in a ship across the Atlantic Ocean.

Liquid Helium (LHE)

Laboratory-Scale Production

After Kamerlingh Onnes at the Leiden laboratory liquefied helium in 1908, one might have expected many low-temperature institutions to follow suit. However, 15 years were to pass before John M. C. Lennon repeated the performance in Toronto in 1923. During the following years this was followed by plants in Berlin (1925) and Kharkov, USSR (1930), and later at Cambridge and Oxford in England and in Berkeley, California.

Thus for 15 years the superconductivity discovered at Leiden could only be studied there, and similarly there was no competition for a long time in the race to reach lower temperatures after Leiden reached 0.83 K in 1922. When Leiden's monopoly was broken, things happened very fast. In Berlin in 1926 Franz Simon (later Sir Francis Simon at Oxford) developed a desorption cryostat for production of liquid helium. Helium gas was desorbed from activated carbon under supercooled LHY (15 K).

In 1930 Martin Ruhemann developed a miniaturized helium cryostat with Joule–Thomson expansion. Until then all helium liquefaction had been done according to this principle, but in 1934 Pjotr Kapitza in Cambridge succeeded in liquefying helium with the help of a Claude system (expansion engine). This principle was also used in the development of the first commercial helium liquefaction apparatus, which was developed by the American Samuel Collins in 1946, and made it possible even for small laboratories to liquefy helium.

Already in 1911 researchers in Leiden had discovered a density maximum for helium at 2.2 K. The phenomenon was studied during the 1920s, but remained unexplained until 1931, when Onnes's successor Willem Keesom and the German Klaus Clusius published studies which suggested that there were two phases in liquid helium. These were called He I and He II, with a transition temperature of 2.19 K.

Large-Scale Production of LHE

Industrial production of helium from natural gas started in Canada the United States in 1917. The process was based on liquefying and separating the other components of natural gas from the helium gas (see Chapter 3.6).

A Success Story that Ended in a Personal Disaster[6]

In reviewing the successful industrial air liquefaction development in Europe by Linde and Hampson, the accomplishments of Charles E. Tripler in the United States should also be mentioned. An article in the February 12, 1898, issue of *Scientific American* reports: "The economical liquefaction of air in large quantities has been recently accomplished by Mr. Charles E. Tripler of New York after several years of experimental work. Two and a half gallons of the liquid were recently sent from his laboratory to Prof. Barker, Professor of Physics at the University of Pennsylvania . . ." The article reflects an interesting scientific development. By 1898, Tripler in New York had, independent of his European colleagues, constructed a similar but much larger air liquefier, driven by a 75-kilowatt steam engine, which produced literally gallons of liquid air per hour (Figs. 4.1.12 and 4.1.13).

Tripler was born in New York in 1849. From his early years he showed a great interest in mechanics and experimenting. In the early 1870s he tried to produce a motor that would be driven by gas. He experimented on an engine driven by ammonia. His aim was to find a continuous cycle, with Carnot's cycle providing his inspiration. He also tried gasoline and naphtha, but these did not work to his satisfaction because of troubles with gas-tight joints at high pressures. At this time Thomas Edison was beginning his work, and it was an era in which electricity was being applied in chemistry. For a few years Tripler left the field of mechanical and scientific work and took up art, painting and exhibiting. When he returned to scientific work about 1884 he studied gold extraction and amalgamation and then returned to his first love: gas experiments. At this point he discovered the principle on

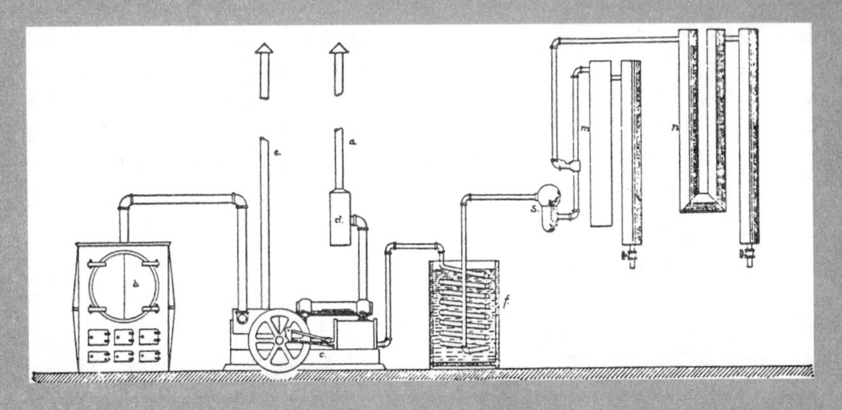

Figure 4.1.12. Tripler's air liquefaction plant, ca. 1897. From Sloane (1899).

Continued

Figure 4.1.13. Charles Tripler. From Sloane (1899).

which his work on the liquefaction of air would be based. He liquefied various gases, and during the 1890s began his attempts to liquefy air.

In 1897 he succeeded in liquefying up to 22 liters per hour of air in a completely home-made liquefier, at that time the largest in the world. The most striking feature of the Tripler process, apparatus, and plant was its efficiency and simplicity. Although his European colleagues were assisted in their expensive efforts by liberal gifts from governments and private individuals, Tripler's plant was erected at his own expense.

In order to preserve air, Charles Dewar in England developed the vacuum bulb. Tripler took common tin cans, lined them with felt, filled them with liquid air, and sent them off hundreds of miles by rail. Whereas Dewar excited great enthusiasm by public lectures on liquid air and liquefied gases (see Fig. 3.1.11), Tripler showed interested scientists his experiments with liquid air in his private laboratory, along with his boiler, air compressor, and simple liquefying apparatus. Tripler used no extra sources of cold in his laboratory. All the liquefaction was done by its own powers within its own system. A steam boiler was used to supply steam to a compressor at about 6 bars (Fig. 4.1.14).

Tripler's enthusiastic demonstrations drew attention from the media and investors, but he lacked sufficient theoretical scientific knowledge, which led to his downfall. In 1899 he began announcing that he had discovered a practical means of perpetual motion. Ten million dollars was raised by investors on Wall Street to start his General Liquid Air and Refrigeration Company. Inevitably his ideas regarding perpetual motion were, of course, found to be wrong and in 1902 Tripler was declared bankrupt. The interest in investing money in cryogenics and air liquefaction from U.S. financiers was lost for a long time. For this reason Carl von Linde ran into great difficulty in establishing his subsidiary in the United States in 1907.

Figure 4.1.14. Tripler's liquefaction laboratory. See also Fig. 4.1.12. From Sloane (1899).

The Liquid Helium Race[7]

The liquefaction of gases, that is, their transformation to liquid form by cooling, is used to achieve the lowest of temperatures. Therefore the sought-after honor of being the first to liquefy helium, the gas with the lowest liquefaction temperature, started a race at the turn of the century [see the book by Mendelssohn (1966) cited in the References to this subsection.]

James Dewar, professor at the Royal Institution in London, stated in an interview in *The Times* in 1894 that he was to engage in the "last possible" stage toward absolute zero by attempting to liquefy hydrogen (the properties of helium had not been discovered at that time). Much earlier, the Dutchman J. D. van der Waals had theoretically determined the boiling point of hydrogen. In Krakow, there were other contestants, Szygmunt Wroblewsky and Karol Olszewsky. They had been working toward the same goal as Dewar, and at the University of Leiden, Heike Kamerlingh Onnes had just founded a cryoinstitute directed toward studying the low-temperature characteristics of different materials.

One year later, relations between Olszewsky and Dewar became strained as a result of the Pole's having publicly accused Dewar of stealing results from Krakow without referring to their source. At one of Dewar's public presentations of his work on the condensation of hydrogen, Ramsay, who had close contact with the Poles, erroneously stated that Olszewsky had already achieved the goal. Since no contradiction from Ramsay was forthcoming, the eccentric Dewar was bitter, to say the least. At the same time, Dewar was engaged in a contest of priorities with Hampson, who had patented a method for the large-scale condensation of air.

Continued

Ramsay discovered helium and other noble gases in 1895–1898, and Dewar's condensation of hydrogen was made public in 1898. The goal now became that of liquefying helium. Dewar needed Ramsay's helium and neon for his low-temperature experiments, whereas Ramsay needed liquid hydrogen in order to separate helium and neon. Cooperation between the two men was unthinkable. Ramsay solved his problem by appointing Morris Travers as his assistant. Travers was clever at building apparatus, and he produced liquid hydrogen only 2 years after Dewar had first been successful. Now Dewar had three competitors, although two were to fall by the wayside in 1905, leaving only Onnes in Leiden, who first condensed hydrogen in 1906.

Dewar was an odd genius, somewhat eccentric and impulsive, but he was also a brilliant physicist and capable craftsman, who had even made a name as a violin maker. He loved to shine as a scientist, but unfortunately it was sometimes for the results of others. In many ways Onnes was his opposite, in that he was a scientific administrator and founder of institutions who surrounded himself with the brightest of technicians and made them work hard. His experiments were planned down to the minutest detail and his method of working was open, even to his competitors. Methodically he built up a plant which, step by step, would enable him to condense helium.

In 1908, in an article entitled "*The Nadir of Temperature*," Dewar wrote about the hardships that had beset him in his attempts to liquefy helium. He claims he would have succeeded had he had 100–200 liters of helium gas. The gas was there, just a few miles away, but it was with Ramsay and was therefore unavailable. Not only that, but a young technician had released Dewar's collected helium into the air by turning the wrong tap.

A footnote was later added to the article: "*Helium liquefied by Dr. Kamerlingh Onnes in Leiden on the 9th July 1908.*" The race was over; Dewar never got over the defeat. He deserted low-temperature research and began, among other things, to study thin films of liquid in soap bubbles. Not until 15 years later was any other institution in the world able to liquefy helium.

Cryoliquids

It is relatively easy to achieve cooling, that is, to remove heat. All that is required is a heat sink at a lower temperature; the heat is transferred to it automatically.

Unfortunately, nature only has one heat sink that reaches down to the cryogenic temperature range (defined as $<120\,K$). It is very remote and keeps the earth in heat balance at about 300 K. In most cryogenic applications, a liquid having a low boiling point or a cooled gas is used as a heat sink. The first success in producing a cryoliquid (a gas liquefied at a cryotemperature) was in 1877, when Louis Cailletet liquefied a few drops of oxygen.

The highest temperature at which a gas can be in liquid form at any pressure is called the critical temperature. The temperature at which a cryoliquid boils at normal pressure (1 bar absolute) is called the normal boiling point. The highest temperature at which a gas can be in solid form at any pressure is called the triple point. The temperature regions of cryoliquids are shown in Fig. 4.1.15. Cryoliquids produced on a large scale are used in industry as cooling agents. During the last 50 years, however, efficient cooling machines, usually called cryogenerators, have been developed.

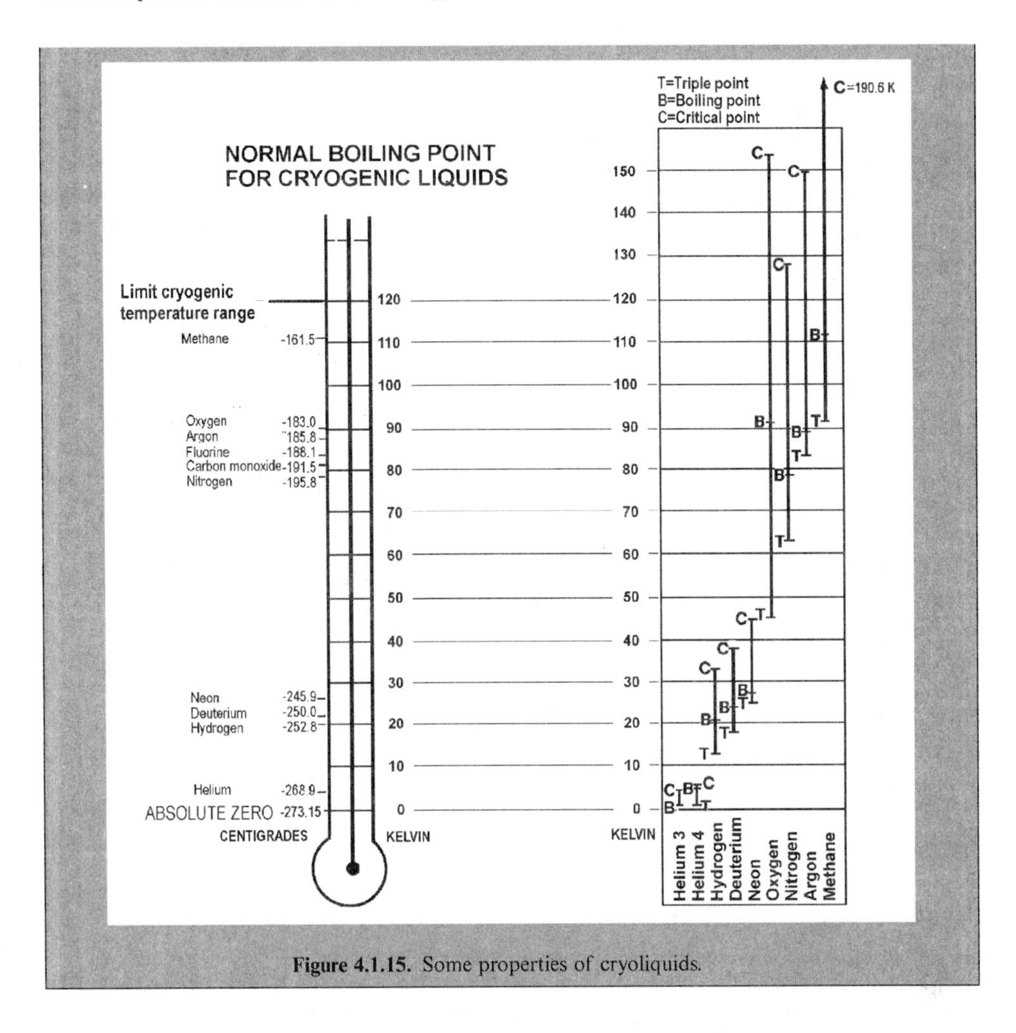

Figure 4.1.15. Some properties of cryoliquids.

References

1. Claude, G. (1926). *Air liquide, oxygène, azote, gas rares.* Paris: Dunod 1926.
2. Almqvist, E. (1988). Kryoteknikens historik [The history of cryotechnology]. *Kosmos (Swedish Physical Society annual)* **1988**:137–160.
3. Brickwedde, F. G., Hammel, E. F., and Keller, W. E. (1992). The history of cryogenics in the USA. Part I: Cryoengineering. Part II: Cryogenic research. In *History and Origins of Cryogenics* (R. G. Scurlock, ed.) (Chapters 11 and 12), pp. 357–519. New York: Oxford University Press.
4. Tripler, C. E. (1899). Liquid air—newest wonder of science. *The Cosmopolitan* **25**:119–126.
5. O'Conor Slone, T. (1899). *Liquid air and the liquefaction of gases*. London: Sampson Low, Marston.
6. Morton, H. (1899). The liquid air fallacy. *Scientific American* **80**(6):404–405.
7. Mendelssohn, K. (1966). *The quest for absolute zero.* London: Weidenfeld & Nicholson.

4.2. THE PATH TO ABSOLUTE ZERO. SUPERCONDUCTIVITY[1,2]

There must of necessity exist a limit to the degree of cold; this state must be manifested by a complete cessation of the rotary movement of particles.

Mikhail V. Lomonosov (1711–1765)

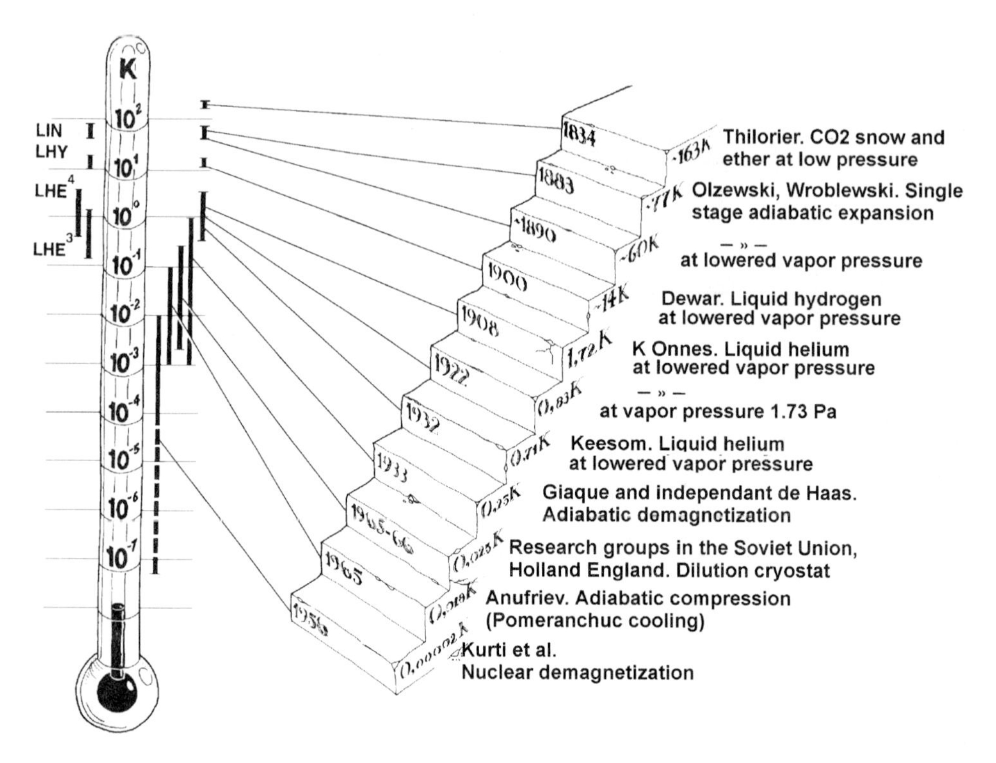

Figure 4.2.1. Important steps toward absolute zero. Courtesy of Teknisk Illustration, Umeå, Sweden.

Toward Absolute Zero[3,4]

Following Kamerlingh Onnes's successful liquefaction of helium in Leiden 1908, repeated attempts were made to reach lower temperatures (Fig. 4.2.1). The method of choice was to reduce the vapor pressure of liquid helium with several and more effective pumps. By the time Kamerlingh Onnes died in 1926 it was thought that the last steps had been taken toward absolute zero. Using a Langmuir pump, W. Keesom in 1922 reached 0.83 kelvin with a vapor pressure of 1.73 pascals. A month or so after Onnes's death, the Canadian Francis Giaque in Berkeley, California, and Peter Debye in Leipzig, Germany, working independently of each other, presented a new method of cooling using adiabatic demagnetization. A paramagnetic substance was used in place of a gas or liquid and a magnetic field was used instead of expansion to reach lower temperatures. An external magnetic field has a powerful effect on the amplitude of the electron lattice vibration. The first successful attempts using this method were carried out by Francis Giaque and Duncan McDougal (Figs. 4.2.2 and 4.2.3) in Berkeley and at the same time by Wander de Haas, E. C. Wiersma, and Hendrik Kramers in Leiden. During those first attempts 0.25 kelvin was reached. Using the same method in 1976, 1.2 millikelvins was reached. Due to the magnetic cooling a breakthrough had been achieved toward temperatures where classical science no longer applied. The implications of the third law of thermodynamics as set out by Walther Nernst in 1906 were now clarified by, among other things, the theoretical work of Giaque.

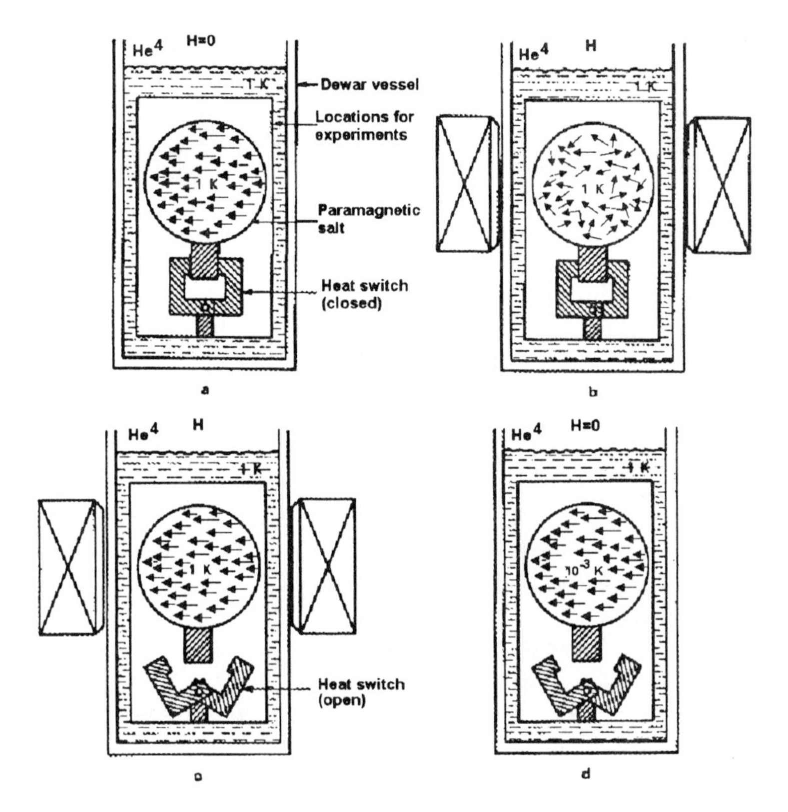

Figure 4.2.2. Magnetic cooling principle. Courtesy of Dr. T. Lindquist.

Figure 4.2.3. William Giaque, Nobel Laureate in Chemistry, 1949. Courtesy of the Nobel Foundation. ©Nobelstiftelsen.

Nuclear cooling, that is, adiabatic demagnetization carried out by affecting nuclear spin, was presented as a theory by Cornelius Gorter in Leiden in 1934 and independently by Nicholas Kurti and Franz Simon, in Kurt 1935. Kurti, Simon, and Kurt Mendelssohn had fled to Oxford and continued the development of the nuclear spin method, analogous in principle to the electron spin method, but experimentally very much more difficult. In 1956, starting from a temperature of 0.012 kelvin, a research team working under Kurti reached approximately 0.00002 kelvin for a very short period. Since then, temperatures of <0.1 kelvin have been reached for longer periods.

The superfluid properties of He II were discovered and studied during the 1930s by groups at places including Leiden, Toronto, Cambridge, and Oxford. The phenomenon was at the same time reported independently by Jack Allen and Donald Misener in Cambridge and Pjotr Kapitza, then in Moscow. The thermomechanical effect was described by Jack Allen and Harold Jones in 1938, and the converse, the mechanocaloric effect, was observed by John G. Daunt and Kurt Mendelssohn in Oxford.

A new platform for low-temperature studies down to 0.2 kelvin was created when the helium 3 isotope became commercially available at the end of the 1950s. Helium 3 (^{3}He), which was first liquefied in 1938, is very scarce in nature, but can be obtained through decomposition of tritium in fusion reactions.

The first ^{3}He cryostat with reduced pressure appeared in 1955, and the cooling method was fully developed around 1965.

The ^{3}He/^{4}He dilution cryostat is based upon the fact that below 0.8 kelvin the helium isotopes cannot be mixed in all proportions. An ^{4}He-enriched, heavier phase and a lighter, ^{3}He phase are created. By means of a pumping system the ^{3}He can be made to migrate during the absorption of heat. Heinz London suggested the use of this method in 1951. The first apparatus was designed by London, Eric Mendosa, and Geoffrey Clark in England in 1962 and was built 3 years later by Dutch (Leiden), Russian (Dubna), and English (Manchester) groups. A stationary 3 millikelvins has been achieved with this method. Superfluid ^{3}He was discovered in 1972 in the United States by Douglas Osherhoff, Robert Richardson, and David Lee at Cornell University.

The Pomeranchuk cooler, as invented by the Russian Isaak Pomeranchuk in 1950, uses adiabatic compression of a two-phase mixture of liquid and solid ^{3}He. This cooler was first constructed in 1956 by the Russian Yuri Anufriev in Moscow. He reached 18 millikelvins (Fig. 4.2.4).

Most low-temperature research has been aimed at the study of mechanical, electrical, and magnetic properties of solid materials at very low temperatures. Fundamental properties can be studied much more easily at these temperatures because the thermal vibration movements of the atom are considerably reduced.

Superconductivity[5–8]

Discovery

This phenomenon was discovered by Kamerlingh Onnes and his assistant Gilles Holst in 1911 when, during the study of the low-temperature properties of mercury, they found that the external resistance disappeared (Fig. 4.2.5). Attempts were made

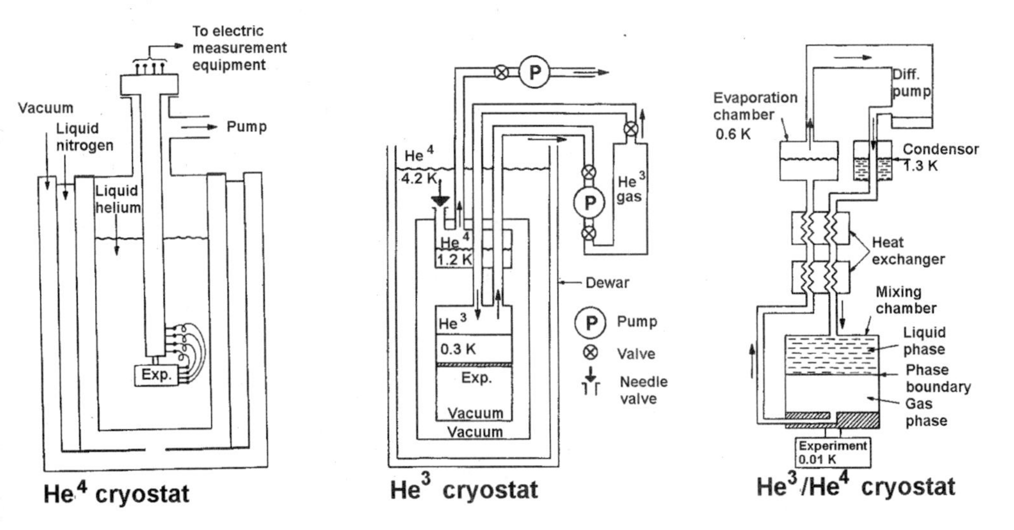

Figure 4.2.4. Cryostats for very low temperatures. Courtesy of Dr. T. Lindqvist.

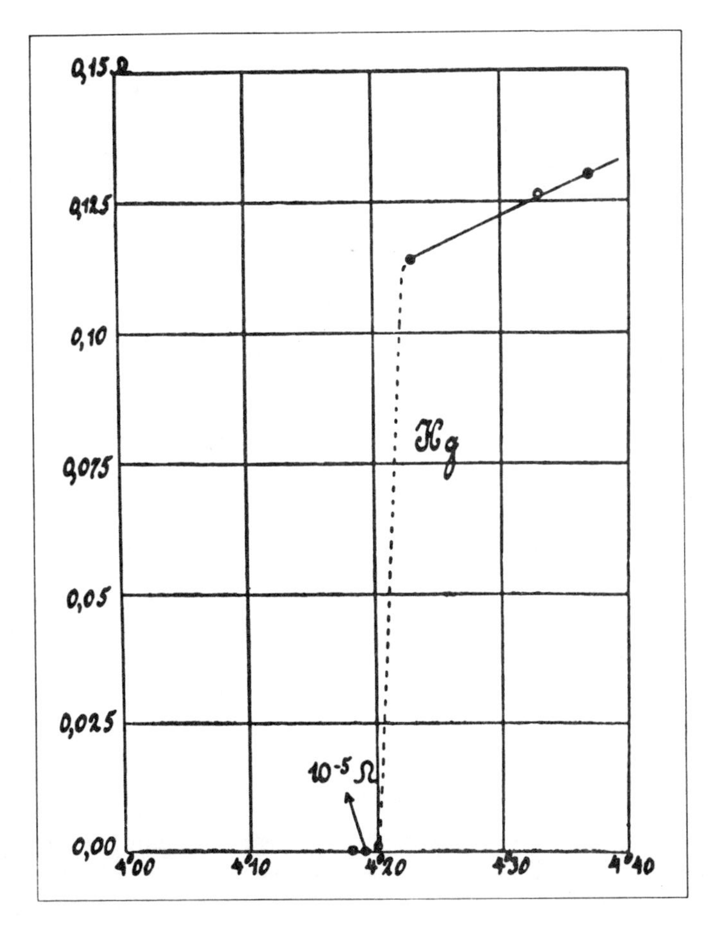

Figure 4.2.5. Kamerlingh Onnes' result illustrating the discovery of superconductivity in 1911. Mercury shows a sudden drop in resistance at liquid helium temperature.

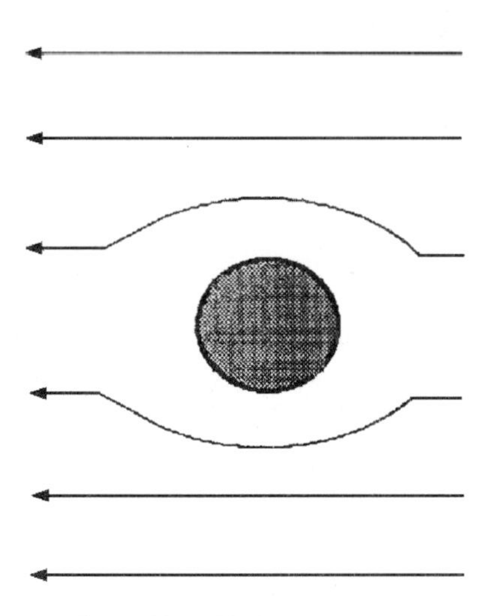

Figure 4.2.6. The Meissner effect.

to utilize the discovery in electromagnets, but failed with the metals available at that time. The superconductivity disappeared when a magnetic field was applied. The next great discovery took place in Berlin in 1933, when Walther Meissner and Robert Ochsenfeldt found that superconductors showed an almost perfect diamagnetism, that is, an apparent repulsion of lines of magnetic induction from a superconducting material. This phenomenon is known as the Meissner effect (Fig. 4.2.6). Superconductors showing the Meissner effect are known as type I. The others, type II, are gradually penetrated by an external magnetic field in quantized eddies. Superconductors of type II were discovered in Kharkov in 1937 by Lev Shubnikov. The phenomenon was explained by Alexej Abrikosov 20 years later.

During the 1930s different theories about superconductors were presented. The two-liquid theory was presented by Cornelius Gorter and Hendrik Casimir of Leiden in 1934. The phenomenological theory of the Germans Fritz and Heinz London in 1935 stated that superconductivity is a quantum phenomenon. The theory of the Londons was generalized in 1950 by the Russians Vitaly Ginzburg and Lev Landau, with others (the GLAG theory). The theory was tested and adapted by Lev Gorkov, who showed it to be in agreement with the microscopic theory and Abrikosov's theory of type II superconductors. Flow quantization in accord with the 1935 ideas of the Londons was shown experimentally in 1961–1962. In 1957 John Bardeen, Leon Cooper, and Robert Schrieffer presented the BCS theory of superconductors.

Development

In 1960 Ivar Giaever demonstrated the so-called tunnel effect, a quantum process whereby quasiparticles passes through barriers where classical physics says they should have been stopped. In 1962 in Cambridge, Brian Josephson discovered a special case: According to the BCS theory, paired electrons could pass through a thin insulation barrier which separated two superconductors. The so-called Josephson effect led to developments in superconductive electronics and extremely sensitive detectors.

After World War II many laboratories had access to liquid helium, and systematic work began to find materials that showed superconductivity at higher temperatures. The American Bernd Matthias has been a noted pathfinder. The work of his group led to the discovery in 1954 of Nb_3Sn, with a critical temperature of 18 kelvins, and in 1973 of Nb_3Ge, with a critical temperature of 23 kelvins.

A new era began in 1964 when Vitaly Ginzburg studied laminated materials having alternate metal and dielectric layers and William Little examined long organic molecular chains. These systematic studies of materials resulted in a breakthrough in April 1986 when, after 4 years of work, Georg Bednorz and Alex Müller at IBM in Zürich discovered superconductivity at 35 K in a ceramic metallic oxide belonging to the Ba–La–Cu–O system. This discovery led to a new race to find breakthroughs in superconductor technology. In February 1987 Paul Chu in Houston successfully achieved superconductivity at 90 K by replacing the lanthanum with yttrium in the above-mentioned ceramic system.

Applications

The first practical application of superconductivity came in 1956 with the cryotron, an ultrarapid switch discovered by Dudley Buck. The first superconductive magnets were made at the beginning of the 1960s and in 1963 they were used for the first time in the field of high-energy technology as a replacement for LHY-cooled magnets. In 1976 superconductive magnets were introduced into CERN's proton accelerator. A superconductive motor (the Fawley motor) was constructed in 1969 by Anthony Appelton at IRD in England.

Also in the 1960s superconducting quantum interference devices (SQUIDs), the most sensitive detectors of magnetic fields available, were developed. They came into practical use about 1976. At that time a prototype levitating train track was built in Japan, and later one was built in West Germany. Another application for superconducting magnets is magnetic resonance imaging, introduced in 1981, which is now a common diagnostic tool in medicine. Other electronic uses of superconductivity are in infrared detection and in signal processing devices.

Superconductors are also of potential use for, for example, very large scale applications in the generation, storage, and transmission of electricity and thermonuclear fusion reactors. The crucial point is the development of reliable materials that are superconducting above liquid nitrogen temperature (77 kelvins).

References

1. Kurti, N. (1978). From Cailletet and Pictet to microkelvin. *Cryogenics* **18**:451–458.
2. Mendelssohn, K. (1966). *The quest for absolute zero.* London: Weidenfeld & Nicholson.
3. Giaque, W. F., Johnston, H. L., and Kelley, K. K. (1927). *Journal of the American Chemical Society* **49**:2367.

4. Lounasmaa, O. V. (1979). Towards the absolute zero. *Physics Today* **1979**(12):32–41.
5. Bardeen, P. J., Cooper, L. N., and Schrieffer, J. R. (1957). *Physical Review* **108**:1175.
6. de Nobel, J. (1996). The discovery of superconductivity. *Physics Today* **1996**:40–42.
7. de Bruyn Ouboter, R. (1997). Heike Kamerlingh Onne's discovery of superconductivity. *Scientific American* **1997**(3):84–89.
8. Dahl, P. F. (1992). *Superconductivity.* New York: American Institute of Physics.

4.3. DEVELOPMENT OF THE PRODUCTION OF AIR GAS (1910–CA. 1975)

La liquéfaction industrielle de l'air n'est pas seulement une révolution scientifique, c'est aussi et surtout une révolution économique et sociale.

Arsène d'Arsonval 1903

Air Gas Liquefaction and Separation[1–3]

The industrial production of air gases by distillation of liquid air began in England, France, and Germany in 1902, after liquefaction on a large scale had been pioneered at the end of the 19th century by William Hampson, Georges Claude, and Carl Linde in Europe and Charles Tripler in the United Sates. Several of the great industrial gas companies of our day descend directly from these developments.

The problem remained of separating the valuable oxygen from the other components of air. Linde began studying devices for obtaining pure oxygen by means of rectification. A successful practical solution, based on an idea of the Canadian LeSueur in 1900, was developed by Carl Linde's son Friedrich in 1902. Linde used a high-pressure distillation column to produce nearly pure oxygen. A feed air pressure of about 50 bars was required to create and maintain liquid in steady-state conditions. The gas leaving the top of the column contained about 7 percent O_2. Only 70 percent of the oxygen from the feed air could be recovered. Linde's company Linde Eismachinen, introduced a single-column system for rectifying liquid air in 1903.

With the development by Rudolph Wucherer and Linde of a double-column system in 1910, the cryogenic air-separation industry really began. Two columns are needed to separate oxygen and nitrogen from air if no special recycle system for reboil and reflux is used. The lower column operated at about 5 bars and the upper one at just over 1 bar. The development of this device solved the requirements of reflux and reboil in the two parts of one column.

Claude also studied the distillation problem. By 1905 his plant could produce about 100 cubic meters per hour of oxygen of 93 percent purity. He used a reflux condenser (dephlegmator) in place of the Linde high-pressure column. Under steady-state conditions, for a plant producing gaseous products, the feed air pressure could be dropped to 18 bars. This apparatus was easier to handle from the thermodynamic point of view, but the efficiency of fractionation was not very great. Nitrogen purity and oxygen recovery were not as high as with the Linde double column.

A further process for oxygen production of higher purity was developed in this period by the electrolysis of water. This, however, was unable to match the existing

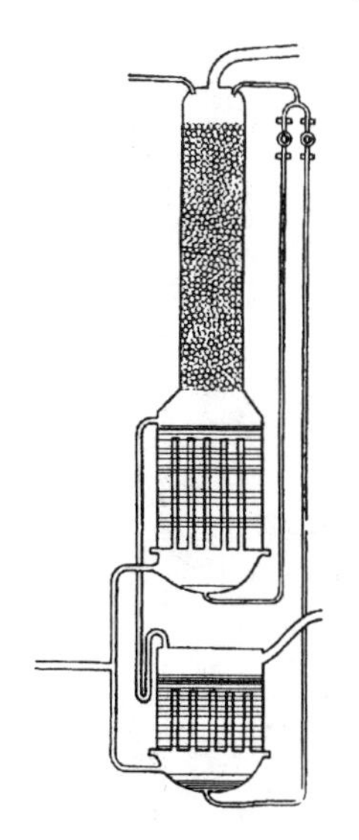

Figure 4.3.1. Linde single column, 1902. From Claude (1926).

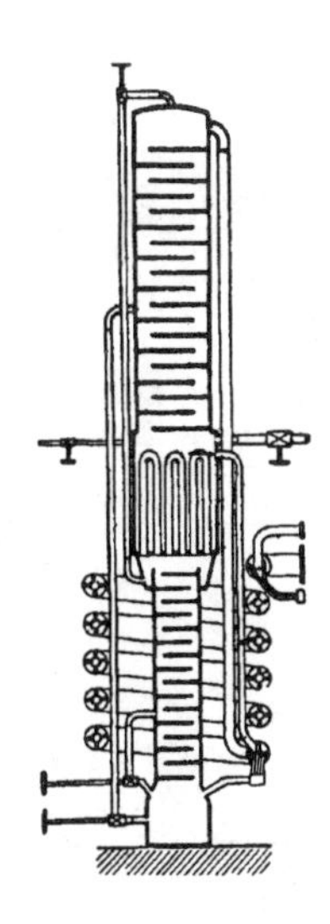

Figure 4.3.2. Linde double column, 1910. From Claude (1926).

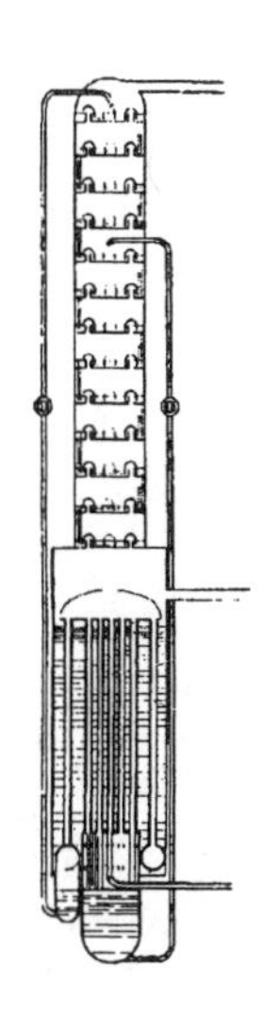

Figure 4.3.3. Claude column, 1905. From Claude (1926).

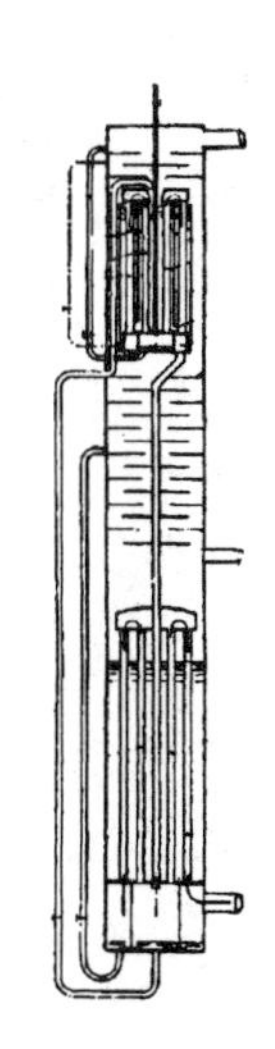

Figure 4.3.4. Messer column, ca. 1912. From Claude (1926).

processes due to the high cost of power required and the lack of demand for the hydrogen by-product at that time.

Early Air Gas Applications[4]

During the first years of the 20th century the market demand for oxygen grew as a result of new welding and cutting methods. The production of pure oxygen was of great importance for the development of the oxyacetylene torches used in metal-working. Pure nitrogen was needed for the large-scale production of calcium nitrate, ammonia, and saltpeter.

In 1903, Linde cofounded the Vereinigte Sauerstoffwerke GmbH to market oxygen. He also developed apparatus for producing pure nitrogen and hydrogen from water gas by means of the partial condensation of carbon monoxide (the Linde–Frank–Caro process, 1910). Linde's development of the double column laid the basis for the modern air separation plant. The decade after 1910 saw great demand for the other main components of air: From 1913 nitrogen was used in ammonia production according to Haber's principle. Other chemical industries needed a new source of nitrogen because the importation of Chilean potassium nitrate was stopped at the end of World War I. From about 1912 the newly discovered rare gases argon (Ramsay and Rayleigh, 1894) and neon, krypton, and xenon (Ramsay and Travers, 1898) found technical uses, among other things, for filling electric light bulbs.

During the 1914–1918 war the production of oxygen was greatly increased due to demands particularly for welding and cutting. In the postwar period the discovery of a safe method of storing acetylene in a porous mass in steel cylinders enhanced the use of oxygen together with acetylene in welding activities. From the late 1920s it became apparent, especially in Germany, that there was potentially a large demand

Figure 4.3.5. Early Linde plant with three units each producing 300 cabic meters of oxygen per hour. From Claude (1926).

for oxygen in various industries. In particular, the steel industry, for example, in the bottom-blown Bessemer process, could accelerate production and improve its product by using oxygen instead of air. A reduction in cost was necessary, however, to make these new uses feasible.

Further Development of Air Separation[5,6]

In the 1920s and 1930s oxygen production increased throughout the world and new developments arrived such as rare gas recovery, production and transport of liquid oxygen, and the emergence of tonnage-scale gas plants. In so-called tonnage plants production figures were at least 100 tons per day and were related to new demands in iron- and steelmaking and in fuel conversion. Required changes to the air separation plant saw the introduction of more efficient methods for heat exchange, purification and gas expansion.

Heylandt and Liquid Production

Up to the 1920s oxygen was distributed under pressure in steel cylinders. As the demand increased from some customers, this became increasingly unsatisfactory due to the high weight of metal that had to be transported. The development in Germany

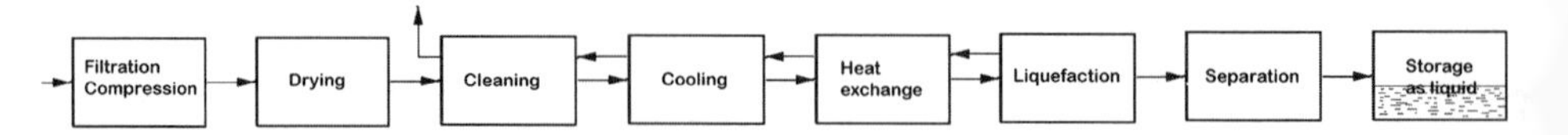

Figure 4.3.6. Process steps, air gas plant.

of the Heylandt' process for producing liquid oxygen in the 1920s was a great step forward. Paul Heylandt was quick to recognize the advantages of distributing and handling air gases in the liquid state. In 1924 he presented a design for oxygen production, which was realized in 1928 (Fig. 4.3.7). Liquid plants came into use in the 1930s, but did not become common until after World War II, when the demand grew for both oxygen and nitrogen in the liquid state.

In the postwar years the Heylandt process was superseded because large gaseous oxygen (GOX) plants could produce part of the output as liquid. Operating large liquefiers using turbocompressors and expanders significantly reduced the relative capital and operating costs and improved reliability and efficiency.

The design solutions which improved air separation technology both economically and in terms of quality include the following:

1. Regenerators (Fig. 4.3.8a), which offered better efficiency than previously used counterflow heat exchangers and potassium hydrate towers. These are large, cylindrical vessels which contain material with a large surface and heat capacity. They are used in pairs. In one, the incoming air is cooled, so that water vapor and carbon dioxide are frozen out, while the filling in the other is cooled with a backflow of cold gas from the distillation column. The disadvantages of the regenerator are that residual gases that remain after changeover affect the purity of the end product.
2. The revex (reversing exchanger) unit (Fig. 4.3.8b), which replaced the regenerator as a combined heat exchanger and water/carbon dioxide separator in the air gas process. The revex unit has a number of separate passages through a compact heat exchanger in which carbon dioxide and water are frozen out. The direction in the unit is alternated so that the impurities are continually flushed out with the air that cooled the heat exchanger.
3. The molecular sieve, which removes carbon dioxide and hydrocarbons through adsorption in the warm state. It was introduced in the beginning of the 1960s, and allows 80 percent of the incoming air to be used in the process, compared to 50 percent when using the revex unit.

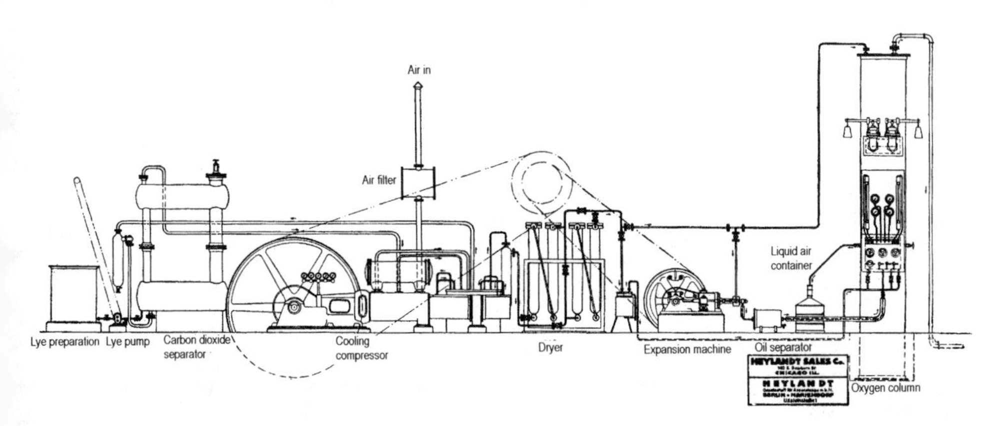

Figure 4.3.7. Heylandt plant, 1928. From Lashin (1929).

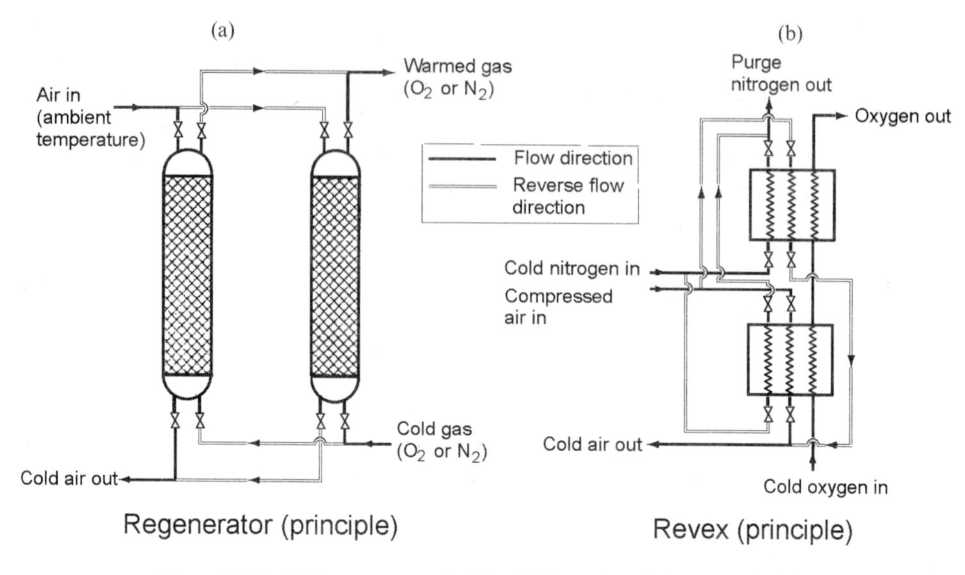

Figure 4.3.8. (a) Regenerator principle. (b) Reversing exchanger principle.

4. The expansion turbine, which replaced piston engines and offered higher thermodynamic efficiency and reduced working pressure down to 6 bars (600 kilopascals) (Fig. 4.3.9).

Fränkl and Regenerators

Mattias Fränkl began working on regenerators in 1925. He developed regenerators with a very efficient "pancake" packing design, using corrugated aluminum ribbons, and incorporated many heat transfer surfaces, which presented only a small pressure drop. The alternating flow streams and rapid switching meant that the precipitated moisture and carbon dioxide could be reevaporated into the outgoing product streams.

Collins and the Revex Unit

In the early 1950s, based on wartime experience with mobile air gas plants, Prof. Samuel Collins at MIT, supported by a research team of Arthur D. Little Corporation, developed a very efficient small reciprocating expander and various designs of reversing exchanger. The expander was suitable for relatively low inlet pressures, say 6 bars. The reversing exchangers, in contrast to regenerators, allowed uncontaminated high-purity oxygen to be produced in a very compact unit. The system was taken up by the M. W. Kellogg Company, which designed plants to produce 1000 cubic feet per hour (30 cubic meters per hour). Together with Hydrocarbon Research, they developed the reversing heat-exchanger system for industrial air gas plants.

Milton and Molecular Sieves

Adsorption processes with alumina, silica gel, or zeolite molecular sieves as adsorbents have been used for many years to remove water vapor, CO_2, and hydrocarbons from the air feed before it is cooled. Adsorptive removal by molecular sieves was first considered to be too expensive due to the high energy requirement for generating pressure and heat. The heating requirement for this type of adsorption is very modest, however, since in combination with pressure swing it is only necessary to drive a temperature pulse through the beds rather than raise the whole bed to regeneration temperature.

In 1949 Robert M. Milton of Union Carbide, Linde Division, successfully synthesized a molecular sieve; this sieve was followed by many others in short order. Many new commercial applications for molecular sieves were then developed, optimizing their use in cryogenic air separation plants.

Kapitza and the Expansion Turbine

The first expansion turbines for use in oxygen plants seem to have been designed and used by the German Zerkowitz at Linde AG around 1935. They were perfected by Pjotr Kapitza of the USSR, who used a high-speed turbine (40,000 rpm). The Russians continued experiments in the field in the mid-1950s, using turbocompressors and expansion turbines for liquefaction of air and later for liquefying other gases, especially helium. Turbomachines have also been used in other countries to liquefy helium, mainly since 1970.

Due to the high density of air and the resulting lower gas velocities, it is best to use a radial flow design, thus capitalizing on the large centrifugal contribution. The size of the required turbine is very small. For an air flow rate of 1000 cubic meters per hour, Kapitza's turbine had an outside diameter of only 80 millimeters and ran at 40,000 rpm. Most expansion turbines in industrial cryogenics after the 1940s have been based on Kapitza's design, which has substantially increased the overall thermodynamic efficiencies of large air gas separation plants.

Tonnage Development

The throughput of plants has increased and reached values as high as 1000 tons of oxygen per day. Almost all components are fabricated from aluminum and most joints are welded. These components are built into "cold-boxes" designed as rigid structures into which all the plant components are assembled and pressure-tested in the manufacturer's workshop. Then they are transported and finally erected on site. The slag-wool which was the traditional cold-box insulant has been replaced by powdered perlite, which has a lower thermal conductivity and can be produced on-site and handled pneumatically.

The Linde double column was dominating air separation when BOC introduced a new type of single-column cycle with nitrogen recycling, the Rescol (recycle single column) cycle for tonnage plants. The Rescol cycle was devised by P. M. Schuftan in an attempt to reduce power costs by using a thermodynamically more reversible

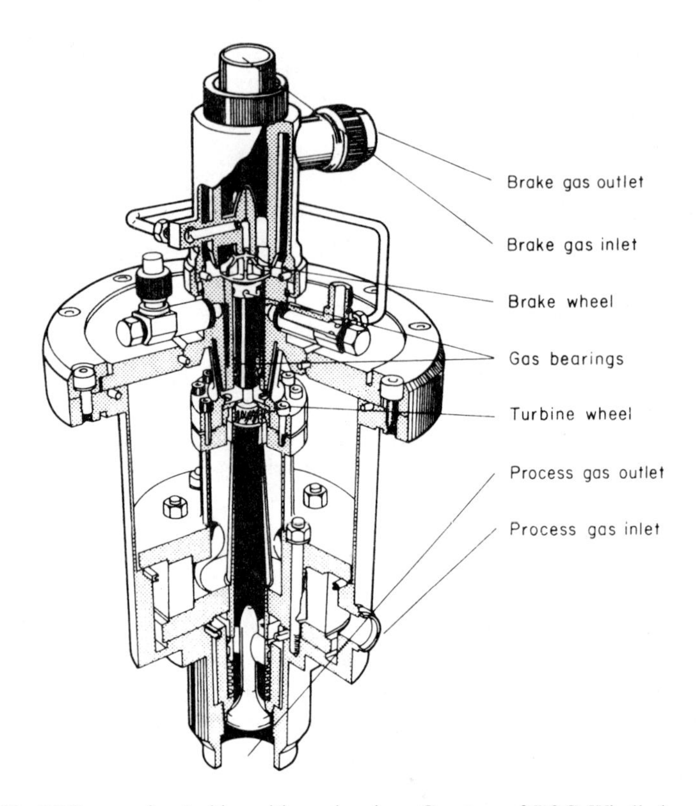

Figure 4.3.9. BOC expansion turbine with gas bearings. Courtesy of BOC, Windlesham, England.

operation. A number of such plants (capacity about 200-tons per day, with an oxygen purity of 95–98 percent) were installed in the United Kingdom, but did not appear to be as flexible as the double column for the complex product mix being demanded. In many places air separation plants should be able to produce oxygen both as gas and liquid (with a purity of 99.5 percent). Some of the nitrogen is also required as high-purity product, part of which is needed as liquid. There is also a growing market for argon as well as for krypton and neon. So additional columns have been attached to the double column and additional refrigeration provided by recirculating nitrogen in a special cycle with its own expander (Fig. 4.3.10).

Noble Gases from Air

Argon boils between liquid oxygen (LOX) and liquid nitrogen (LIN) temperatures and can be recovered in the air separation process (Fig. 4.3.11). Helium and neon have boiling points far below those of the main constituents of air, and krypton and xenon have higher boiling points. The following table shows some of the parameters important in the separation process (the yield refers to a plant producing 1000 tons of oxygen per day).

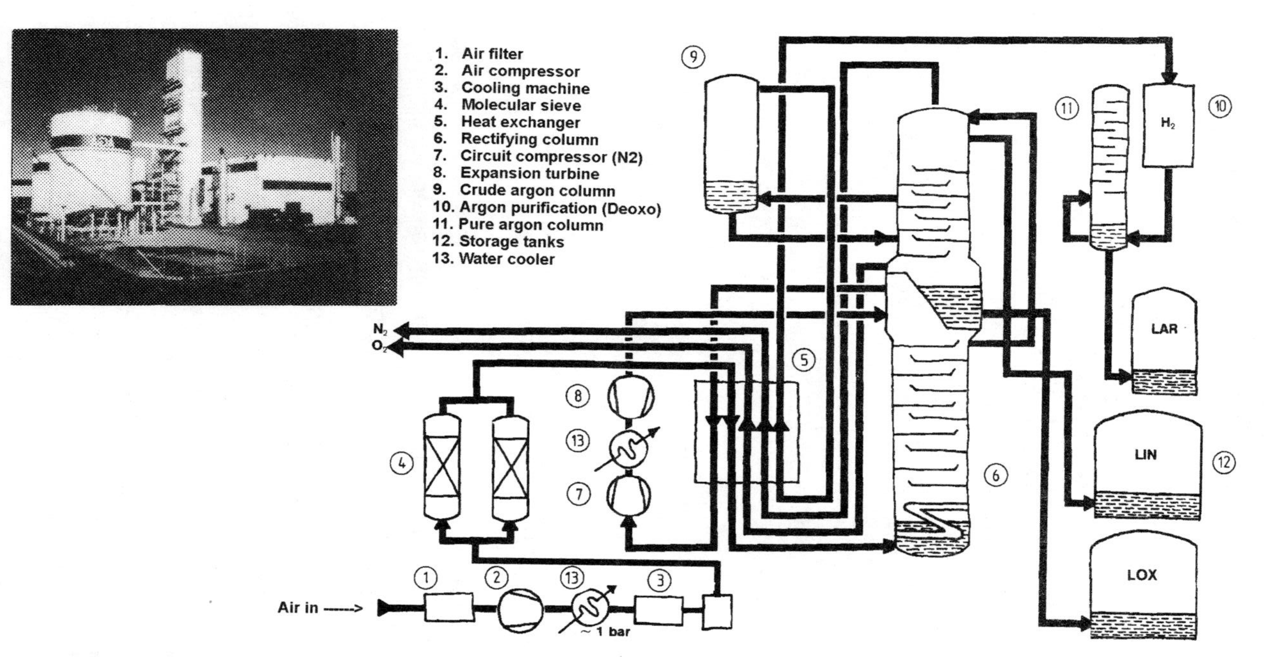

Figure 4.3.10. Principles of an air separation plant. Courtesy of Teknisk Illustration, Umeå, Sweden.

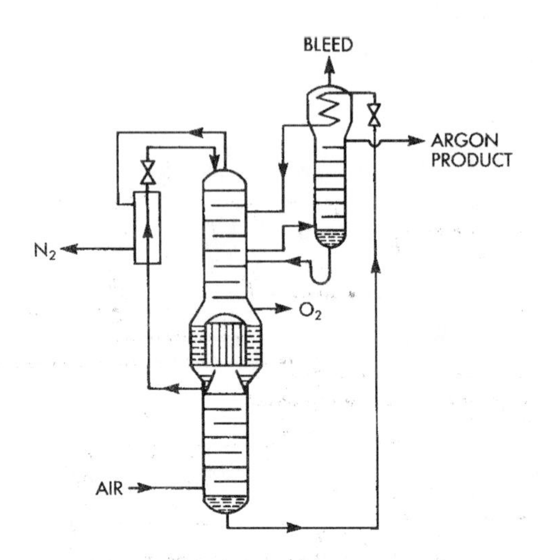

Figure 4.3.11. Argon separation.

Constituent	Boiling point (°C)	Concentration in air (% by volume)	Typical yield (%)
Oxygen	–182.96	20.9	
Nitrogen	–195.80	78.1	
Argon	–185.86	0.93	55
Helium	–268.94	0.0005	60
Neon	–246.00	0.0019	60
Krypton	–153.40	0.0001	30
Xenon	–108.12	0.000009	30

Argon

Argon may be produced either in the air separation process or from synthetic ammonia processes. Linde first separated argon from air in 1913 shortly after the double column was developed. The classical method of argon recovery, which was introduced in the 1930s, is to add a side refraction column to the double column in an air separation plant. Because argon boils at a temperature just below that of oxygen, a high concentration builds up (about 15 percent) just above the oxygen product. At this point the enriched argon is tapped and further enriched into a side (crude argon) column. Reflux and reboil rates are selected to accumulate argon in the lower section of the crude argon column to reach 95–97 percent argon concentration. From there the argon stream passes a separate purification plant, where oxygen is removed by addition of hydrogen followed by catalytic combustion and drying. Argon with less than 10 parts per million of impurities is achieved.

Helium and Neon

Helium and neon are not "condensed" even at the lowest temperatures in a double-column system. They accumulate in the condenser of the lower column. In large modern plants using plate-fin-type condenser/evaporators the helium and neon are dissolved in liquid nitrogen and some recovery is achieved by expanding the liquid into a separator en route to the top of the upper column and collecting the "flash" gas. Recovery is mainly for neon.

Krypton and Xenon

Krypton and xenon are relatively nonvolatile, and collect in the liquid oxygen in the upper column sump. In early (prewar) plants, recovery was achieved by feeding a small side column with LOX from the sump in which the krypton and xenon were concentrated to about 0.5 percent. The krypton and xenon were further concentrated and separated outside the main plant.

In modern plants producing gaseous oxygen an equilibrium amount of krypton and xenon builds up in the sump of the upper column. Where recovery is required, the gaseous oxygen product is scrubbed in a special unit prior to passing on to the heat-exchanger system.

Short Biography

Pjotr L. Kapitza (1894–1984) had a profound influence on the development of condensed matter physics and became a dominant figure in Soviet science.

He came to England in 1921 after a dreadful family tragedy. After the Russian Revolution and Civil War he lost his father, his wife, and two small children to epidemics. In Cambridge he met Rutherford, who invited him to spend a few months at the Cavendish working on a problem concerned with alpha rays. Rutherford was greatly impressed by Kapitza's ingenuity and encouraged him to extend his visit. He continued at the Cavendish for 13 years and turned his attention to studies of the behavior of metals in high magnetic fields. It soon became apparent that the behavior became more interesting at lower temperatures and he developed original methods of hydrogen and helium liquefaction.

During a routine visit to the Soviet Union in 1934, he was refused permission to return to Cambridge and was told he must set up a new institute in Moscow. After a year or so of negotiations at the highest level, the Soviet authorities still would not allow him to return. Therefore Cambridge sold the USSR the essential parts of Kapitza's special equipment for installation in the new Institute for Physical Problems in Moscow.

In the new laboratory he took up a quite different problem, the behavior of liquid helium. In 1938 he discovered the remarkable property of superfluidity and made experiments on the connections between the liquid flow and the flow of heat.

Kapitza also continued his interest in gas liquefaction and developed a turbine method of liquefying oxygen on an industrial scale. This became important for the steel industry during and after World War II. His achievements were recognized by prestigious government awards.

References

1. Claude, G. (1926). *Air liquide, oxygène, azote, gaz rares* (2nd ed.). Paris: Dunod 1926.
2. Haselden, C. G. (1986). The story of low temperature gas separation. *Proceedings of the 4th BOC Priestley Conference*, pp. 201–217.

3. Gardner, J. (1986). One hundred years of commercial oxygen production. *BOC Technology Magazine* **5**:3–15.
4. Scurlock, R. G. (ed.). (1992). *History and Origins of Cryogenics*. New York: Oxford University Press.
5. Laschin, M. (1929). *Der flüssige Sauerstoff*. Halle, Germany: Carl Marhold.
6. Appelmann, F. (1930). Flüssigsauerstoff-Verfahren System Heylandt. In *Proceedings Zehnter Internationaler für Acetylen, Autogener Schweissung und verwandte Industrien*, pp. 220–222. Zurich.
7. Ruhemann, M. (1949). *The Separation of Gases*, pp. 154–162 Clarendon Press, Oxford 1949:
8. Hagenbach, G. F., Wheeler, H. P., Jr., Stern, S. A., and Cook, G. A. (1961). Separation, purification, storage, and distribution. In *Argon, Helium and the Rare Gases* (G. A. Cook, ed.), Chapter XII. New York: Interscience.
9. Wagner, N. (1980). Development of air separation technology. *Proceedings of the Linde-symposium: Air separation plants*, pp. 1–28.
10. Thorogoud, R. M. (1991). The development of air separation. *Gas Separation and Purification* **1991**(5):83–94.

Further Reading

Hands, B. A. (ed.). (1986). *Cryogenic Engineering*. New York: Academic Press.
Hausen, H., and Linde, H. (1985). *Tieftemperaturtechnik*. Berlin: Springer-Verlag.

4.4. EQUIPMENT FOR DISTRIBUTION AND STORAGE OF GAS

Background: Distribution of Gases

When gases first were produced and studied in laboratories in the 18th century they were collected in animal bladders, usually from ox or pig, or in specially manufactured gas bags.

The first hydrogen balloons used in 1783 by the Charles brothers in Paris were made from silk painted with a rubber solution. Later, gas bags in oiled textile or silk with gilt paint were common. When gases, mainly oxygen, began to be used for medical purposes and for limelight in theaters commercial generation started. The gas could be purchased from a pharmacist and was usually distributed in large leather bags. In use the flow and pressure of the gas was controlled by applying weights on the gas bag. These gas bags were later replaced with steel cylinders to contain the gas and with regulators to give a steady pressure and flow (Fig. 4.4.1).

From Gas Bags to High-Pressure Cylinders

The first small compressed gas tank was invented around 1810 and was used to store dry coal gas (0.5 kilogram each). About 50 years later, the German Julius Pintsch developed a product called Pintsch Gas, which was formed by cracking oil at high temperatures to form an oil gas, which was compressed at about 8 bars and was stored in cylinders.

In 1888 the founder of Liquid Carbon Company, Jacob Baur, began to store liquefied carbon dioxide in cylinders (Fig. 4.4.2). The first seamless steel high-pressure cylinders were introduced by the Mannesmann Company in Germany. At the same time Brin's Oxygen Company began distributing oxygen in high-pressure cylinders.

Figure 4.4.1. Transportation of coal gas in Paris around 1860. From Ålund (1875).

Vacuum-insulated Vessels[1–4]

In 1882 the French physicist Jules Violle announced to the French Academy of Science that he had worked out a way to isolate liquid gas by using vacuum in a double-wall glass vessel. However, his work was forgotten and in 1887, Arsène d'Arsonval designed an efficient insulation vessel to contain methyl chloride, made of a double glass wall with an intermediate vacuum. This vessel was improved in 1892 by Dewar. His vessels became known as "Dewar flasks," or, simply, dewars.

Dewars are double-walled vessels with a vacuum in the annular space to minimize heat transfer by conduction and convection. The walls are silvered to reflect radiant heat. Dewar's design was a very significant contribution to the storage and transportation of very cold liquefied gases such as oxygen, nitrogen, air, hydrogen, fluorine, and helium. After his successful liquefaction of hydrogen, Dewar became very confident about storing and transporting it in his vacuum vessels, predicting that it could be handled as easily as liquid air.

Dewar's efforts to improve the thermal insulation by filling the vacuum space with different substances, however, achieved no positive results (Fig. 4.4.3). In 1910 the Pole Maryan Smoluchkovski showed that the thermal conductivity of porous substances is reduced at lower pressure. It is upon his work that today's evacuated powder

Figure 4.4.2. Liquid Carbonic salesmen in the 1890s carried filled gas cylinders with them in order to demonstrate their new product. From Stiegerwald (1988). Courtesy of Praxair, Danbury, Connecticut.

insulation is based, where perlite and silica gel are used as powders. It was not until the 1940s that evacuated powder insulation began to be used in cryotanks.

The ability to achieve a high-vacuum state was improved in 1915–1916 by use of the diffusion pump, which was developed by the German Wolfgang Gaede and the American Irving Langmuir. During the 1950s, ion and cryosorption pumps arrived, with which it was possible to achieve ultrahigh vacuum ($<10^{-11}$ bar, 10^{-4} pascal).

Multilayer insulation, so-called superinsulation (Figs. 4.4.4 and 4.4.5), was first presented by the Swede Peter Petersen in 1951. In this design, aluminum and fiberglass foils (about 1 and 10 micrometers, respectively, in thickness) are alternated in a matting to fill the vacuum space, which is then evacuated to a high vacuum (Fig. 4.4.6).

Storage and Transport Tanks for Liquefied Gases[5–9]

The distribution of oxygen in liquid form, the advantages of which were stated as early as 1930 (Fig. 4.4.7), was applied on a small scale in Europe 10 years after World War I. It was not until 1950 that a considerable development of the distribution of air gases in liquefied form occurred. The first tanks for the storage and transport of liquid oxygen were constructed around 1920, for the supply of small vessels

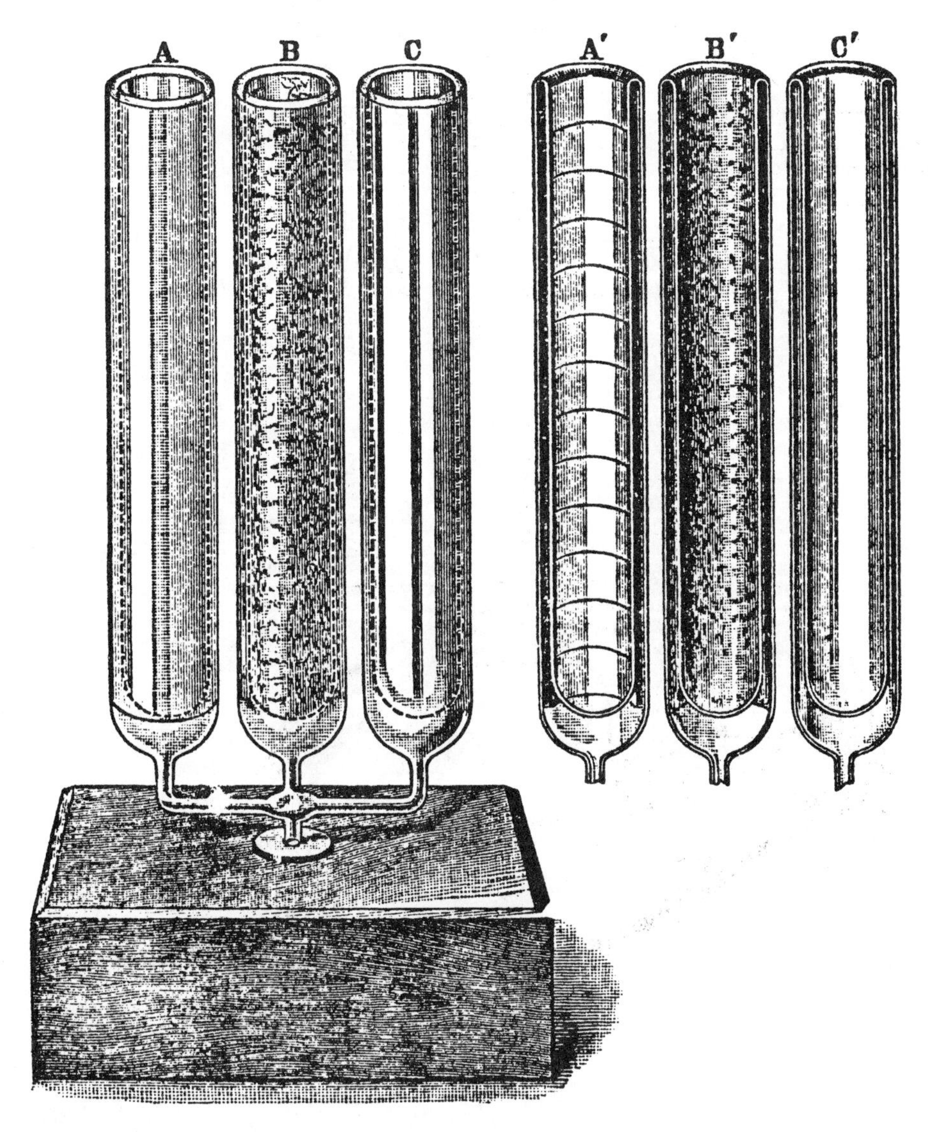

Figure 4.4.3. Dewar's experiments with layer insulation in an evacuated vessel. From Dewar (1911).

to places where explosives were prepared and used. This was the first case of distribution of oxygen in liquid state. These vessels, with a capacity of 500 or 600 liters, had a spherical shape and included an inner brass or copper sphere suspended by means of chains inside a steel outer shell. The insulation was achieved by filling in the inner space between the walls with powdered materials such as magnesium carbonate, diatomite, cork, or animal hair. A branch pipe starting from the gaseous phase and coiled around the internal vessel inside the insulant was intended to keep the insulation cool.

The capacity of transport tanks was restricted by their maximum diameter, which was dependent in practice upon the width of the platform of the covered trucks

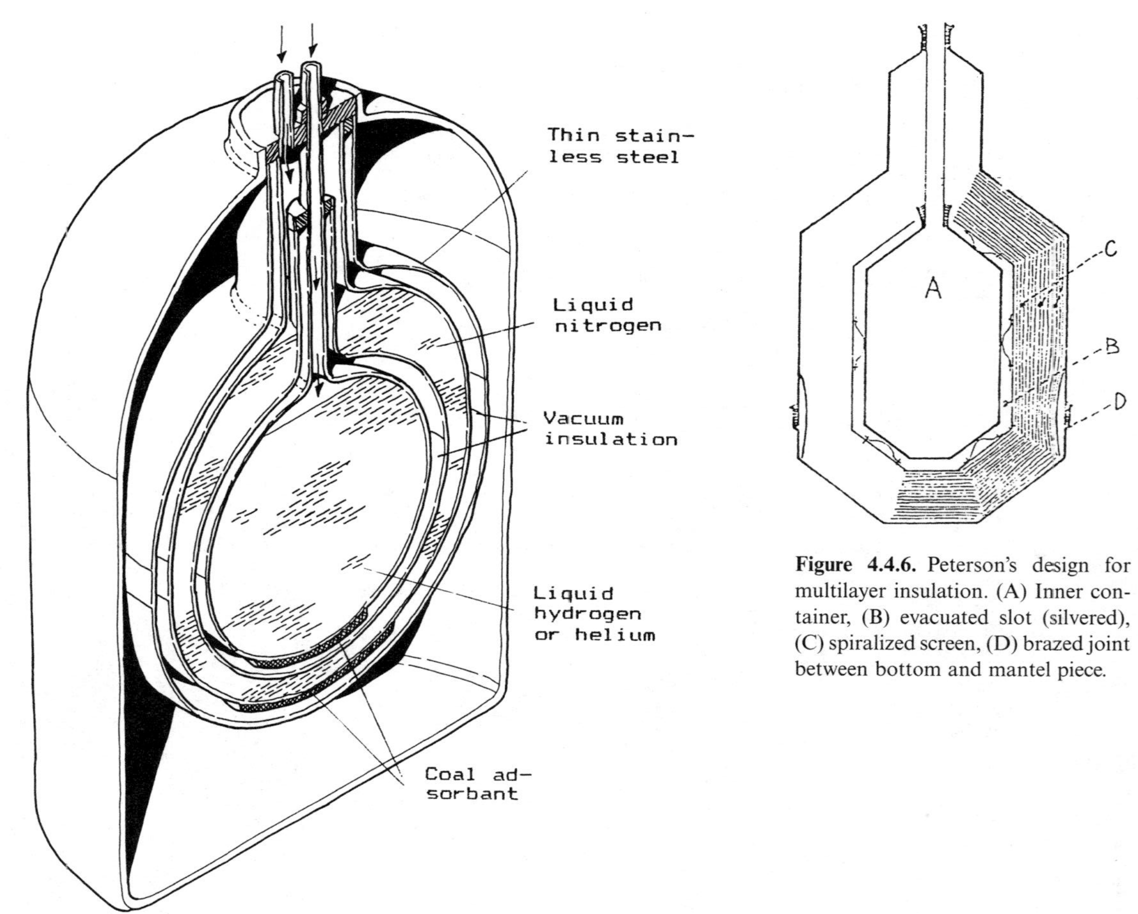

Figure 4.4.4. Superinsulated vessel. Courtesy of Teknisk Illustration, Umeå, Sweden.

Figure 4.4.5. Small vessel for liquid helium. Courtesy of Teknisk Illustration, Umeå, Sweden.

Figure 4.4.6. Peterson's design for multilayer insulation. (A) Inner container, (B) evacuated slot (silvered), (C) spiralized screen, (D) brazed joint between bottom and mantel piece.

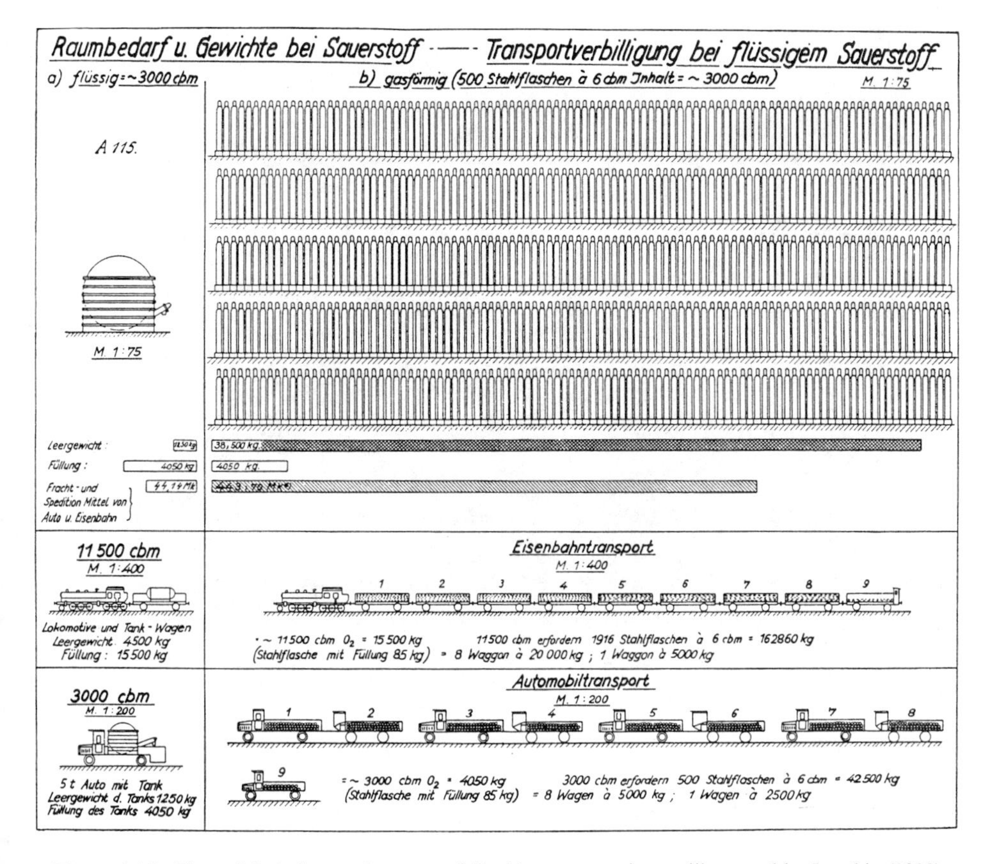

Figure 4.4.7. The weight/volume advantage of liquid transportation as illustrated by Laschin (1929).

transporting them, that is, 1200 liters and then 2500 liters. Fixed storage tanks constructed according to that same principle soon reached a capacity of 8000–10,000 liters.

The economic advantages of liquid distribution were stated by Paul Heylandt in Germany about 1922. He demonstrated the possibility of storing oxygen in the liquid form, whereas its bulk storage in gaseous form was practically impossible. The first equipment designed according to his ideas for the production, storage, and transportation of oxygen in liquid form was the real revolution in the transport of compressed gases to market (Figs. 4.4.7 and 4.4.8). It occurred in Germany around 1925 when the C. W. Heylandt Company, now owned by Linde, introduced the system. In 1930 the first vacuum-insulated truck and railroad tanks were built to carry gases in their refrigerated liquefied state at cryogenic temperatures. In the United States the Linde Division of Union Carbide began using liquid oxygen trucks from about 1931.

Small Vessels

The first industrial design of small cryovessels with high insulation efficiency was developed in the form of polished concentric spheres in copper or brass brazed

Figure 4.4.8. Early tank car for liquid oxygen. Courtesy of Linde.

with silver or welded with tin, with a small-diameter neck. Between these spheres a vacuum as high as possible was produced. These vessels were made in different sizes with a capacity ranging between 5 and 25 liters. The vacuum between the walls was improved at low temperature by the presence of charcoal. During World War II important research work was undertaken in the United States and Europe to improve the first generation of equipment. Aluminum alloys were sometimes used to reduce the weight of spherical transport tanks. High-pressure pumps were also improved and vacuum-insulated horizontal cylindrical vessels, still insulated with powdered material, were built.

Pipelines

Already before World War I, Air Liquide had participated to the first industrial tests using oxygen in steel making in Belgium. As early as 1925, a pipeline was installed in Italy between the SIO plants and the steel mill owned by Fiat in Turin. Today, large steel mills in many industrialized countries are supplied through pipeline systems connected to oxygen-producing facilities. During the 1950s industrial gas pipeline networks were constructed by UC Linde and Big Three in the United States and by Air Liquide in the Benelux countries and France in Europe (Fig. 4.4.9).

The first large U.S. industrial gas pipeline went in into service 1951 from the UC Linde plant in East Chicago, Indiana, to the Inland Steel steel mill at Indiana Harbor.

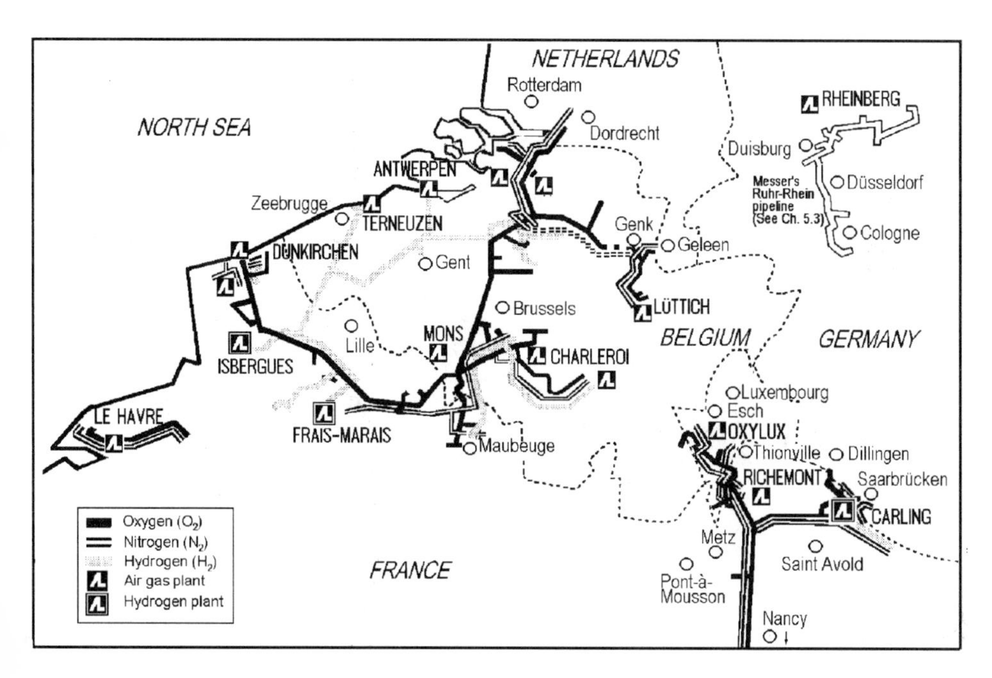

Figure 4.4.9. Industrial gas pipeline systems in Europe. The Air Liquide Grand Masse and the Messer Ruhr-Rhein pipelines (ca. 1994). Adapted from Air Liquide brochure with permission.

Big Three Industries built a multicustomer pipeline supplied with 11 oxygen plants from Lake Charles, Louisiana, to Bay City, Texas.

In Europe from 1955 onward Air Liquide built the world's largest industrial gas pipeline system in the Benelux countries and France to transport oxygen, nitrogen, and hydrogen from several plants. Several plants are interconnected to a ensure a reliable supply. Since 1964 Messer Griesham has used pipelines to supply customers with nitrogen in the Ruhr area of Germany.

New Distribution Processes Change the Structure of the Industrial Gas Business

From about 1950 strong efforts were made to design distribution processes which could keep down losses and also minimize the energy consumption for the production and conversion of liquid into gas. Developments in metallurgy made it possible to find commercially available aluminum alloys and weldable steels with satisfactory low-temperature impact properties to substitute for the copper and brass used until then for low-temperature vessels. Reliable processes for welding these new low-temperature materials were also developed. The evaporation losses were considerably reduced by combining a high vacuum (less than 1 millimeter of mercury) with new developed insulants.

Figure 4.4.10. An AGA tank car in 1962. Courtesy of the AGA Historical Picture Archive.

From 1960 onward, production was centralized and the structure of the industry of industrial gases was modified. Liquid distribution came into general use and plants producing several thousand liters per hour were erected; the smaller plants that had been producing gaseous oxygen were shut down or operated only as cylinder filling stations.

Around 1965, "superinsulants," multilayer insulants under high vacuum (10^{-4} millimeter of mercury), allowed the improvement of small vessels for the transport of liquid, and also the design of vessels for the transport of liquid hydrogen and liquid helium.

References

1. Dewar, J. (1911). Thermal transparency at low temperatures. *Proceedings of the Royal Institution*, (London) **15**:820–826.
2. Petersen, P. (1951).The heat-tight vessel. University of Lund, Lund, Sweden.
3. Petersen, P. (1958). Some means to improve te vacuum insulation of the Dewar vessel. *Teknisk Vetenskaplig Forskning* **29**(4):151–168.
4. Soulen, R. J., Jr. (1996). James Dewar, his flask and other achievements. *Physics Today* **1996**(3): 32–37.
5. Laschin, M. (1929). *Der flüssige Sauerstoff.* Halle, Germany: Carl Marhold.
6. Jamault, M. (1979). Historical past, present and future transport and distribution of cryogenic liquids. In *Proceedings of the IGC Symposium on Safety in the Transport and Distribution of Cryogenic Liquids.*
7. Bräutigam, M., and Huppertz, P. H. (1987). Road transport tanks for carrying liquefied gases—Yesterday and today. *Linde Reports on Science and Technology* **42**:39–44.
9. Sloop, J. L. (1978). *Liquid hydrogen as a propulsion fuel. 1945–1959.* Washington DC: NASA.

4.5. PRODUCTION: ALTERNATIVE GAS SEPARATION METHODS

Air Gas Separation Methods[1–3]

The predominant method for the production of air gases is liquefaction of air followed by separation of oxygen, nitrogen, and argon in rectification columns. Where large quantities or liquid products are required for their cold value, cryogenic separation is advantageous economically, but when scaling down to small sizes other principles come into play. Theoretically there are at least four different technically feasible alternatives to cryogenics for gas separation: chemical reaction, absorption, adsorption, and membranes. Since the 1980s there have been two efficient alternative processes for the production of air gases: the molecular sieve or pressure-swing adsorption (PSA) process and the membrane process.

Chemical Reaction Techniques

These methods of gas separation originated long ago. The Brin process was based on chemical oxidation of heated barium oxide under pressure with subsequent liberation of the oxygen under vacuum. Many attempts have since been made to find a better chemical reagent than barium oxide, but no feasible chemical reaction process has been found for separating oxygen and nitrogen.

Absorption

Some successful chemical absorption processes for the CO_2 removal from natural gas have been developed, such as the Rectisol process and the Benfield process.

Adsorption

Like absorption, this is a physical process. In adsorption, one component of a gas mixture is preferentially adsorbed on a porous solid surface and is thus separated from the rest. Pressure-swing adsorption (PSA) processes for separation of oxygen were developed in the early 1960s in which nitrogen is adsorbed by zeolite. PSA nitrogen plants were than developed that use a carbon molecular sieve to adsorb oxygen from the inlet air (Fig. 4.5.1).

Membrane Separation

This involves transport of one constituent of a liquid or gas mixture across a relatively thin barrier. Membranes can be made of metal, glass, or polymer in tube or sheet form. Membranes have been involved in separations in cells since life itself

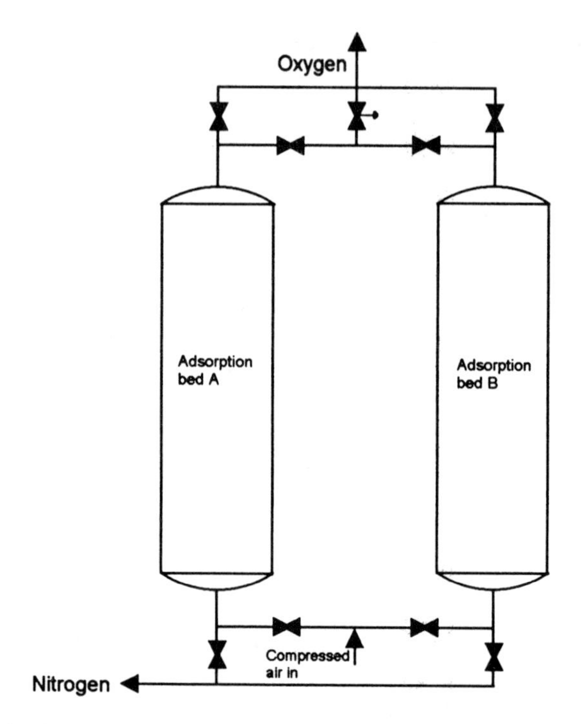

Figure 4.5.1. Principle of presure-swing adsorption: a typical two-vessel oxygen system. Compressed air is fed to the first adsorber vessel. As the air passes over the adsorber, the nitrogen is adsorbed, so that gas leaving the top is oxygen. While adsorption occurs in one vessel, the second vessel is regenerated by venting to ambient pressure. After about 1 minute, the vessels switch roles. The adsorption–regeneration cycles are repeated indefinitely. This basic process is also used to produce nitrogen, with the zeolite adsorber replaced with activated coal.

began: In biological tissues many natural functions involve the transport of fluids across membranes. These processes have been copied in industrial processes such as desalination and hydrogen recovery in ammonia plants.

Pressure-Swing Adsorption[4–6]

In the 1940s it was found that naturally occurring molecular sieves could separate certain mixtures; in some cases, they also could separate molecules on the basis of molecular size. Thus, the name "molecular sieves" was given to this class of minerals. In the late 1940s, the Linde Division of Union Carbide Corporation became interested in the properties of these materials in connection with a possible method of air separation. Because the naturally occurring minerals of greatest interest were rare and rather impure, Linde attempted to synthesize molecular sieves in the laboratory.

In 1956 D. W. Breck, W. G. Eversole, and R. M. Milton reported that synthetic zeolite, which is chemically different from, but structurally similar to the naturally occurring mineral faujasite, can be used to separate nitrogen from mixtures with oxygen and argon (Fig. 4.5.2).

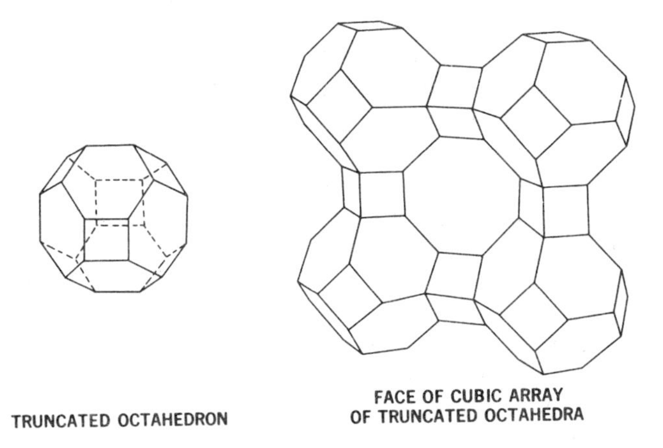

Figure 4.5.2. Structure of synthetic zeolite, containing a cubic array of truncated octahedra.

In 1963 it was found that a material with a precise pore system consisting of channels of molecular dimensions of 3 micrometers or just below could separate air into its components. The critical diameters of the air gases are 2.8, 3.0, and 3.83 micrometers for oxygen, nitrogen, and argon, respectively.

In 1949 Milton synthesized as a molecular sieve "crystal A" (Fig. 4.5.3), which has turned out to be one of the most commercially important molecular sieves. Many

Figure 4.5.3. Robert Milton showing a molecular sieve model. Courtesy of Praxair, Danbury, Connecticut.

other molecular sieves were synthesized in short order. They were all characterized and tested for potentially useful properties, including air separation. Unfortunately, the attempt to develop an economically attractive air separation process at that time was unsuccessful. However, other potentially useful applications were found, particularly in the dehydration of gases and liquids. These applications led to the development of the molecular sieve business.

Industrial-scale PSA units are used in many areas of chemistry as well as for producing nitrogen or oxygen from air. Besides Union Carbide's Linde Division, Esso and Aerojet-General in the United States and Air Liquide in France began to study the possible applications of molecular sieves to air separation and for separating H_2, CH_4, CO_2, and a variety of other gases from industrial process streams. Dehydration of industrial gases was the first widely accepted application in the industrial gas industry.

These efforts resulted in the establishment of the Molecular Sieve Department at Union Carbide. The first commercial-scale separation by molecular sieves was oxygen removal from an argon stream. This process operated successfully for about 20 years.

In the 1970s a small PSA O_2 concentrator for home care was introduced by Mountain Medical Equipment and Xorbox.

For production of 99.9 percent pure nitrogen in a PSA process a special form of activated carbon is used. This activated carbon process was first developed by Bergbau-Forschung in Essen, Germany, in the mid-1960s. PSA units for this type of material were first licensed on the market by Mahler in Germany and BOC/Airco world-wide. In 1982 the C. M. Kemp Manufacturing Company in Maryland started the production of PSA units for nitrogen for the U.S. market.

Work with zeolites in the early 1980s showed that a 5-angstrom molecular sieve material could adsorb nitrogen at elevated pressure and release it at reduced pressure. This discovery allowed the development of PSA systems capable of producing oxygen from air with a purity of 90–95 percent. Engineering developments in the use of zeolites for adsorptive separation led to the use of regeneration pressures of less than atmospheric pressure. Lower overall power requirements are the result of regeneration at subatmospheric pressure by so-called vacuum swing adsorption (VSA) or vacuum–pressure swing adsorption (VPSA). VPSA apparatus operates at a lower pressure and with reduced operating power costs. Early designs utilized three adsorption beds in the VSA process, but improvements in the zeolite material and process innovations have reduced modern systems to a two-bed design. Such plants are capable of producing either enriched O_2 (93–95 percent) or N_2 (to 99.5 percent). VPSA plants feature a blower to introduce air at a positive pressure and vacuum pump for the regeneration cycle.

Membranes[7]

In 1866 the Scot Thomas Graham, based on the treatment of diffusion in John Dalton's theories of mixed gases, became the first to recognize that a selective gas

Figure 4.5.4. PSA on-site plant. Courtesy of Nippon Sanso, Tokyo, Japan.

separation can be performed by means of different permeation velocities of various gases through a membrane. It was, however, not until the early 1940s that gas separation through membranes became technically interesting as a means to enrich uranium 235. This procedure was extremely costly, but at that time there were no alternative enrichment processes available.

In the 1950s and 1960s, experiments were performed on membrane units for gas separation. There was, however, a lack of membranes with adequately high permeability and separation efficiency for commercial use. The solution to this problem became available in 1960 through the invention of an asymmetrically structured membrane consisting of a very thin, nonporous, selective skin layer resting on a highly porous, nonselective support layer. The American Monsanto Company, after development and in-house tests, made the first commercial breakthrough in 1977 with its composite Prism-Module membrane for H_2 separation from hydrogenation and NH_3 production processes (Fig. 4.5.5). Monsanto was followed by the Separex Company, Spectrum Division, and Dow Company Generon Systems. In 1988 Air Liquide joined with DuPont to form Medal.

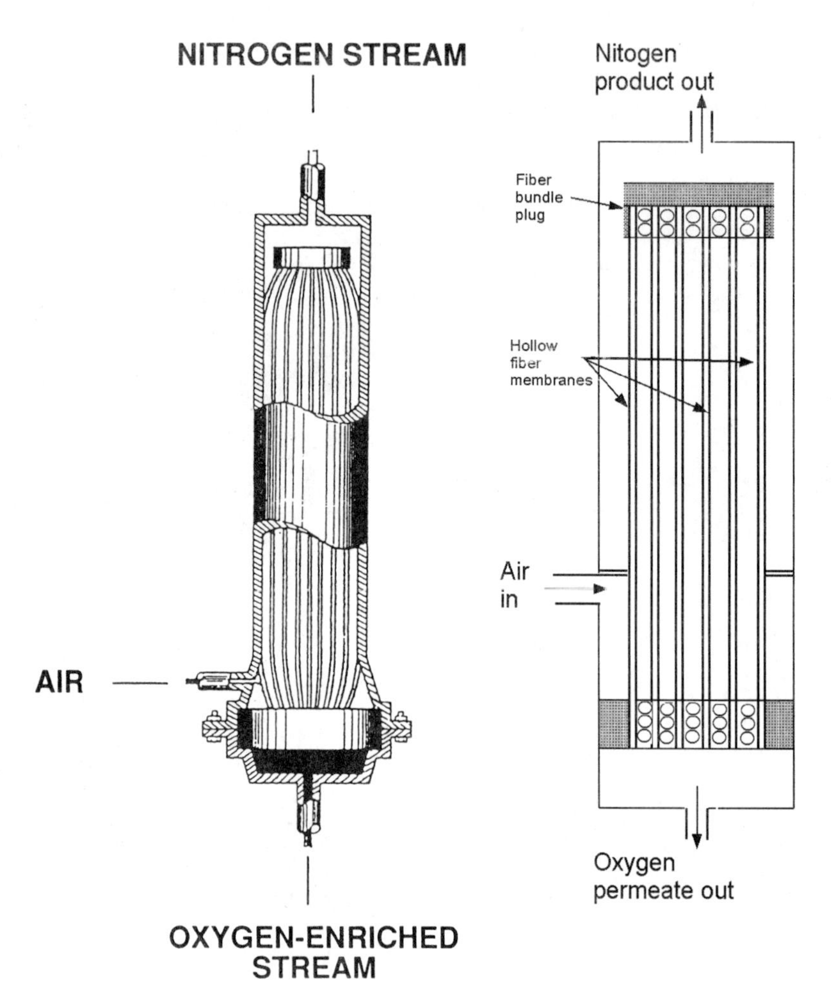

Figure 4.5.5. Principle of membrane separation. Shown on the left is a Prism separator from Air Products Courtesy of Air Products and Chemicals, Inc., Allentown, Pennsylvania.

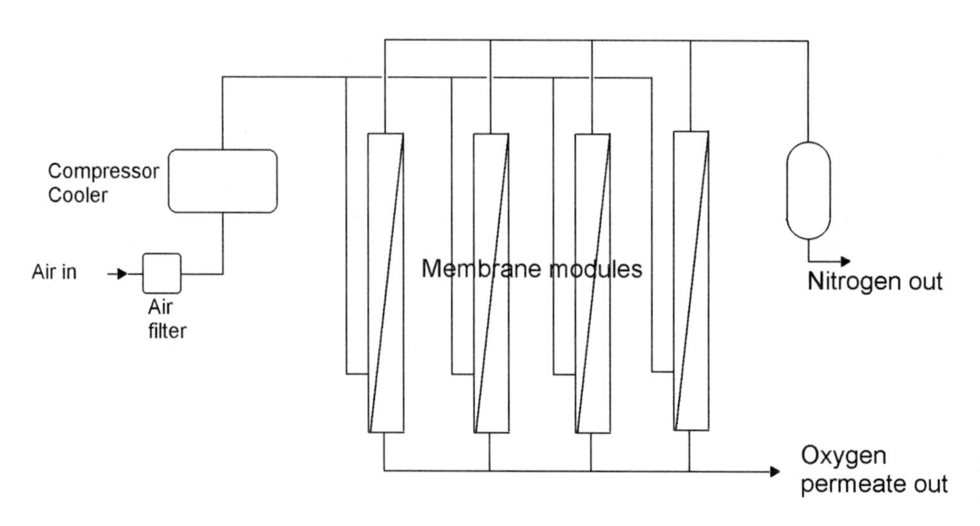

Figure 4.5.6. Membrane separator principle.

The Membrane Process

Figure 4.5.6 shows a membrane air separator. Packed in a cylindrical pressure vessel are many fine tubes each made of a mesh of hollow polymer fibers with the width of a human hair. One end of the tubes is open and the other is closed. Pressurized air is introduced in the space between the tubes. Oxygen diffuses into the tubes and flows toward the outlet, while nitrogen flows in the other direction outside the tubes.

References

1. Foeng, W. (1987). Non-cryogenic air and gas separation. *IOMA Broadcaster* **8**:9–14.
2. Stanley, M. (1986). Thomas Graham and gaseous diffusion. In *Proceedings of the 4th BOC Priestley Conference*, pp. 161–170.
3. MacLean, D. Advances in non-cryogenic gas separation technologies. BOC Technology Magazine **5**:38–45.
4. Breck, D. W., Eversole, W. G., and Milton, R. M. (1959). *Journal of the American Chemical Society*, **78**:2338.
5. Kerr, G. T. (1989). Synthetic Zeolites. *Scientific American* **1989**(7):82–87.
6. Jain, R., and Tseng, J. K. (1998). Production of high purity gases by cryogenic adsorption. *BOC Technology*. **1998**(November):29–33.
7. L'Air Liquide. (1998). 10 ans de succès de la technologie membrane. Paris: L'Air Liquide.

5

DEVELOPMENT OF THE INDUSTRIAI GAS BUSINESS

5.1. LIQUID CARBONIC[1]

The Liquid Carbonic company was founded in 1888 in Terre Haute, Indiana, and for over 50 years concentrated on the production and supply of carbon dioxide (CO_2) and related equipment to the soda fountain and soft drink bottling trades. After that time, new technology in the use of CO_2 showed that the beverage industry was only a portion of the market Liquid Carbonic could serve. In 1939, it entered the industrial/medical gases business. An aggressive acquisition program changed Liquid Carbonic from one of the world's largest carbon dioxide suppliers into a major industrial gas company during the 1940s and 1950s.

In 1957 Liquid Carbonic merged with General Dynamics and in 1969 it became part of Houston Natural Gas Corporation. In August 1984 Houston sold Liquid Carbonic Industries Corporation to CBI Industries Inc., a metal plate construction company (formed in 1889). In 1996 Praxair took over CBI and integrated Liquid Carbonic into their industrial gas organization.

Background

The last decades of the 19th century, during the American industrial revolution, were a good time for new ideas for making carbon dioxide available for commercial use. With the concept of a process for manufacturing CO_2 and distributing it under pressure in cylinders, Jacob Baur, a chemist from Terre Haute, Indiana, moved to Chicago and started development. Together with his brother Charles and seven other friends and relatives, he established the Liquid Carbonic Acid Manufacturing Company in November 1888.

Baur's Concept

Gas from combustion was used as the source, compressed and liquefied, and contained in steel cylinders. In May 1889, the first cylinder of liquid compressed gas was delivered for commercial use. The term "liquid" may not have been correct because the gas was kept under conditions where it could exist only in the vapor phase. It has, however, persisted as the common term for the compressed carbon dioxide

Figure 5.1.1. Interior of a Liquid Carbonic plant, ca. 1895. From Stiegerwald (1988) with permission.

cylinder gas. The business developed rapidly and other producers soon entered the market. In 1898, competitors existed in Boston, Chicago, Cleveland, and Detroit among other places.

Jacob Baur's method of manufacturing and distributing CO_2 gradually replaced the common practice of on-site gas generation, which was used by druggists and bottlers to make carbonated beverages. At that time, nearly every soft drink manufacturer had its own gas generating outfit and dissolved CO_2 into its beverages by agitation within closed vessels. This was dangerous, costly, and inefficient. When compressed carbon dioxide cylinders were introduced in 1886 the cost was about 25 cents per pound, which was over three times what it cost beverage manufacturers to make it for their own use in their own plants. In spite of that fact the Red Diamond cylinders of Liquid Carbonic from 1889 began a new era in the beverage industry.

Market Development

Carbonators

Baur's concept brought with it new ideas for improved carbonating equipment. New business opportunities and products developed. Manufacture of carbonators became the company's second business with the release of what probably was the first automatic carbonator in 1895 (Fig. 5.1.4). During the period 1894–1905 Liquid Carbonic introduced equipment with names such as Geyser, Perfection, Magic, Excelsior, Niagara, Little Giant, and Columbia.

Liquid Carbonic also began making dispensing apparatus for beer and other beverages, and after a visit to Europe in 1894 Jacob Baur and his chief engineer

Figure 5.1.2. Liquid Carbonic salesman, ca. 1900. From Stiegerwald (1988) with permission.

purchased plant equipment to improve their product, which was installed in CO_2 plants in Chicago, Pittsburgh, St. Louis, and Milwaukee.

New Product Line

Bottlers throughout the country who had started using gas in cylinders and the automatic carbonator also needed dependable flavorings for their beverages. In 1896 flavors and extracts were added to the Liquid Carbonic line of products. In 1897 the company felt they were "forced into" manufacturing syrups for soda fountain users as well as soda fountain equipment. By the turn of the century, Liquid Carbonic was

Figure 5.1.3. Early cylinder transport 1902, Milwaukee. From Stiegerwald (1988) with permission.

Figure 5.1.4. Liquid Carbonic developed a new "rocking" carbonator in the 1890s. From Stiegerwald (1988) with permission.

supplying bottlers with all the equipment they needed and had 12 large factories in Chicago, New York, Pittsburgh, St. Louis, Milwaukee, and Cincinnati.

In 1914, Liquid Carbonic introduced a low-pressure filling system for bottling carbonated beverages, which increased production and gave economic savings and thus revolutionized the industry. The first Liquid Low Pressure Filler was purchased by the Coca-Cola Bottling Company of Mobile, Alabama, in July 1914, 2 weeks before World War I.

War-Time

At first general business declined during the war, but soon the company found new production possibilities. In 1917 the company's woodworking and cabinet department began the production of airplane propellers for the U.S. Government. At the same time Liquid Carbonic began a changeover from the by-product process of manufacturing carbon dioxide to the "coke process," which was more dependable and economical; the purity of the gas increased to about 99.8 percent average. The first CO_2 plant using the coke process was built in Minneapolis, Minnesota. In 1918, Liquid Carbonic moved abroad for the first time, into Canada, establishing a branch in Toronto. In 1920, Havana, Cuba, became the next foreign site to carry the company's complete line of products.

In 1926 its name changed from the Liquid Carbonic Company to the Liquid Carbonic Corporation and the company expanded its network of CO_2 gas plants in Detroit, Indianapolis, and Louisville.

Dry Ice

In 1926 Liquid began producing carbon dioxide snow and so-called "dry ice" on a commercial scale. In 1928 Liquid entered into an agreement with the Dry Ice Corporation to provide CO_2 for Dry Ice's manufacturing equipment, which was to be installed in Liquid's plants. Liquid Carbonic also acquired a 25 percent interest in the Dry-Ice Holding Corporation. In 1931 Liquid Carbonic made a deal with the Dry Ice Corporation and began direct production and sale of dry ice. Marketed under the trade name of Carbonic Ice (Fig. 5.1.5), the product was manufactured in 15 of the

Figure 5.1.5. First dry ice transport in 1931. From Stiegerwald (1988) with permission.

company's CO_2 plants. A major competitor, the General Carbonic Company, based in New York, was acquired by Liquid in 1929. This added eight CO_2 plants to Liquid's network. In addition, two labeling machine companies were acquired.

Business Expansions

In 1930 the next major acquisition was made despite the stock market crash and the beginning of the Great Depression. A Canadian subsidiary was formed by the acquisition of Canadian Carbonate Company (founded in 1898), which from 1911 was based in Montreal, and operated eight CO_2 plants in Canada. In the following year further expansion in Canada was accomplished by acquisition of Dominion Carbonic Company Ltd. In the eastern United States, Liquid increased its market with the purchase of Keystone Carbonic Company of Pennsylvania.

Liquid Carbonic West Indies, Ltd., was acquired in 1931, and CO_2 production was later begun in Trinidad. At its 50-year anniversary in 1938, Liquid owned 28 plants and warehouse properties in the United States and 7 in Canada.

Industrial/Medical Gases

1939, Liquid Carbonic entered the industrial/medical gases business through the acquisition of Wall Chemicals, Inc. (Fig. 5.1.6), a producer of oxygen, acetylene, and other compressed gases, with plants in Chicago, Detroit, and Buffalo. At about the same time, Liquid also acquired Acetylene Gas & Supply Co. (Toledo) and began construction of an oxygen plant at its Chicago headquarters facility. Next year Liquid

Figure 5.1.6. Acquisition of Wall Chemicals 1939, Liquid Carbonic's start in industrial gases. From Stiegerwald (1988) with permission.

purchased Independent Oxygen Company (Cincinnati), and Wall Chemicals, Ltd., Canada, which had plants in Montreal, Toronto, and Windsor, marking its entry into the industrial gases market in Canada.

With the buildup of America's defense program, a survey of customers in 1941 indicated that about 70 percent of the products sold by Liquid's oxyacetylene division were used for defense purposes. In 1942, Liquid Carbonic started a medical gases division with the acquisition of Cheney Chemical Company (Cleveland), a manufacturer of nitrous oxide, ethylene, and other medical gases.

Expansion of the Industrial Gases Business

In 1947 the Stuart Oxygen Company of San Francisco, the fourth largest producer of oxygen in the United States and a major producer of acetylene, suggested to Liquid that it considered purchasing its company. This was accomplished in 1948, with the addition of 10 oxygen and 10 acetylene Liquid Carbonic-owned plants in the United States and Canada.

Two years later Pennsylvania-based Paschall Oxygen Company was acquired, opening a new industrial gases market for Liquid Carbonic on the East Coast. In 1953 the purchase of Bird Gas Corporation of Detroit, a producer of oxygen and acetylene and a major distributor of welding machines and supplies, was a further step in Liquid's development of its industrial gas business. This transaction widened Liquid's supply network to a total of 26 plants in the United States and Canada. The first liquid oxygen distribution started with the purchase of a plant in Detroit in 1955.

Industrial Air Products's industrial gas facilities in Louisiana and Mississippi and Western Oxygen, Inc., of Seattle were acquired in 1956, and in the same year a hydrogen plant was built in San Carlos, California, to serve the rapidly expanding industries in the Bay Area.

In 1957 Liquid Carbonic merged with General Dynamics. The next acquisition was Hench Associated Gas Enterprises of New Jersey, which had 16 industrial/medical gas plants in eight East Coast states. An electrolytic hydrogen plant was acquired in Toronto. During the 1960s Liquid's international operations increased in Canada and South America and new operations were started in Spain.

Figure 5.1.7. Liquid Carbonic Trinidad, 1952. From Stiegerwald (1988) with permission.

Figure 5.1.8. Liquid Carbonic 16-tube trailer for transporting hydrogen. From Stiegerwald (1988) with permission.

Pioneering Applications of Liquid Nitrogen

In 1965, the company began test marketing of "a revolutionary new food freezing machine." It was the first of its kind in the world. The Cryotransfer freezer (Fig. 5.1.9) used liquid nitrogen to "flash-freeze" food products without sacrificing flavor and texture, which had not been possible to do effectively with conventional blast freezing. This development continued in 1971 with liquid CO_2 freezers (Cryo-Shield) and liquid nitrogen freezers (Cyro-Shield, Flavor-Shield), which considerably increased sales of liquid CO_2 and liquid nitrogen.

New Owners

In 1969 Liquid Carbonic became part of Houston Natural Gas Corporation, and in the next year entered the specialty gas business.

Figure 5.1.9. Cryogenic liquid nitrogen food freezing equipment, the Cryotransfer freezer. From Stiegerwald (1988) with permission.

In 1970 Liquid Carbonic acquired Canada National Oxygen, Ltd., and in 1971 entered into a joint venture (50:50) with Mitsui Toatsu Chemicals, Inc., of Tokyo. The Korea Liquid Carbonic Company, Ltd., was established with headquarters in Seoul.

During the 1970s and 1980s Liquid Carbonic's geographic expansion continued, and by 1981 it had a total of 85 carbon dioxide plants world-wide. In 1984 Houston National Gas sold Liquid Carbonic to CBI.

In 1986 Liquid Carbonic entered the electronic gases market with a high-purity specialty gas laboratory in Orlando, Florida.

In 1996 Praxair purchased Liquid Carbonic's owner CBI, and after some years fully integrated Liquid Carbonic into Praxair.

Short Biography

Jacob Baur (1856–1912) was born in Louisville, Kentucky (Fig. 5.1.10). His father, John Jacob Baur, came to the United States from Zürich, Switzerland, and opened a large retail drug establishment in Terre Haute, Indiana.

Jacob graduated from the Philadelphia College of Pharmacy in 1881 and returned to Terre Haute to help his father manage the drugstore. Having worked in the store since he was 14, Jacob had long been interested in "soda water," or carbonated drinks, which led him to make a scientific study of them. He developed his retail drug business to the point where he soon began to see its larger possibilities.

In the early 1880s he perfected a process for compressing and liquefying carbon dioxide and placed it on the market. He went to Chicago to form a company for that purpose called the Liquid Carbonic Acid Manufacturing Company. In 1903 it changed its name to the Liquid Carbonic Company, and was operating plants in 10 cities as well as marketing carbonator equipment. During 1903–1905 the business expanded and increased its equipment production capability by acquiring three specialized machine manufacturing companies.

After Jacob Baur's unexpected death in 1912, Charles Minshall, who was Vice President and a director of the company, served as Liquid's chief executive officer. Jacob's widow, Bertha Duppler Baur, was elected Vice President.

Bertha Duppler Baur (1872–1967) had a long life filled with achievement and was involved in politics for most of her 95 years. Though politics may have been her

Figure 5.1.10. Jacob Baur. From Stiegerwald (1988) with permission.

life-long calling, her years in the business world were remarkably successful. Having earned a law degree in 1908—the same year she married Jacob Baur—she was fully capable of defending her stakes in Liquid Carbonic when her husband died. She assumed a leading role in the business and doubled the worth of Jacob's initial investment.

Milestones

1888	Liquid Carbonic Acid Manufacturing Company formed by Jacob Baur in Chicago
1889	Liquid Carbonic's Red Diamond cylinders introduced
	Carbonator developed
1895	First automatic carbonator developed
1896	New product lines introduced of flavorings for beverages and soda fountain manufacturing
1898	Onyx Soda Fountain Company acquired after a fire destroyed Liquid's new factory
1900	Business expansion to 12 large factories in Chicago, New York, Pittsburgh, St. Louis, Milwaukee, and Cincinnati
1914	Low-pressure filling system introduced
1917	Manufacturing of very pure carbon dioxide by the "coke process"
1918	First foreign plant investment, in Toronto, Canada
1920	Manufacturing and marketing of full product line in Havana, Cuba
1926	Dry ice introduced on a commercial scale
1928	Agreement to install Dry Ice Corporation's manufacturing equipment in 15 of Liquid's plants
1929	General Carbonic Company, New York, acquired
1930	Dominion Carbonic Company, Toronto, acquired
	Keystone Carbonic Company of Pennsylvania acquired
1931	Liquid Carbonic West Indies formed in Trinidad
1939	Wall Chemicals, Inc., acquired, marking Liquid Carbonic's entry into the industrial gas market
	Acetylene Gas & Supply Co, Toledo, acquired
1940	Independent Oxygen Company, Cincinnati, acquired
1942	Medical gases division formed by acquisition of Cheney Chemical Company, Cleveland
1948	Stuart Oxygen Company, San Francisco, acquired
1949	Paschall Oxygen Company of Pennsylvania, acquired
1953	Bird Gas Corporation, Detroit, acquired, producer of oxygen and acetylene, and a major distributor of welding machines and supplies
1955	Start of liquid oxygen distribution
	Fully automated CO_2 plant built in Oakland, California
	Industrial Air Products industrial gas facilities in Louisiana and Mississippi, acquired

1956 Western Oxygen, Inc., Seattle, acquired
Construction of hydrogen plant in San Carlos, California
1957 Liquid Carbonic merged with General Dynamics
1959 Hench Associated Gas Enterprises of New Jersey, acquired
An electrolytic hydrogen plant in Toronto, Canada, acquired
1963 Liquid's international operations include 32 carbon dioxide plants and 13 industrial gas plants in Latin American countries
Liquid Carbonic introduces the Mono-Trak flame cutting machine
1965 Introduction of the Cryotransfer freezer using liquid nitrogen to "flash-freeze" food products
1966 Liquid Carbonic's largest air separation plant (200 tons per day) built in Calumet City, Illinois
Purchase of a site in Scarborough, Ontario, Canada; industrial gases complex built including an electrode plant, a hydrogen plant, and a tonnage air separation plant
The largest commercial hydrogen plant in Canada built at Rexdale, Ontario
1967 New oxygen plant in La Paz, Bolivia, built at an altitude of 4300 meters
1969 Liquid Carbonic becomes part of Houston Natural Gas Corporation
1970 Liquid Carbonic enters the specialty gas business
Canada National Oxygen, Ltd., Oakville, Ontario, acquired
1971 Japanese joint venture between Liquid Carbonic and Mitsui Toatsu Chemicals, Inc., Tokyo
Production and marketing of CO_2 in Japan under the name Mitsui Toatsu Liquid Carbonic Inc.
Korea Liquid Carbonic Company, Ltd., established in Seoul
1971 Fabrication of liquid CO_2 freezers (Cryo-Shield) and liquid nitrogen freezers (Cyro-Shield, Flavor-Shield)
1976 Startup of Canada's largest commercial hydrogen plant near Sarnia, Ontario
1979 Air separation plant operating at 300 tons per day at Irwindale near Los Angeles
1981 Liquid Carbonic has a total of 85 carbon dioxide plants world-wide
1984 Houston Natural Gas Corporation sells Liquid Carbonic to CBI
1986 Air separation plants in La Porte County, Indiana, and Stockertown, Pennsylvania, operating at 600 tons per day
High-purity specialty gas laboratory in Orlando, Florida
1996 Liquid Carbonic owner CBI purchased by Praxair; Liquid Carbonic becomes a fully integrated part of Praxair

Reference

1. Stiegerwald, D. (1988). *The Liquid Story 1888–1988. A History of Liquid Carbonic's First Hundred Years.* Liquid Carbonic.

5.2. LINDE AG[1–5]

Background

Linde AG was founded in 1879 in Wiesbaden under the name Gesellschaft für Linde's Eismaschinen to develop and exploit Carl Linde's inventions in the refrigeration field. In 1890 Linde left the executive management of his company to resume teaching at the Technical University in Münich. There he began his study of the air liquefaction problem and made a pioneering invention: In 1895 he made an air liquefier using a counterflow recuperative heat exchanger and a Joule–Thomson expansion valve, for which he was granted a patent. For his inventions he was also awarded an honorary doctorate by Braunschweig College of Technology and was raised to the nobility in 1897. In October the same year von Linde contracted with Gesellschaft für Linde's Eismaschinen to exploit his air liquefaction process. The first use of the air liquefaction process was for research in university institutes and as a coolant in the separation and production of chlorine by freezing a chlorine/air mixture. By 1901 the Linde workshop in Lothstrasse in Munich had built about 60 air liquefiers for laboratory use. At the 1900 World's Fair in Paris the liquefier (Figs. 5.2.1 and 5.2.2) was awarded the Grand Prix.

Pure Oxygen and Nitrogen from the Air

The next task, which Linde had begun to studying as early as 1895, was to find ways to separate oxygen from nitrogen, and in 1902 he found how oxygen could be separated by means of rectification.

Partial vaporization of liquid air was used to produce air with up to 50 percent concentration of oxygen, so-called "Linde air." It found some applications in the

Figure 5.2.1. The Linde air liquefier at the World Exhibition in Paris, 1900. From Claude (1926).

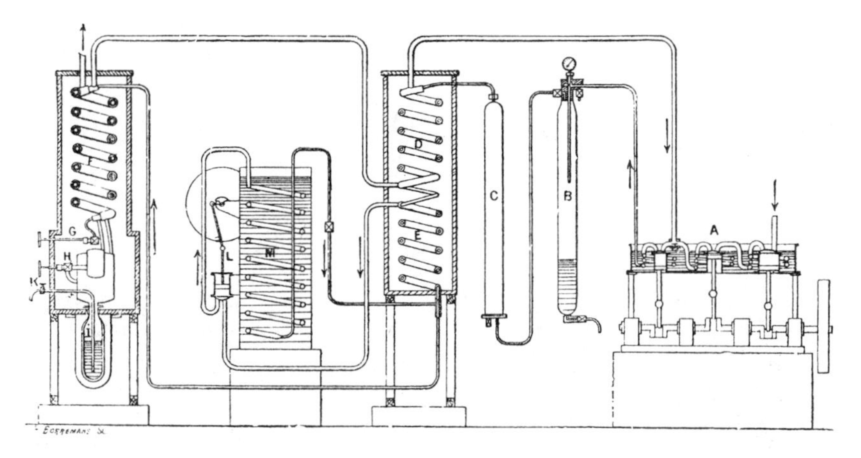

Figure 5.2.2. Principle of the Linde air liquefier shown at the World Exhibition in Paris, 1900. From Claude (1926).

chemical industry. Another use was as an explosive, Oxyliqvit, containing "Linde air" combined with oxidizable materials (as charcoal, sulfur, and petroleum). This explosive was used for blasting the Simplon Tunnel about 1900 and also played a role during the shortage of explosives during World War I.

"Linde air" did not have much of a future because newly developing large-scale applications such as metalworking required oxygen of higher purity. An experimental air liquefier for research and development was installed in the Linde Eismaschinen works at Höllriegelskreuth in 1901 (Figs. 5.2.3 and 5.2.4).

Figure 5.2.3. Interior of Werk Höllriegelskreuth, 1900. Courtesy of Linde AG, Wiesbaden, Germany.

Figure 5.2.4. Exterior of Werk Höllriegelskreuth, 1901. Courtesy of Linde AG, Wiesbaden, Germany.

The demand for a method of manufacturing oxygen with up to 99 percent purity increased. In 1903, a rectification process was found capable of producing oxygen at a rate of 100 cubic meters per hour. The gas could then be compressed in steel cylinders for sale.

A demand for pure nitrogen arose at the same time in the chemical industry. Linde showed in 1904 that it was possible to reach a purity of 99.7–99.8 percent by volume, which was sufficient for the time being. The first plant of this kind started operation in 1905 for nitro lime (calcium cyanide) manufacturing in Italy according to a new process developed by Adolf Frank and Nicodemus Caro.

Market Development

Initial Sales

Linde Eismaschinen decided in 1903 to construct its first test plant in Münich to sell pure oxygen locally. Steel cylinders were purchased for distribution of gas in the Münich area. For more complete coverage of the country, it was decided to form the first oxygen joint marketing organization together with two existing sellers of chemically produced oxygen who already had a distribution system in place. This led to the formation of Vereinigte Sauerstoffwerke GmbH, Berlin, in 1903, a company which handled the distribution of oxygen for Gesellschaft für Lindes Eismaschinen, Sauerstoff-Fabrik Berlin (earlier Elkan), and Kohlensäurewerke C. G. Rommenhöller AG of Herste, Westphalia. They made oxygen via the barium route and the Kastner lead process, respectively. Linde was obliged to provide the gases to the joint venture, and did so at a profit. He also shared in the profits of the joint venture. Oxygen was first produced in the facility at Höllriegelskreuth, which had expanded from the pilot stage into a larger, true manufacturing plant, and from a second factory at Unterbarmen. Next, a plant was constructed in Berlin in 1904. During the next 5 years seven additional oxygen plants were built in Germany. Von Linde contributed to further progress by improving the rectification process in 1903 to make possible the production of pure nitrogen from air as well and in 1910 by designing the double-column rectifier, which gave a greatly increased yield of oxygen without using more power and made it possible to produce pure oxygen and pure nitrogen in the same plant.

The first Linde acetylene plant started production in Düsseldorf in 1910 and experiments to produce rare gases from air led in 1913 to a process for argon manufacture on an industrial scale. At the end of 1910, Vereinigte Sauerstoffwerke was dissolved as an oxygen distributor and was replaced by Sauerstoffwerke GmbH, Berlin, in 1911.

Foreign Subsidiaries

Outside Germany, gas manufacturing companies were also established by Linde or with Linde's participation. An important example was the founding, jointly with Société d'Applications de Acétylène in Paris, of Internationale Sauerstoff-Gesellschaft AG in Berlin in 1906. This in turn formed companies to manufacture gases in several other European countries and the United States. The most important subsidiary of Internationale Sauerstoff-Gesellschaft at the time was the Linde Air Products Company, founded in 1907 in Cleveland, Ohio, which in 1917 became a part of the Union Carbide Corporation, the forerunner to Praxair.

On the European continent, Linde became established in almost all countries, in the Austro-Hungarian empire under the name Oesterreichisch-Ungarische Sauerstoffwerke GmbH, later Hydroxygen Ges.mbH, and in Italy under the name Società Italiana Ossigeno ed Altri Gas (SIO), Milan, both founded in 1909. During the same period, Linde invested in subsidiaries or affiliates in Switzerland (Sauerstoff- & Wasserstoff-Werk Luzern AG, Lucerne), Spain (Abelló Oxigeno-Linde, S. A., Barcelona), and France (Bardot, Paris; Duffour, Igon & Cie., Bordeaux). New companies were formed in Scandinavia: Dansk Ilt & Brintfabrik in Copenhagen, Norsk Surstof & Vandstoffabrik in Oslo, and Nordiska Syrgasverken in Stockholm. Gas plants were also licensed by Internationale Sauerstoff-Gesellschaft between 1906 and 1912 in Belgium, Russia, Romania, Argentina, Chile, Brazil, China, and Japan. By the time Internationale Sauerstoff-Gesellschaft was dissolved in 1912 it had air gas production activities in 29 countries.

International Competition

Linde and Air Liquide were the first in the business of cryogenically produced oxygen and very soon each formed its own subsidiaries. A significant difference in marketing approach was that Air Liquide aimed at selling gas, whereas Linde concentrated on selling plants to customers, together with a license to make and market

Figure 5.2.5. Oxygen transport lorry for Linde Air Products, 1914. Courtesy of Linde AG, Wiesbaden, Germany.

gas for a defined geographic area. In this way Linde could keep direct control over the production of oxygen in the company's interests.

After some disputes Linde and Air Liquide made an agreement whereby they licensed each other's patents in various countries, depending on the stage of subsidiary formation. In Germany, for instance, Air Liquide had transferred its patent rights to Griesheim Electron, so Linde and Griesheim came to a patent agreement in 1908. By that time there were many other European countries in which Linde and Air Liquide had started gas businesses. In Scandinavia, South America, and some other countries the market was now divided between Linde and Air Liquide.

In Great Britain, Linde joined with Brin Oxygen in 1906. Brin had been producing and selling oxygen made by the barium process since 1890. Linde contributed its British Linde patents and a plant owned by Linde Refrigeration of Great Britain. Brin Oxygen was renamed the British Oxygen Company. After World War I, Linde lost its subsidiaries in the United States and Great Britain because German assets were confiscated.

Other Gas Separation Applications

Hydrogen Production

From about 1904 on, von Linde began studying the possibilities of separating other technically interesting gases and gas mixtures. At his request, Adolf Frank came up with the idea of producing hydrogen by separation of water gas. At that time there

Figure 5.2.6. Early ASU 1914 (before World War I). Courtesy of Linde AG, Wiesbaden, Germany.

was a great need for hydrogen for filling airships. Some years later other large-scale hydrogen applications arose in chemical industries for fat hardening and ammonia synthesis, which became very important for the manufacture of fertilizers, fuels, and ammunition in Germany during World War I.

The Linde–Frank–Caro Process

Linde developed the first plant where hydrogen was produced by means of partial condensation of carbon monoxide (the Linde–Frank–Caro process, 1910). A pilot plant was built in Höllriegelskreuth in 1909. Because hydrogen has a much lower boiling point than the other gases in water gas, separation was effected by condensing the other, heavy components. Using a bath of liquid nitrogen under reduced pressure, a content of carbon monoxide of about 1 volume percent could be reached.

The removal of carbon dioxide, which solidifies at low temperatures, was done by using soda lye.

Separation of Coke Oven Off-Gas

It was thought that the large supplies of coke oven off-gas might be used in the same way as water gas for hydrogen, methane, and ethylene production, but coke oven gas contains many more impurities with high boiling point, so the risk for clogging was considered to make this process impossible in practice.

In 1921, however, a process for separation of coke-oven gas into different fractions such as hydrogen, methane, ethylene, and propylene was developed. In 1924, the production of a hydrogen–nitrogen mixture with an extremely low content of carbon monoxide (poison for catalysts) and suitable for ammonia synthesis was developed. This stimulated a growing demand for very large low-temperature plants from the chemical industry all over the world. Although Linde as a manufacturer of air separation equipment was subject to increased competition, it completed its 500th oxygen–nitrogen plant in 1928.

Ammonia

About 1912 the Linde–Frank–Caro process and the pure nitrogen process acquired a special use for the synthesis of hydrogen and nitrogen to obtain ammonia by the Haber–Bosch procedure. With the outbreak of World War I, Chilean saltpeter was no longer available and new nitrogen and hydrogen sources for production of ammonia (NH_3) were particularly important. The construction of very large nitrogen plants for the production of calcium cyanamide was also essential. These nitrogen plants had at least four times the throughput of those built hitherto. The oxygen plants also were built much larger.

Argon Production

In 1912, the lightbulb industry recognized that the gas filling in lamp bulbs must contain a heavy rare gas that also was a bad heat conductor. This led von Linde to start work in the field of rare gas production. The increasing demand for argon for bulb filling led to the installation of equipment for the production of argon in some of the larger oxygen plants.

Plant Construction

In 1920 Linde took over Maschinenfabrik Sürth, Berlin, as a part of Deutsche Oxhydrique, which Linde and Griesheim Elektron acquired in 1916. Sürth had manufactured equipment for carbon dioxide production and applications since 1885 and sold dry ice production units from the 1930s. In 1922 Linde purchased part of the Heylandt Gesellschaft für Apparatebau and acquired patents for liquid distribution of industrial gases, a very important investment acquisition for the future. Paul Heylandt later took a very active part in the German development of liquid oxygen-fuelled rockets.

Figure 5.2.7. One of the first tank cars, 1930s. Courtesy of Linde AG, Wiesbaden, Germany.

In 1925, the Vereinigte Sauerstoffwerke GmbH (VSW) was formed once again in Berlin, owned 50 percent each by Gesellschaft für Lindes Eismaschinen and Chemische Fabrik Griesheim Elektron. VSW was a cooperative effort to make gas distribution cheaper and more efficient. Its most important customers were the iron and steel industry, the chemical industry, and the metal-working industries. It was founded at the time when the first steps were being taken to achieve liquid distribution of air gases and to build tonnage plants. VSW remained in existence until 1945.

Regenerators

1925, Matthias Fränkl, an engineer and businessman from Augsburg, got the idea of using regenerators in air separation plants instead of continuously working recuperators. In 1926, he told Linde about his idea, but was first met with some scepticism. Fränkl built a first air separation plant equipped with regenerators in his workshop, using corrugated aluminum strips as heat reservoir. His plant suffered problems which had to be studied in a thorough research program. In 1930 Linde decided to build a pilot plant in Hannover to master the problems. This expensive engagement turned out to be a very good investment. Through this technology large tonnage quantities of atmospheric gases could be produced economically. Soon large Linde–Fränkl plants were erected for metallurgy and chemistry customers. In 1932, a breakthrough was achieved with the first plant for the liquefaction of helium.

Krypton and Xenon Separation from Air

From about 1930 air separation production of krypton and xenon from the oxygen in large Linde–Fränkl plants was feasible. The process idea originally came from Georges Claude in Paris. For development of the separation procedure a cooperative company called Krypton was set up by Linde, Air Liquide, and Griesheim Elektron.

Expansion Turbines

The piston expander was found to be the bottleneck in attempts to build air gas plants with high throughput. The expansion turbine was thought to be a more efficient refrigerating machine, but the first turbines installed from about 1934 had too low an efficiency. It was not until 1950, after the Russian scientist Pjotr Kapitza developed this equipment further, that the required increased efficiency was obtained.

Molecular Sieves

Water vapor and carbon dioxide can be removed at ambient temperatures with the so-called molecular sieve. About 1960 this procedure was developed for air separation plants and it is still applied today. The first Linde air separation plant equipped with a molecular sieve came into operation in 1968.

Applications

In 1937 Linde started a welding equipment department, Ellira, based on the Electro-Linde-Rapid submerged arc welding method. This department developed and introduced new electric welding and cutting techniques. The welding business of Linde was sold in 1972 to Messer Griesheim, which discontinued the construction of air separation plants.

A Linde contribution to environmental conservation was the development of the Lindox method for the biological purification of sewage and waste waters using oxygen. The first Lindox plant has been in operation since 1974.

During World War II Linde plants were used to meet the great oxygen needs of the military. By 1945 many plants (like large parts of the Höllriegelskreuth facilities) had been destroyed by Allied troops and production figures were back to those of 1921. During 1948–1951 the main industrial gas plants were reconstructed, and Linde's plant manufacturing business grew rapidly. In 1953 the then largest plant for air separation (ca. 300 tons per day of oxygen and 540 tons per day of nitrogen) was built for export to the United States. In 1964 Linde constructed the world's largest air separation unit to that date (1130 tons per day of gaseous oxygen, 30 tons per day of gaseous nitrogen, and 30.7 tons per day of liquid argon) at Linz, Austria.

In 1965 the name of the company was changed from Gesellschaft für Linde's Eismaschinen to Linde AG, and in 1999 Linde Gas was formed following the acquisition of AGA.

International Market Developments

In 1956 Technische Gase Ges.mbH (TEGA) was founded in Obereggendorf, Austria, as a cooperative venture between AGA and Linde. This was Linde's first major postwar company investment outside of Germany. In 1972 Linde and MG started a joint venture, Likos AG, in Zurich, Switzerland, for development of their international industrial business. Manufacturing and marketing companies were started in Belgium, France, The Netherlands, and South Africa. This joint venture remained until 1989, when the European Economic Community forbid

Figure 5.2.8. Linde ASU, Trieste. Courtesy of Linde AG, Wiesbaden, Germany.

Figure 5.2.9. Salzgitter plant. Courtesy of Linde AG, Wiesbaden, Germany.

this type of joint activity in Europe. In 1974 Aeroton Gases Industrial Ltd. in Rio de Janeiro became a new subsidiary in Brazil and Linde Gas Pty. Ltd. was started in Sydney, Australia.

After the Berlin Wall fell in 1989 Linde soon expanded into Eastern Europe through cooperative ventures and acquisitions. A cooperative venture for delivering industrial gases to the Leuna-Werke (Fig. 5.2.10) in the former East Germany began in 1990, followed by a full takeover by Linde in 1998. In 1991 a majority stake was taken in the leading Czech gases company, Technoplyn a.s. In 1995 Technoplyn became a Linde subsidiary. In 1999 Linde acquired the Polish gas subsidiary of the U.S. gas company Airgas, and became the largest supplier of industrial gases in Poland. On 16 August of that year Linde announced its takeover of AGA.

In 1992 Linde increased its ownership in nv W. A. Hoek's Machine en Zuutstoffabriek in the Netherlands to 60 percent. The Italian gas supplier Caracciolosigeno s.r.l. was purchased in 1994, and in 1997 Linde acquired the Austrian hydrogen business of Air Products GmbH, also in that year extending its carbon dioxide business in Austria by further two acquisitions.

Linde's industrial gas plant and systems manufacturing business in Asia took a major step forward by means of a joint venture with a Chinese manufactrurer in Dalian, China, in 1994. During the 1990s Linde also took an active part in the German solar hydrogen project (Fig. 5.2.11).

Figure 5.2.10. Linde hydrogen plant, Leuna. Courtesy of Linde AG, Wiesbaden, Germany.

Figure 5.2.11. Hydrogen storage tanks for the solar hydrogen project. Courtesy of Linde AG, Wiesbaden, Germany.

Short Biography[6]

Carl Linde (1842–1934) was born at Berndorf, Upper Franconia, Germany, the third of nine children. His father was a Protestant minister. Raised in Kempton, Germany, he had cultural interests together with great enthusiasm for technical matters. After studying machine construction, he went to the famous Polytechnicum in Zürich in 1861, where he studied for 4 years under R. Clausius, among others, and developed a passion for thermodynamics. In 1868 he joined the newly opened Munich Technical University and became a full professor in 1872. Here he developed his first compression refrigerating machines using methyl ether and studied the difficulties of working with this refrigerant. In 1876, he built his first ammonia compressor. Brewery owners asked Linde to set up a refrigeration company under his own management.

Linde left his teaching position in 1879 and founded the Gesellschaft für Linde's Eismachinen in Wiesbaden to develop his refrigeration process on an industrial scale. Contracts were signed with prominent companies such as MAN and Sulzer for production of the Linde-designed refrigeration machines. The first foreign company to be set up was the Linde British Refrigeration Company, London, in 1885.

By the time Linde left Wiesbaden in 1891 to resume his teaching and research, 1200 of his refrigeration machines had been installed. At that time nobody had succeeded in liquefying oxygen in other than very small quantities. Linde's aim was to work on a method to liquefy air for the separation of oxygen. He combined two known effects. One was to use a countercurrent heat exchanger arrangement like the one used by William Siemens in 1857 to save heat by using the waste heat of a process, which produced hot gas. Siemens, however, had not thought of the reverse: the accumulation of cold and air liquefaction.

The other effect Linde used was that of achieving refrigeration by expanding air. The work performed by such an expansion has to come out of the heat content

Figure 5.2.12. Carl Linde. Courtesy of Linde AG, Wiesbaden, Germany.

of the gas, thereby cooling the gas. Such cooling could be produced by expanding air in a piston expansion engine.

Early in 1894 Linde asked Sulzer Brothers in Switzerland, who manufactured compressors for the Linde refrigeration business, to design a suitable piston expansion engine. He soon realized, however, that no lubrication material could be found for the extremely low temperatures he wanted to reach, and the insulating losses would be too large. Therefore, in December 1894 he canceled his request to Sulzer and began studying the Joule–Thomson effect. In 1863 Joule and Thomson had published results of their experiments on expanding air through a porous plug. They found that the air cools on expansion, but to a very small extent—only 1/4 degree Celsius for every 1 bar of pressure drop. No one paid much attention to this minimal cooling effect.

Linde found that a simple valve which works exactly like the porous plug could be used. The valve employed was far more reliable than an engine at very low temperatures. Linde now designed and ordered a spiral heat exchanger consisting of a 100-meter-long, 4-inch pipe with a $1\frac{1}{2}$-inch pipe inside it. It was wound in a coil (33 turns, diameter ca. 1 meter, weight ca. 1.3 tons). Today, a heat exchanger weighing 20 kilograms would be able to do the job. (See Fig. 4.1.8)

The process now consisted of only three components: a compressor for use at near-ambient temperature, a spiral heat exchanger, and an expansion valve. In May 1895 the arrangement was started up. After 3 days to cool the heavy coil it produced 3 liters per hour of liquid air on the very first try.

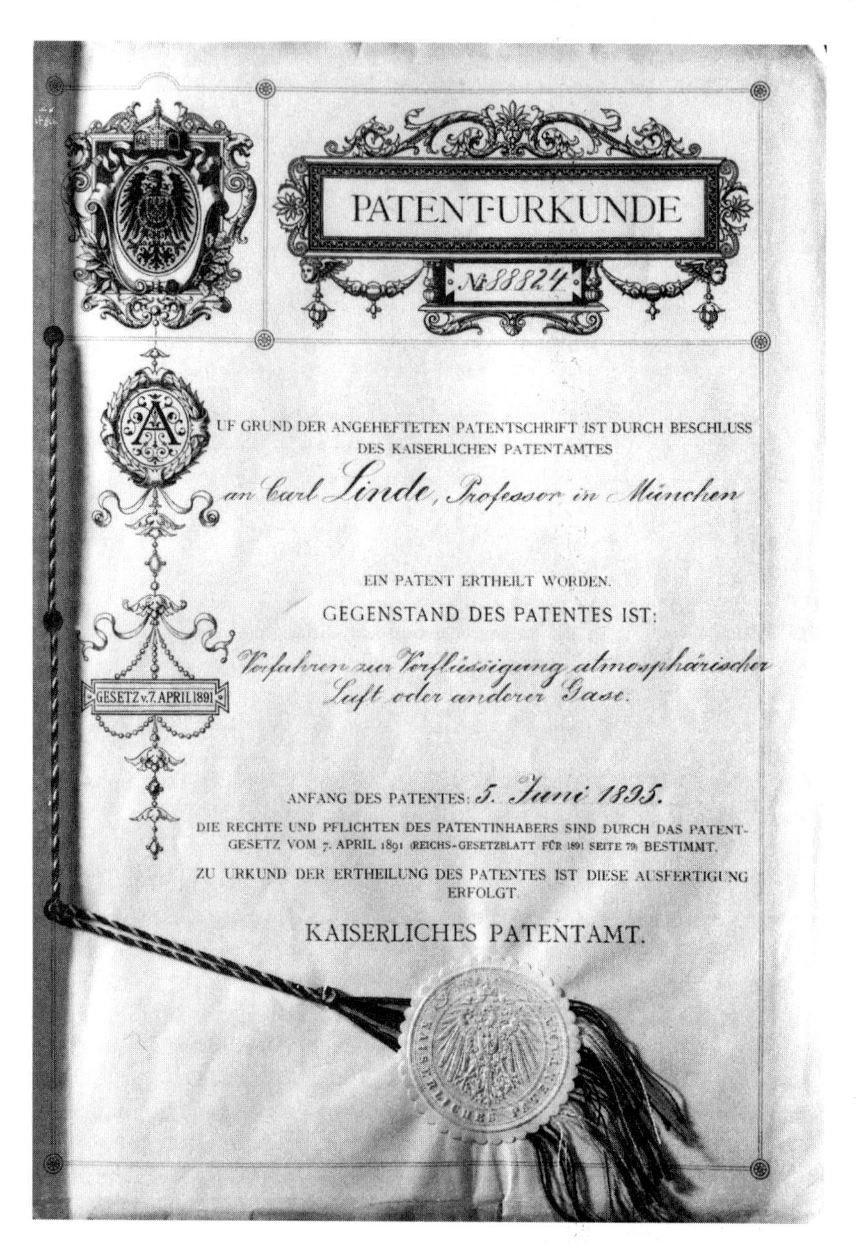

PATENT-URKUNDE

№ 88824

GESETZ v. 7. APRIL 1891

AUF GRUND DER ANGEHEFTETEN PATENTSCHRIFT IST DURCH BESCHLUSS DES KAISERLICHEN PATENTAMTES

an Carl Linde, Professor in München

EIN PATENT ERTHEILT WORDEN.

GEGENSTAND DES PATENTES IST:

Verfahren zur Verflüssigung atmosphärischer Luft oder anderer Gase.

ANFANG DES PATENTES: 5. Juni 1895.

DIE RECHTE UND PFLICHTEN DES PATENTINHABERS SIND DURCH DAS PATENT-GESETZ VOM 7. APRIL 1891 (REICHS-GESETZBLATT FÜR 1891 SEITE 79) BESTIMMT.

ZU URKUND DER ERTHEILUNG DES PATENTES IST DIESE AUSFERTIGUNG ERFOLGT.

KAISERLICHES PATENTAMT.

Figure 5.2.13. The Linde air liquefaction patent, 1895. Courtesy of Linde AG, Wiesbaden, Germany.

Linde's inventions had immediate effects on industrial gas development. Laboratory experiments using very low temperatures could now be done. By 1897 the company had sold 14 laboratory liquefiers.

Linde was chairman of the supervisory board of the company until 1931 and remained a member of the board until his death in 1934.

Milestones

1879	Gesellschaft für Linde's Eismaschinen founded on June 21 in Wiesbaden, Germany
1890	Carl Linde leaves the executive management of his company to be professor at the Technical University of Munich; he keeps his position as chairman of the board in the company until 1931
1895	Patent granted to Carl Linde for the "process for liquefaction of atmospheric air or other gases"
1897	Production started on air liquefaction machines
1900	Building of a trial plant for development of air liquefaction in Höllriegelskreuth, near Munich
1902	First air separator employing rectification process installed at Höllriegelskreuth
1903	Cofounding of Vereinigte Sauerstoffwerke GmbH in Berlin by Linde and two other partners as a company for distribution of gaseous oxygen
	Additional Linde gasworks opened in Germany in Barmen, Berlin, Düsseldorf, Mülheim, Altona, Nuremberg, and Dresden, and abroad in Paris, Toulouse, and Antwerp
1906	Holding acquired in British Oxygen Company, London
	Cofounding of Internationale Sauerstoff-Gesellschaft AG in Berlin for organization of oxygen gas plants and gas distribution abroad
	New companies formed in Barcelona, Stockholm, and Vienna
	Investment in affiliate at Lucerne and licenses for oxygen production in other countries
1907	Linde Air Products Company launched in Cleveland, Ohio, by Linde and other partners.
1909	First acetylene plant constructed at Düsseldorf–Reisholz, marking Linde's start in the acetylene business
1910	Patent granted to von Linde for the double-column rectifier for production of pure oxygen and nitrogen gases from liquid air
1912	Two large plants for nitrogen and hydrogen production supplied to BASF for ammonia synthesis. Property of Internationale Sauerstoff-Gesellschaft AG taken over by Linde
1913	First trials on manufacture of rare gases from air
1914	Patent for argon production process
1918	Loss of numerous industrial rights, important branches, and company holdings abroad following the end of World War I

1920	Maschinenfabrik Sürth assimilated by Linde as part of Deutsche Oxhydrique
1922	Investment in Heylandt Gesellschaft für Apparatebau mbH in Berlin, acquiring patents for activities still in effect in the field of cryogenic and process engineering
1928	First contract concluded with Matthias Fränkl for exploitation of his inventions in low-temperature technology, including the replacement of heat exchangers by regenerators
	The 500th air separation plant delivered by Linde
1934	Death of Carl von Linde at the age of 92
1937	Introduction of the Unterpulver submerged arc welding method in the newly formed Ellira (Electro-Linde-Rapid) welding engineering department
1945	Destruction of plant facilities and renewed loss of industrial rights, investments, and subsidiaries.
1947	Participates in Sauerstoffwerk Wilhelmshaven GmbH, its first postwar company investment
1949	Rebuilding of destroyed works
1953	Completion of largest air separator built hitherto in Europe for delivery to United States, capacity 300 tons per day of oxygen and 540 tons per day of nitrogen
1955	Construction of plant to separate heavy hydrogen at process temperatures of –252°C
1956	Cofounding of TEGA, Technische Gase Ges.mbH, Obereggendorf, Austria, the first major postwar company investment outside of Germany and starting point for reconstructing business abroad
1965	Name change to Linde AG
1969	Construction of air separator for BASF, one of world's largest, generating oxygen, high-purity nitrogen, and argon at 1300, 3400, and 50 tons per day, respectively
1971	Commission of first tonnage plant for liquid natural gas storage built by Linde; geometrical tank capacity 30,000 cubic meters
	Construction of Linde's own tonnage oxygen plants initiated with completion of Salzgitter works
1972	Together with MG, sets up Likos AG in Zurich, Switzerland, for development of their international industrial gases business, leading to organization of manufacturing and marketing companies in Belgium, France, The Netherlands, and South Africa; at the same time an agreement is made resulting in Linde taking over air separation plant production and MG taking over welding equipment manufacturing
	Acquires 50 percent of the acetylene producer Industriegas GmbH & Co., Cologne
1974	Development of industrial gas business in Brazil and Australia by starting Aeroton Gases Industrial Ltd. in Rio de Janeiro and Linde Gas Pty. Ltd. in Sydney
	The first LINDOX plant for sewage treatment delivered

1990 Cooperation with Leuna-Werke, which relinquishes its industrial gases section to Linde

1991 Majority stake in the leading Czech gas company Technoplyn a.s.

1992 Increase of participation held since 1974 to 60 percent in the largest supplier of industrial gases in the Netherlands, nv W. A. Hoek's Machine- en Zuurstoffabriek

1994 Inauguration of Linde's largest industrial gases center, in Leuna

Purchase of Italian gases supplier Caracciolossigeno s.r.l. and founding of joint venture in Dalian, China, with a Chinese manufacturer of system units for the design and construction of air separators in Asia

1995 Construction of further acetylene plants and filling units in Portugal, Czech Republic, and Hungary

Takeover of Technoplyn a.s. in Czech Republic

1996 Purchase of the U.S. engineering company the Pro-Quip Corporation (TPQ), Oklahoma, world market leader in hydrogen plants

1997 Joins an international consortium awarded a 15-year nitrogen supply contract by Pemex, the national oil company of Mexico; Linde builds the world's largest air separation plant, which delivers 40,000 tons per day of nitrogen transported in pipelines for increasing oil flow from offshore wells 130 kilometers away

Acquires the Austrian hydrogen business of Air Products GmbH, Salzburg, and extends its carbon dioxide business in Austria by gaining two majority stakes

1998 Acquires the complete industrial gas supply of the Mitteldeutsche Erdoelraffinerie (MIDER) in Leuna

Takes over the air separation plant from Thyssen Krupp Stahl AG in Duisburg-Ruhrort

Builds and commissions a new air separation plant in Warsaw, Poland

Acquires from Millennium Petrochemicals Inc., Cincinnati, Ohio, a synthesis plant at La Porte, Texas, which produces merchant gas, sold via Holox, Inc., a subsidiary of Hoek Loos and part of the Linde Group

1999 Takes over the Polish industrial gas business of Airgas, and becomes the largest supplier of industrial gases in Poland

Announces takeover of AGA

Announces intention to separate the gases section from Linde AG to form a new company, Linde Gas

References

1. Baldus, W., and Scurlock, R. G. (1992).The story of Linde AG. In *History and Origins of Cryogenics* (R. G. Scurlock, ed.), Chapter 4, pp. 122–140. New York: Oxford University Press.

2. Gesellschaft für Linde's Eismachinen AG (1954). *75 Jahre Linde*. Wiesbaden Germany.
3. Linde AG. (1979a). *Linde 1879–1979. Technische Gase- vorsprung der bleibt*. Wiesbaden Germany.
4. Linde AG. (1979b). *Linde 1879–1979. Jubiläumsausgabe anlässich des 100jährigen Bestehens der Linde AG.*: Wiesbaden Germany:
5. Weiss, O. (1975). Cryogenics worldwide-Germany. Linde AG and cryogenics. *Cryogenics* **1978**(12):635–640.
6. Linde, C. (1916). *Aus meinem Leben und von meiner Arbeit*; reprint: Oldenbourg Verlag, Munich 1979.

5.3. MESSER GRIESHEIM[1]

In 1965, after near 70 years of separate, but almost parallel history, Messer Griesheim GmbH combined the results of work in industrial gas and welding technology pioneered by two former German competitors, Ernst Wiss (Fig. 5.3.1) in Griesheim and Adolf Messer (Fig. 5.3.2) in Frankfurt, who both started in 1898. Messer concentrated on making welding equipment, acetylene generators, and air and gas separation plants, using cryogenic technology, whereas Griesheim Elektron concentrated on the gases business and welding technology.

Messer Griesheim GmbH was founded in 1965 in a merger between Adolf Messer GmbH (formed in 1898) and the Knapsack-Griesheim AG, owned by Farbwerke Hoechst AG. Hoechst bought two-thirds of the Messer Griesheim shares and the Messer family kept the rest of the capital stock.

Knapsack-Griesheim was founded in 1952 when Chemische Fabrik Griesheim-Elektron (CFGE) (formed in 1898) joined AG für Stickstoffdünger, Knapsack (formed in 1911), owned by Hoechst.

Background: Messer

In April 1898, Adolf Messer, a student at the Technische Hochschule in Darmstadt, Germany, started business in a small workshop in Hoechst, near Frankfurt-am-Main. There he produced acetylene generators and acetylene lighting plants, based on skills acquired in gas technology during his apprenticeship at a gas engine factory. The years around the turn of the century were a golden time for acetylene gas, which was used widely throughout industry and in private households. Adolf Messer's business grew rapidly. In 1899, he moved to larger premises in Frankfurt and started Frankfurter Acetylen-Gas-Gesellschaft Messer & Cie. In 1902, he filed his first patent, for an "air vent pipe on acetylene generators."

At that time, Messer foresaw future problems for his business: Gas light and electric lighting were technically and economically superior to acetylene lighting, and because of numerous accidents acetylene had acquired a bad reputation; its market success appeared to have come to an end for households.

Figure 5.3.1. Ernst Wiss.

Figure 5.3.2. Adolf Messer. Courtesy of the Messer Group, Frankfurt-am-Main, Germany.

Welding and Cutting

From 1902 onward, Messer looked for additional sales potential abroad, but at the same time he started to adapt his business operations to a new and promising field of activity: oxy-fuel welding. The change to a new product line was completed in 1904. Now the company faced new challenges, improving welding and cutting tools. In the manufacturing and metal-working industries the demand for improved products increased and created stiff competition among established companies in this branch of business. Adolf Messer knew that he could only compete by offering customers equipment of superior quality and safety. From 1908 his range of torches, acetylene generators, pressure regulators, and other accessories for oxy-fuel welding and cutting were marketed under the trade name Original Messer.

Internationalization

By the outbreak of the World War I in 1914 Adolf Messer had increased the reach of his business activities both to foreign business and to the growing market for the production of oxygen for oxy-fuel applications. The industrial demand for oxygen was enormous, but consumption was held in check by its relatively high price. Messer chose to build and sell oxygen plants, not to sell the gases himself. His strategy was to build air separation plants directly on the customer's premises.

The first Messer-constructed air separation plant was sold to the Spanish firm Rodrigo de Rodrigo, Madrid, in 1910. In the same year, Messer opened its first

Figure 5.3.3. Interior of Messer air separation plant. From Koch (1993) with permission.

representative office abroad, in Oslo, Norway. The company expanded. In 1911 the company, now named Messer & Co. GmbH, employed 180 persons (Fig. 5.3.4). Messer established the Messer Company, Philadelphia, in the United States, using as original capital 5000 dollars from the sale of 10 welding and cutting torches with some accessories which he brought with him and sold in New York. In the United Kingdom he started Messer Engineering Co. Ltd., London. These foreign investments were lost because of World War I.

Business Development

After the war Messer resumed his normal business activities. His plants in Germany were relatively undamaged and he was able to restart production immediately. The old sales markets outside Germany, however, no longer existed. Messer began again, traveling abroad to form companies or joint ventures. In the 1920s great advances were made in the development of welding and cutting equipment and in the construction of air separation plants. In 1924, the first electric welding trials were carried out at Messer. In 1930, the company started to produce coated welding electrodes and in 1932 to manufacture extruded electrodes for arc welding.

The company's air separation plant business was initially restricted to markets outside Germany because of patent disputes with Linde, the major supplier of air separation plants in Germany. In 1929 these problems were resolved and Messer was then able also to supply the domestic market. With the construction of a large separation plant for coke oven gas in the Netherlands in 1930, the company fulfilled the role of general contractor for the first time.

Figure 5.3.4. Messer advertisment. Courtesy of the Messer Group, Frankfurt-am-Main, Germany.

From 1933, Messer and his firm had to deal with National Socialism. Messer refused to compromise politically and gave personal support to employees who were prosecuted by the regime At the same time, however, he benefited from the market demand stimulated by the state.

World War II destroyed much of Messer's production plants. In summer 1945 Messer acquired from the U.S. military government the necessary operating license for Adolf Messer GmbH and started the task of cleaning up. Once again he had to travel abroad to reactivate old business and to create new ones. Hans Messer, the 20-year-old son of the company founder, took an active part in the reconstruction of the firm, especially the arc welding electrode works.

Hans Messer was appointed Managing Director in 1951. In 1953, he became head of the company; 1 year later Adolf Messer died, at 76 years of age.

Figure 5.3.5. Messer head office, 1948. Courtesy of the Messer Group, Frankfurt-am-Main, Germany.

From 1955 onward, the market for cutting and welding equipment from Adolf Messer GmbH continued to grow, benefiting from advances in the metal industry. The market for cryogenic air separation plants also grew. In the late 1950s Messer started to construct spot welding machines for the automotive industry to meet the trend toward automation.

At the beginning of September 1964 negotiations started with Hoechst AG on a possible cooperative arrangement. The result of this was the formation of Messer Griesheim GmbH in January 1965.

Background: CFGE

The Chemische Fabrik Griesheim Elektron (CFGE) was the result of a merger in 1898 between Chemischen Fabrik Griesheim, which had developed the chlorine–alkali-electrolysis method in 1884, and the Chemische Fabrik Elektron, which originated from the Electrochemical Works founded in 1894 by AEG, Berlin. The CFGE electrolysis plant produced large amounts of hydrogen as a by-product from its chlorine cells.

Industral Gas Production

In December 1898 Ernst Wiss joined the Chemische Fabrik Griesheim-Elektron AG. The 28-year-old Wiss was mainly interested in the application and marketing potential of the company's products. His interest was focused on hydrogen, which was used as a lifting gas for balloons.

In his search for further applications of hydrogen, Wiss became involved with lead soldering, an oxy-fuel welding process in common use at that time. He developed a newly designed torch in which the hydrogen gas enters under high pressure carrying air with it. The necessary hydrogen was pressurized and supplied in steel cylinders. To increase the flame temperature, he used oxygen in combination with hydrogen. In 1903 the first oxyhydrogen torch from Griesheim was launched, built by Dräger. Within 2–3 years the metal cutting technology was developed using hydrogen and oxygen. Griesheim bought a license from Cöln-Müsener Bergwerk, a small steel company. They used and patented a hydrogen–oxygen lance, developed in 1901 by Ernst Menne for cutting away the plug in their blast furnace.

Wiss's work showed the CFGE management the growing importance of industrial gases. As a result, a special department was formed in Griesheim in 1904 in order to intensify the search for further applications for these products. Soon the development of welding and cutting led to increased demand for oxygen, and Griesheim decided to produce it. It became a licensee of Air Liquide on patents for the construction of air separation plants. In 1908, the first CFGE-constructed air separation plant went on-stream in Griesheim. As part of a patent dispute settlement between Linde and Air Liquide in the same year it was agreed that the market for oxygen in Germany would be divided 50:50 between Linde and Griesheim/Air Liquide.

The special industrial gas department of CFGE was called the Hydrogen and Oxygen Department. Between 1909 and the outbreak of World War I, this department expanded its market activities. Experience showed that it was not cost-effective to supply oxygen over a wide area due to the high cost of transporting gas cylinders. For this reason, air separation plants were constructed and spread over Germany. By 1913 CFGE had achieved a 44 percent share of Germany's entire oxygen production.

Business Development

In 1916 the Hydrogen and Oxygen Department became a separate business unit, Autogen Works, headed by Ernst Wiss. Concentrating on the German market, his aim was to exploit new market opportunities in order to secure a leading position in cutting and welding technology and in the gas business. In 1920 CFGE, together with Linde, took over Deutsche Oxhydric AG, adding eight more oxygen plants to the Autogen Works. Ernst Wiss became a member of the management board of CFGE in the same year.

Because of continuous growth in the 1920s, a separate business unit had to be set up to handle sales activities. This led to the formation of the Griesheimer Autogen-Verkaufs-GmbH in 1923, with responsibility also for the sale of welding carbide for the AG für Stickstoffdünger (Knapsack).

In 1925 CFGE joined the Intressengemeinschaft der deutschen Teerfarbenindustrie, which was converted into the I. G. Farbenindustrie A.G. The latter took over the ownership of the Autogen Works. The Works divided into two departments, Maschinenfabrik and Gases. The Gases Department was made up of 18 oxygen plants and 8 hydrogen plants. Ernst Wiss remained responsible for the Autogen Works until his retirement in 1932. In 1942 a cooperative arrangement was started between the Autogen Works group and Messer for the marketing of the hardening machines

Figure 5.3.6. Advertisement, Ernst Wiss welding (Griesheim). Courtesy of the Messer Group, Frankfurt-am-Main, Germany.

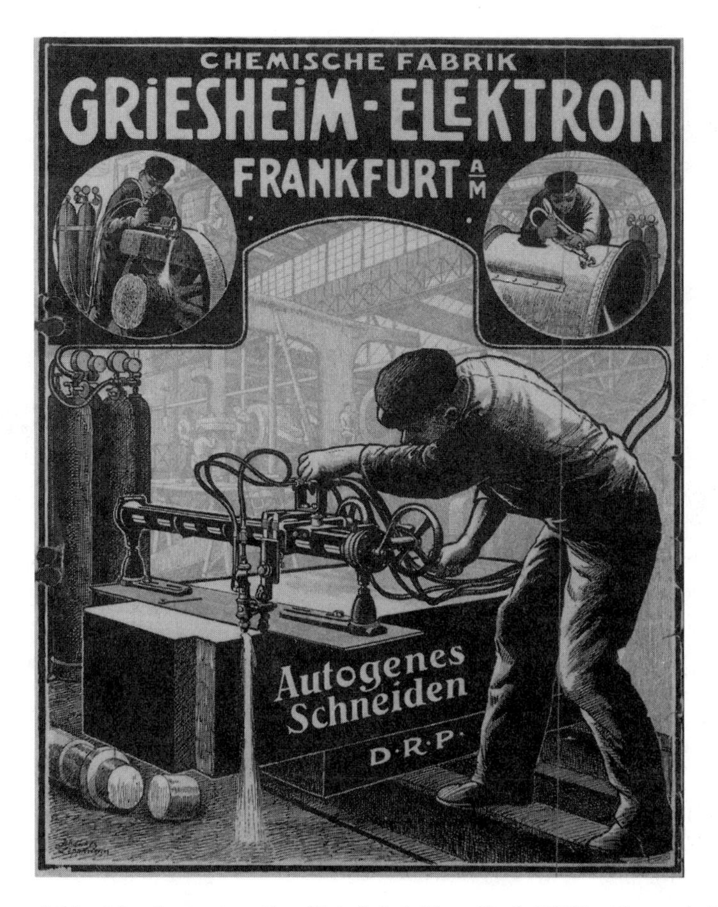

Figure 5.3.7. Advertisement, cutting (Griesheim). From Koch (1993) with permission.

manufactured by the two companies. This agreement led to further contacts between the Messer firm and the Griesheim works, ultimately leading to the formation of Messer Griesheim GmbH.

In 1952, Griesheim-Autogen with 76 oxygen works merged with the AG für Stickstoffdünger, which belonged to Hoechst AG, to form Knapsack-Griesheim AG. The Autogen Works departments Oxygen (plants) and Griesheim-Autogen (equipment) was responsible for about two-thirds of the new company's total sales. During the next 10 years these departments made strong progress in the German market, benefitting from economic developments, increased demand, and the ability of the new company to seek and exploit new opportunities.

Formation of Messer Griesheim GmbH

In January 1965 Messer Griesheim GmbH was formed and has since developed into a world-wide group in the field of industrial gases and related technologies. In 1968 the company was divided into three divisions, Industrial Gases, Cutting and

Figure 5.3.8. Horse-driven cylinder transport, ca. 1905. From Koch (1993) with permission.

Welding Technology, and Cryogenic Equipment. That year the Messer Group (MG) also began operating what is now a 550-kilometer pipeline net for LOX/LIN in the Cologne, Rhein-Ruhr, and Saarland areas in Germany (Fig. 5.3.9).

Market Development

In 1972 corporate management took the structural decision to lay off the Cryogenic Equipment Division (the construction of on-site plants) in order to concentrate on the industrial gases business. In 1973 MG discontinued the manufacturing of air separation plants and sold the cryogenic equipment patents to Linde AG, which in turn sold its welding business to MG. In the 1990s MG went back to the plant business with developments in on-site supply.

International Expansion

In 1972 MG also began to enter the international gas business with the formation of Likos AG, a joint venture with Linde in France, Netherlands, Belgium, and South Africa. In North America MG acquired the Burdett Oxygen Company in 1974. Further opportunities were exploited when Liquid Air acquired the Chemetron Company in 1978. U.S. antitrust laws restricted Liquid Air to the purchase of only 70 percent and MG bought the remaining 30 percent. In 1976 MG formed a subsidiary for welding equipment in Venezuela.

In 1986 MG opened two new markets for industrial gases by the acquisition of SIGAS in the United Kingdom and Società Torinese Ossigeno in Italy. Its strategic

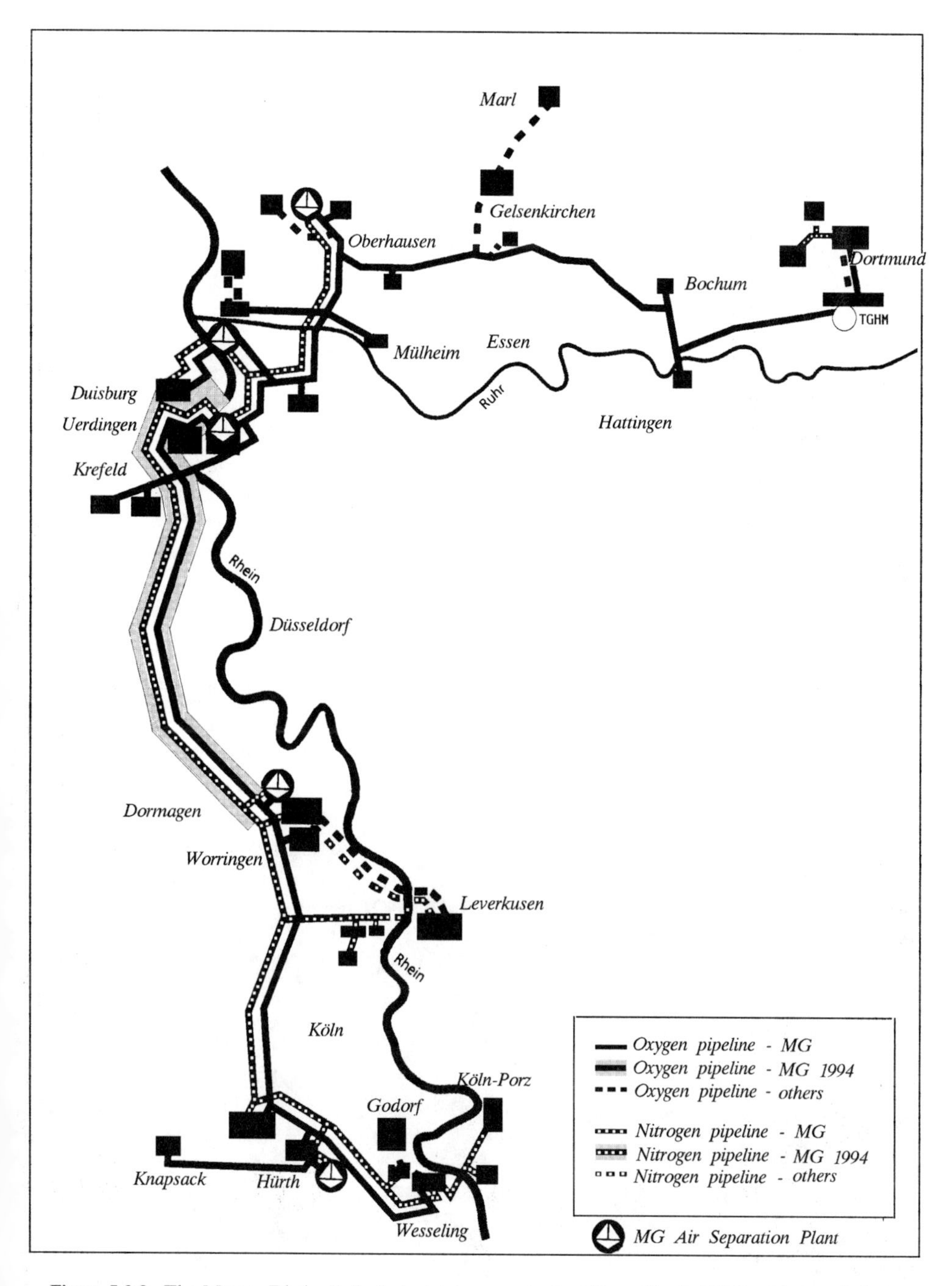

Figure 5.3.9. The Messer Rhein–Ruhr industrial gases pipeline. From Koch (1993) with permission.

position in the electronic and specialty gas business was improved by the signing of a contract for a new helium source in Wyoming. The extension of the specialty gas plants in Duisburg and Krefeld in Germany also improved its position.

In 1987 MG strengthened its position in the United Kingdom by the acquisition of Distillers Co. and extended its activities in France through the takeover of SIAC (together with Linde). In 1989, as a result of an European Economic Community directive, MG and Linde had to separate their joint activities in Europe. MG took over most of the activities in Belgium and France, and the business in The Netherlands went to Linde. At the same time MG acquired the rest of the ownership of the Fedgas Company in South Africa.

The opening up of Eastern Europe and the reunification of Germany created new investment and sales opportunities, which the company was quick to seize. The Group formed its own companies or acquired former state-owned companies in Hungary, the Czech Republic, Slovakia, Slovenia, Croatia, and Poland. In the former East Germany, Messer took over a large industrial gases plant in Leipzig.

In 1993, Herbert Rudolf succeeded Hans Messer as the President of the Messer Group. Between 1993 and 1998 the number of countries where Messer had subsidiaries grew from 19 to 60. In 1999, after the divestiture of the cutting and welding business, the Messer Group consisted of 177 companies in 55 countries.

Figure 5.3.10. Multisec K cutting machine in operation. From Koch (1993) with permission.

Figure 5.3.11. Messer equipment development project for using liquid hydrogen as a car fuel. From Koch (1993) with permission.

In 1996, MG acquired the membrane producer Generon Systems from Dow Chemical and has since become one of the market leaders in noncryogenic gas separation.

Milestones

Prehistory of CFGE

1856	Formation of the Frankfurter Actiengesellschaft für landwirtschaftlich-chemische Fabrikate in Frankfurt-am-Main
	Production of sulfuric acid and nitric acid from Chile salpeter
1859	Soda production by the Leblanc method
1863	Name changes to Chemische Fabrik Griesheim
1871	The chemist Ignatz Stroof starts new chemicals and soda production
1881	Formation of subsidiary Chemikalien Fabrik Mainthal's Teerfarben
	New factory built in Griesheim for nitrobenzol and aniline manufacturing
1884	Development of the chlorine–alkali–electrolysis process
1888	Construction of an electrolysis plant at Griesheim, starting production in 1890

1888–1891 Development of the "diaphragm method", in which, with electrical current, chlorine and soda or potash lye are obtained from sodium chloride and sodium and potassium solutions

1892 New plant built for carbon electrode manufacturing from coke
Plans for enlarged electrolysis capacity
Name changes to Chemische Fabrik Elektron AG

1893 Electrolysis development shown at the World Exhibition in Chicago
AEG, Berlin, forms the Elektrochemischen Werke GmbH (EW), Berlin.

1894 AEG takes over of the production plant in Elektronwerk Bitterfeld

1894–1895 Electrochemical plants started in Bitterfeld and in Rheinfelden am Oberrhein, where energy for an electrolysis plant comes from water works

1895–1896 Badischen Anilin- & Soda-Fabrik AG in Ludwigshafen and the Consolidierten Alkali-Werken in Westeregeln, as well as later plants in France, Spain, Russia, and Poland, start licensed production with the Stroof method

1898 Chemische Fabrik Elektron takes over the works of Elektrochemischen Werbe in Bitterfeld und Rheinfelden, and rebuilds of these works for production with the Stroof method.

CFGE and Messer

1898 Chemischen Fabrik Griesheim and Chemischen Fabrik Elektron joined to form Chemische Fabrik Griesheim-Elektron AG (CFGE)
Ernst Wiss employed by CFGE
Adolf Messer starts producing acetylene generation plants for lighting

1902 Messer develops a welding burner

1904 Griesheim develops a hydrogen burner, first using air, then oxygen, for higher temperatures
Hydrogen business started

1906 Messer changes his business to making welding equipment, later including cutting torches and cutting machines

1907 Griesheim buys rights for using hydrogen–oxygen for cutting

1908 First CFGE-constructed oxygen plant in Griesheim
The oxygen market in Germany divided 50 : 50 between
Linde and Griesheim/Air Liquide

1908 Trade name Original Messer introduced

1910 First Messer oxygen plant built; sold to Spain
First Messer foreign representative, in Oslo, Norway

1913 CFGE has 44 percent share of Germany's entire oxygen production

1916 Hydrogen and Oxygen Department of CFGE established as separate business unit, Autogen Works, headed by Ernst Wiss

1920 CFGE and Linde take over Deutsche Oxhydric AG

1923	Formation of Griesheimer Autogen-Verkaufs-GmbH for sale of welding equipment and carbide
1924	first electric welding trials at Messer
1925	CFGE joins I. G. Farben; Autogen Works, divided into two departments, Maschinenfabrik and Gases
	Ernst Wiss responsible for Autogen Works until his retirement in 1932
1930	production of coated electrodes for arc welding at Messer
1945	Reconstruction of Messer after the war; new subsidiaries and joint ventures formed
1952	Griesheim-Autogen merges with the AG für Stickstoffdünger to form Knapsack-Griesheim AG; company now owned by Hoechst AG
1960	Messer Group (MG) in England

Messer Griesheim GmbH

1965	Formation of Messer Griesheim GmbH in January 1965
1968	Establishment of three divisions, Industrial Gases, Cutting and Welding Technology, and Cryogenic Equipment
	MG starts operating a 500-kilometer pipeline network for oxygen and nitrogen in western Germany
1972	Linde sells its welding business to MG, which in turn discontinues the manufacturing of air separation plants
1974	Acquires the Burdett Oxygen Co. in the United States, a company mainly operating on the East Coast
1976	Establishment of a subsidiary in Venezuela
1980s	Begins to enter the new international gas business with the formation of Likos AG, a joint venture with Linde AG, in France, Netherlands, Belgium, and South Africa
1986	Opens two new markets for industrial gases by the acquisition of SIGAS in the United Kingdom and Società Torinese Ossigeno in Italy
1986	Contracts for gas from a new helium source in Wyoming
	Extension of specialty gas plants in Duisburg and Krefeld in Germany
1987	Acquisition of Distillers Co., England
	Together with Linde, takeover of SIAC in France
1989	MG and Linde break up joint activities in Europe due to European Union directives; MG gets most of the activities in Belgium and France, whereas the business in The Netherlands goes to Linde
	Fedgas in South Africa taken over by MG
1990s	Eastern Europe and reunification of Germany create new investment and sales opportunities
	Forms its own companies or acquires participatory interest in Hungary, the Czech Republic, Slovakia, Slovenia, Croatia, and Poland; in the former East Germany, takes over a large industrial gases plant in Leipzig

1993	Herbert Rudolf succeeds Hans Messer as President Subsidiaries in 19 countries
1996	Acquires membrane producer Generon Systems from Dow Chemical
1999	Presence in 55 countries with 177 companies

Reference

1. Koch, E. (1993). *Ein Unternehmen im Wandel der Zeiten: Messer Griesheim 1898–1998*. Frankfurt/Main, Germany: Messer Griesheim.

5.4. AIR LIQUIDE[1–5]

Background

It was based on ideas concerning cheaper acetylene production that Air Liquide was formed in 1902. In 1896, Georges Claude, a 26-year-old engineer, developed the method still used today for safe transportation of acetylene by dissolving it in acetone contained in a ceramic mass. Claude worked for an electric utility and observed that the heat necessary for production of calcium carbide consumed enormous amounts of electricity. His idea was to generate the heat by using oxygen, but at that time oxygen also was expensive, generated by electrolysis or chemically.

Claude therefore looked at new processes for separating oxygen from air. Among others, he studied the centrifuge, membranes, and solvents. The French physicist Arsène d'Arsonval told him about Carl Linde's successful liquefaction of air in 1895. Instead of the simple expansion valve employed by Linde, Claude set out to produce enough cold by using an expansion engine. He started experiments to improve that method based on the idea that separating oxygen from nitrogen could be done more easily by using liquefied air. His search for the lowest production and separation costs was based on the theory by W. Gibbs in 1876 that no energy was needed to separate the mixed elements of air. This theory was almost true for many chemical mixtures, but turned out to be false for air separation.

Studying the cold recuperative exchanger system invented by W. Siemens (1857) combined with the dynamic cold expander developed by E. Solvay (1887) through which a previously compressed gas could be liquefied, Claude found his own way to cool and liquefy air. Solvay's first Belgian refrigerator of 1887 would not cool below 180 K because of deficiencies in thermal insulation, the freezing of lubricants, and gaseous impurities such as water vapor or carbon dioxide carried by the cycle fluid, and because of frictional heating of moving parts.

In the first months of 1900 Claude started work to optimize a Solvay-type machine by using more efficient insulation, improving the purification of the gas, and developing the lubrication of the bearings first by petroleum ether, later by pentane, and finally by liquid air.

From the time of Claude's earliest air liquefaction experiments he received financial backing from a far-seeing group of associates, headed by a friend and former schoolmate, Paul Delorme. Claude and Delorme were both employed by Cie

Figure 5.4.1. Claude and Delorme with the first air liquefying machine, 1902. From L'Air Liquide (1992) with permission.

Thomson-Houston. On the date on which the backing group was to be liquidated after exhausting its capital of 50,000 francs, a final improvement was made to the air liquefaction machine (Fig. 5.4.1), which enabled sufficient liquid air to be produced. On November 8, 1902, L'Air Liquide: Société Anonyme pour l'Etude et l'Exploitation des Procédés Georges Claude (called Air Liquide for short) was incorporated with a capital of 100,000 francs under the presidency of Paul Delorme (Fig. 5.4.2).

The company was formed to exploit the method of dissolving acetylene in acetone and to use and develop air separation technology (Fig. 5.4.3), both inventions made by George Claude. Early products from the new company were welding equipment and air separation plants, and later bulk chemicals were added, such as hydrogen peroxide in 1908 and ammonia produced with a new, improved process in 1917.

Industrial production of gaseous oxygen and nitrogen, each containing less than 10 percent impurities, had been achieved by Linde in 1902 and 1907, respectively. The dephlegmation apparatus developed by Claude from 1905 onward made it possible to produce the two gases with higher purity at the same time in coupled distillation columns.

The first production plant for industrial gases was built at Boulogne, near Paris, in June 1903, but the real success of oxygen production had to wait until April 1905, when Claude collected 280 cubic meters of oxygen with a purity level of 93 percent.

Early Development of Air Liquide[6]

The market for dissolved acetylene was restricted to lighting applications, and the development of industrial oxygen for low-cost carbide was never justified. In 1901

Figure 5.4.2. Paul Delorme. From L'Air Liquide (1992) with permission.

Figure 5.4.3. Operation of an early air separation column. From L'Air Liquide (1992) with permission.

two French engineers, Charles Picard and Edmond Fouché, invented the oxyacetylene welding torch, which developed a need for pure oxygen. In 1904, Picard also invented the first oxyacetylene cutting torch, which almost overnight revolutionized metal-working processes by making obsolete the old methods for cutting steel. Air Liquide began to expand by building plants to produce oxygen and acetylene and by manufacturing the new torches and other equipment for welding and cutting. In 1908 production of hydrogen peroxide was started, and Air Liquide has since become one of Europe's leading manufactures of that product.

The injection of oxygen into a molten steel converter in Belgium was tried as early as 1906. Some 40 years later this became an industrial application, which now by far is the largest consumer of industrial oxygen.

In order to meet the great demand for dissous gas (dissolved acetylene) and oxygen, Air Liquide in 1907 made agreements with other companies. In France Oxhydrique Française was the owner of the first oxygen cutting patent, and Compagnie Française de l'Acétylène Dissous (CFAD) owned the dissous gas patent. La Soudure Autogène Française (SAF) had the patent for the first oxyacetylene torch. Air Liquide also made agreements with other domestic oxygen manufacturers, Bardot in Paris and Duffour et Igon in Bordeaux, both allied with German Linde.

International Activities and Subsidiaries

Internationally Air Liquide had to compromise with Linde after a dispute regarding plant delivery to the English British Oxygen. In 1908 the market was divided in Hungary, Switzerland, Scandinavia, Romania, and South America with two-thirds for Linde and one-third for Air Liquide. In Germany Linde had 50 percent and Air Liquide/Griesheim Electron 25 each of the market.

Figure 5.4.4. Early Japanese cylinder transport.

Figure 5.4.5. Air separation column, Liege, Belgium, 1908.

Air Liquide started its international development by setting up works in Belgium, Luxembourg, Italy, Spain, Greece, Russia, Japan, and Canada. Use of oxygen to cut metal had become a common practice and an important early application of the young firm's products was in cutting sheet metals at shipyards around the world. It quickly became necessary to set up operations in major international seaports. Things moved fast and before World War I Air Liquide also had plants in Vietnam, Hong Kong, Egypt, Poland, and Algeria. At that time six plants had been built in France.

In 1915 a joint subsidiary, Air Reduction Co. (later Airco), was established with the Rockefeller Group in the United States. The operations were, however, more or less neglected by Air Liquide with a decreasing share in ownership. In 1945 the French government forced Air Liquide to sell their part of Airco because of French need for dollars to pay their war debts. Operations in the United States were restarted in 1968 with a number of successive acquisitions.

From 1919 to 1939, operations began in several other countries, including Portugal (1923), Malaysia (1927), Lebanon (1928), Czechoslovakia (1928), Hungary (1929), Yugoslavia (1929), Senegal (1929), Libya (1930), Argentina (1938) and Paraguay (1939). SOAEO (Société d'oxygène et d'Acétylène d'Extrême-Orient), which was formed in 1916, was busy establishing operations in the Far East, and Air Liquide entered mainland China in 1929.

After World War II Air Liquide continued its expansion with moves into Brazil (1946), the Caribbean, South Africa (1948), Australia (1956), and nations of North and West Africa.

Further Developments[7]

Ammonia

Claude's liquefaction process was his greatest achievement in the industrial field, but he also introduced methods for large-scale separation from air of neon for lighting (1908) and argon for inerting (1914).

In 1918 Claude improved the Haber ammonia process by increasing the operating pressure from 200 to 1000 atmospheres in order to raise the heat of reaction and enhance the catalytic reaction. He also combined the nitrogen produced from the atmosphere with hydrogen extracted at low temperature (100 kelvins) from coke ovens, whose other hydrocarbon constituents were thus liquefied before being revaporized as fuel in the plant.

The Société Chimique de la Grande Paroisse was founded in 1919 to experiment on and exploit the Claude ammonia process and other new procedures within the Air Liquide group. The production of ammonia was easily raised from 3 to 80 tons per day. The success of this production scheme in France, Belgium, Spain, Japan, the United States, etc., added the chemistry business as a strong new activity for the Air Liquide group. From that time on the four principal activities of the group were engineering, gases, welding, and chemistry, all of which were developed extensively.

Krypton

In 1928 Eugéne Gomonet succeeded in separating krypton from the other constituent gases of air. Gomonet's new process consisted in bubbling atmospheric air through a small amount of liquid air to dissolve out the relatively involatile krypton and xenon after precooling in a countercurrent heat exchanger. The results led Georges Claude and his nephew André to develop an industrial process. Together with the company Lumière, a new company was formed, called Société Anonyme pour l'Application de l'Electricité et de Gaz. A krypton production plant was built in 1933 and the first xenon lamps were lit in 1936.

Air Liquide during and after World War II

At the outbreak of World War II Air Liquide had to move its headquarters from Paris to Bordeaux and to coordinate its international activities from Montreal, Canada. Some of the French plants underwent severe damage during the war.

Figure 5.4.6. Delivery truck, Oxigeno do Brazil.

After the war, Jean Delorme, the son of Paul Delorme, was appointed chairman of Air Liquide. His main efforts were directed toward getting things moving again and upgrading and modernizing the plants. The Dunkirk and Strasbourg facilities went back into operation and new plants were built in Saint-Chamond, Angoulème, and Dijon. Jean Delorme also promoted the development of new gas applications in industry and medicine. A new subsidiary was created in South Africa in 1948, and in South America Air Liquide took part in the founding of the Oxigenio do Brazil company in Sao Paulo, Brazil (Fig. 5.4.6).

Delorme saw the need for gas application development and started central research and development labs. In 1957 the Technical Center was formed for welding development, in 1965 the Cryogenic Reserch Center in Sassenage near Grenoble opened, and in 1970 the Claude Delorme Research Center was started in Jouy-en Josas near Versailles.

Gas Technology Developments

During the early years Air Liquide had a cryogenic laboratory where many applications with liquid nitrogen were put into operation by Claude, d'Arsonval, and their collaborators. Most of these applications had to wait about 50 years until gases began to de distributed in liquid form. It was not until 1960 that liquid nitrogen applications found a wide market. Some of these cryogenic applications include the following:

- Artificial clouds for theatrical effects
- Deep freezing of perishable goods
- Hardening and embrittlement of materials
- Purification by fractional crystallization
- High-pressure generation by vaporization of liquid nitrogen
- Cryopumping
- Inhibition of chemical reactions
- Solvent recovery

New Developments

Air Liquide has also pioneered such additional developments in gas technology as processes for production and handling of liquefied natural gas, transport of cryogenic liquids, and pipeline distributions of gas, as well as methods for increasing gas quality and purity for food technology, chemistry, metallurgy, and welding.

In 1947 Jacques Cousteau (Fig. 5.4.7) and Air Liquide engineer Emile Gagnan developed a pressure regulator for scuba diving. The Spirotechnique company was formed to market this successful innovation, beginning a new era of diving technology.

Figure 5.4.7. Jacques Cousteau. Courtesy of Corbis Images.

Developments in Production and Distribution

In the 1950s the strategy of Air Liquide was to build its air gas production around tonnage plants instead of a number of smaller so-called merchant market plants. When the use of air gases in liquid form increased, Air Liquide's production network went from 60 small-scale plants scattered throughout France to 12 major facilities.

The first tonnage gas production unit, called Oxyton, built by Air Liquide started service in 1952 at INCO in Hamilton, Canada. The first Oxyton in France (150 tons per day) was built in Richemont in 1959. In 1961 the first large (2400 liters per hour) air separation plant for liquid production was started in Blanc-Mesnil north of Paris.

From 1959, five new, large Oxyton plants were built in eastern and northern France and in Lorraine and a pipeline network for oxygen and nitrogen deliveries several hundred kilometers in length was built to supply the steel industry in those regions. This pipeline net, "the Grand Masse," was later extended into Belgium, The Netherlands, and Luxembourg (Fig. 5.4.8), distributing the gas supplied by tonnage plants over 2400 kilometers (see Fig. 4.4.9). New Oxyton plants (1500 tons per day) were built at the Grand Masse in Mons and Antwerp in 1970 and in Dunkirk, Richemont, and Charleroi in 1975–1976. A new hydrogen plant using PSA separation (3300 cubic meters per hour) was built in Isbergues (Fig. 5.4.9), supplying two customers ultrapure hydrogen by pipeline.

Figure 5.4.8. Grand Masse pipeline expansion.

Figure 5.4.9. Isbergues PSA hydrogen plant.

Further Business Developments[7,8]

By 1960, Air Liquide was operating in some 50 countries and was considering new possibilities for reentering the U.S. market. The U.S. subsidiary Liquid Air was formed in 1971, and in 1978 Chemetron, the industrial gases division of Allegheny Ludlum, was acquired. Some of the world's largest oxygen plants were built in 1979, with a capacity of up to 2500 tons per day. In 1971 a joint venture (50 : 50) with AGA was formed in Belgium, Luxemburg, The Netherlands, and Germany. This joint venture was dissolved in 1986. In 1976 Air Liquide built the Sasol syngas complex in South Africa (Fig. 5.4.10). In the 1980s Air Liquide decided to expand into the carbon dioxide business, where demand was rising fast from soft-drink bottlers and for food preservation. In 1982 Cardox, the third largest carbon dioxide company in the United States, was bought and became a division of Liquid Air. During 1983 Air Liquide acquired the majority of shares in Carboxyque Française, the largest CO_2 producer in France. The Agefko Company of Germany was purchased in 1985. In Italy Air Liquide purchased the CO_2 company Candia in 1987.

Figure 5.4.10. The Sasol syngas plants in South Africa.

In 1986 the U.S. industrial gas company Big Three Company, at that time ranked number four in the United States and number nine in the world, was acquired. Big Three delivered nitrogen and oxygen to customers in Texas and Louisiana with a 2200-kilometer pipeline (see Fig. 4.4.9).

In 1986 Air Liquide signed an agreement with Du Pont et Nemours regarding membrane production and the Medal company was formed in 1988. At the end of the 1980s Air Liquide pioneered the use of membrane and PSA separation systems for on-site gas production. The Floxal system was launched.

Some years later Air Liquide made major investments in East Europe, Japan, Thailand, and China. It improved its welding position in the European market with the 1991 acquisition of Oerlikon's welding division.

Space

In 1986 Air Liquide signed a contract to supply liquid hydrogen for the Ariane V rocket project. This led to the construction of Europe's biggest hydrogen liquefier, in Wazier in northern France, for hydrogen delivery also to other European industries such as the electronics industry. In 1988, Aerospatiale and Air Liquide jointly set up Cryospace to manufacture the cryogenic tanks used by the Ariane IV and Ariane V rockets.

Figure 5.4.11. Georges Claude.

LNG

In 1964 Grenier of Air Liquide patented and introduced a mixed refrigerant cycle whereby the refrigerant components are successively separated by condensation before their vapors are remixed into a single compressor. This natural gas process is very important because it makes possible the daily liquefaction in Algeria of tens of thousands of ton for transport overseas.

Concentration on Gas

At the end of the 1980s Air Liquide started focusing on its core business—industrial and medical gases—and disposed of, among others, the fertilizer manufacturer Grand Paroisse in 1987. A marked strategy of expanding into Far Eastern and Eastern European markets has also been seen.

Short Biography[9]

Georges Claude (1870–1960) was born in Paris in 1870 (Figs. 5.4.11 and 5.4.12). His father worked successively as a primary school teacher, an engineer, and an assistant manager at the Saint Gobain Glass Works, and in his free time was an inventor. He and his wife took care of their children's basic education completely outside the educational system. After Claude completed his engineering course at the École de Physique et de Chemie in Paris in 1889 he developed his skills in science, technology, and economics in different challenging assignments.

Figure 5.4.12. Georges Claude created Air Liquide.

First he became an electrical inspector in a cable factory and later laboratory manager of an electricity works. Then he edited the magazine *L'Etincelle électrique*, in which task he could satisfy his curiosity about scientific news. Here Claude met, among others, the physicist d'Arsonval, who brought him ideas essential for his future work as pioneer and inventor in the industrial cryogenics field.

In 1896 Claude learned about the explosion risks of acetylene stored in cylinders and liquefied under pressure by the process introduced by Raoul Pictet in Berlin. At that time, acetylene was much used for lighting. Claude was asked to solve that problem and soon found a solution. Already in 1883 Marcelin Berthelot had shown that acetylene was explosive at pressures above 2 bars. In spite of this, Pictet tried to increase the quantity of acetylene in a given container volume by compressing the gas to liquid form. In 1896 he started the commercial transport of liquid acetylene, with catastrophic results. Based on a study of seltzer-water-filled siphon bottles, Claude looked for liquids in which acetylene was highly soluble, and soon found acetone, with a solubility figure of 25. The resulting gas was called acétylène dissous (dissous gas). Berthelot and Vieille found this combination much less explosive than the liquid, and considered it safe up to 10 atmospheres. There was still a high risk of explosion, however, because free gas could arise in the gas cylinder during transport.

The means of preventing this problem was found by Georges Claude and Albert Hess, after developing an idea by Henri Le Chatelier, whose advice was to disperse the liquid in the pores of diatomite. The mass created a fine capillary system into which the acetone was absorbed. Because an explosion cannot spread in a space having a diameter of a fraction of a millimeter, the risk of explosion was theoretically eliminated. Claude and Hess launched their invention in 1896, and in the following year the Compagnie Français de l'Acétylène Dissous (CFAD) was founded by members of Laboratoire Volta. Dissous gas, compressed to 10 bars, went onto the market. However, the porous mass used proved to have shortcomings because shaking and vibration could lead to defects in the porosity. This created explosions, which gave acetylene a bad reputation that lasted for a long time.

Claude became an engineer with Laboratoire Volta group in the Thomson–Houston Company. Here he became interested in cheaper ways of producing calcium carbide and began looking at new ways to generate cheap oxygen for creating the enormous amount of heat needed. Claude was told of Linde's invention for air liquefaction and started a project that was financially supported by some of Claude's friends. In 1899 a syndicate was established under the direction of Paul Delorme to support experimental tests on the liquefaction and distillation of air.

Claude was working by day as an engineer for the Thomson–Houston Company and by night was performing experiments in the hall of a streetcar depot, where he could get the necessary compressed air. At long last, the machine began producing liquid air. On November 8, 1902, a business was incorporated under the name of Air Liquide with a capital of 100,000 francs put up by 26 shareholders and started with head offices in Paris, under the direction of Paul Delorme.

Claude developed his air separation technology working day and night at an industrial gas plant built at Boulogne, near Paris, in 1903; again there were funding problems. In April 1905 a large amount of oxygen with a purity level of 93 percent was first obtained. Now the high oxygen demand for welding and cutting could be met. Now more firmly established, Air Liquide began to expand its operations.

Other Developments by Georges Claude

Neon Lighting

When Claude was looking for a cheap way to produce oxygen for use in hospitals he happened to find a method of injecting inert gases into colored tubes. Claude and his nephew André began in 1907 to purify industrially produced neon by cryosorption for use in lighting applications. Claude's neon-filled discharge tubes were first exhibited in 1910 in a fixture to illuminate the Grand Palais in Paris. Two years later Claude sold the first neon advertising sign to a barber in Paris.

Believing that neon illuminating tubes might be able to compete with the Edison light bulb, the two Claudes now tried to reconstitute white light by combining tubes with red discharges in neon and blue in mercury vapor. André Claude coated a lamp bulb wall with a layer of pigments which were fluorescent in ultraviolet radiation and lowered the excitation voltage by adding some traces of krypton or xenon to the internal atmosphere.

Incandescent Lighting

Incandescent bulbs, originally vacuum lamps, were improved by Langmuir, who around 1915 suggested filling them with a gas sufficiently heavy and inert to prevent the evaporation of the filament. Because nitrogen is too reactive, Claude, his nephew André, and Lerouge undertook the extraction of the 1 percent of argon contained in air for use in high-luminosity bulbs.

The slight difference in physical properties between argon and oxygen was industrially exploited by Air Liquide in 1916, 1 year after Linde. In 1918 the first glow lamp with argon filling was released. At that time the idea arose of replacing argon by the much rarer krypton or xenon, the boiling points of which are higher (120 and 165 kelvins, respectively). The first tests, carried out in 1922, with very small quantities of these noble inert gases obtained from the by-products of oxygen manufacture showed that white light could be obtained by incandescence with yields of 10–30 percent.

Military Developments

During World War I Claude took part in the military and industrial war effort and invented among others the cryogenic aerial bomb using liquid oxygen and a method for the detection by sound of the enemy position for artillery batteries. He was also engaged as a civilian in combating the effects of Ostwald's poison gases, and constructed gas masks.

Ammonia Synthezising

The shortage of imported nitrates in Europe during World War I led to the production of substitutes by the synthetic nitrate processes: nitric acid by the method of Birkeland and Eyde, calcium cyanamide by the Frank–Caro method, and the ammonia of Le Chatelier by the Haber process. Having examined production methods for the last of these as used by the Badische Anilin factory, Claude suggested in 1918 to the Air Liquide group that they should undertake research to extend their chemical efforts in order to make France self-sufficient in nitrogen compounds. Claude's high-pressure ammonia process gave a higher yield than the Haber process.

Krypton Gas Lamps

When Eugen Gomonet in the late 1920s succeeded in separating krypton with a process using liquid air, Georges and André Claude set out to develop an industrial process. This resulted in commercial production of krypton in 1933 and the development of xenon lamps in 1936.

This work was the last contribution made by Claude to the development interests of the Air Liquide group. The French inventor gave up his studies of the atmosphere and its gases to devote his efforts until his death in 1960 to the extraction of gold from the sea and to geothermal energy.

Milestones

1902 — Use of dissolved acetylene for lighting; use of liquid air for food storage, medical care, chemistry, and theater effects

1905	Construction of oxygen and nitrogen generators, oxyacetylene torches, air separation units
1906	First oxygen gas metallurgy process for melting of steel, in Belgium
1907	Generation of neon and helium
1908	Production of hydrogen peroxide
	Beginning of oxygenated water production
1910	Use of pure neon for public lighting
1911	Cold recovery of volatile solvents
1912	First cutting machine
1914	First industrial production of argon
1916	Founding of the Air Reduction Company in the United States with J. D. Rockefeller as part owner
1917	Claude process for synthetic ammonia
1918	First incandescent lamp with argon filling; first all-welded ship
1922	First krypton gas lamp
1924	Underwater cutting demonstrated
1925	First Air Liquide pipeline from SIO plant and Fiat in Turin
1930	Teisan KK formed in Japan with the Sumitomo Group
1933	Industrial production of krypton and xenon from air
1935	Development of hydrogen burners
1936	First xenon fluorescent tube
1938	Pipeline to supply Canadian steel mill with oxygen
1945	Jean Delorme (1902–1997) elected chairman of Air Liquide
	Air Liquide sells its minority stake in Airco
1947	Development of first low-weight diving equipment (Jacques Costeau and Emile Gagnan)
	Founding of Spirotechnique
1948	First oxygen burners for the steel industry
1952	First Oxyton tonnage (300 tons per day) plant built in Canada at INCO
1953	New nitric acid syntheses process
1954	First bottom injection of oxygen in steel converter
1955	Subsidiary in Australia
1956	The TIKKO system for transport and storage of liquid carbon dioxide
1955–1959	First part of pipeline network Grand Masse supplies air gases to metallurgy plants in north and northeast France, later extended to Luxembourg, Belgium, and Holland
1961	Participation in tests of liquid oxygen/liquid hydrogen propulsion
	Pilot plant for cryogenic separation of deuterium in Toulouse
	Liquefaction of natural gas by cascade cooling technique
	Cellulose bleaching with oxygen
1962	Formation of Deutsche Air Liquide Edelgas GmbH

1963	Start of cryogenic storage of living cells
1966	The "ammoniac process"
	Deuterium and heavy water production in cooperation with Sulzer and CEA
	First liquid hydrogen plant in Europe
1967	High-vacuum pump for space simulation
	First programmed biological freezer
	First hydrogen and helium turbine with gas bearings
	First production of high-purity gaseous hydrides: silane, arsine, phosphine
1969	First laser welding equipment
	Acquisition of American Cryogenics, Atlanta, Georgia
1970	New method for potassium peroxide production for regeneration of confined atmospheres
	Oxyton plants in Mons and Antwerp (1500 tons per day)
1971	Joint venture with AGA in Germany and Benelux
	Liquid Air is founded
1973	Cryogenic design of the third stage of the Ariane space rocket
1975	Utilization of oxygen in hydrometallurgy, for example, zinc, uranium
1976	Construction of the biggest oxygen production center in the world: the Sasol syngas complex in South Africa.
1977	New procedures for controlled atmospheres in heat treatment of metals
1978	Acquisition of Chemetron, industrial gas division of Allegheny Ludlum
1982	Purchase of the U.S. CO_2 company Cardox
1983	First process for freeze-drying of milk
	Acquisition of Carboxyque Françaíse, leading CO_2 supplier in France
1985	Construction of the biggest Oxyton plant in the world, capacity 2500 tons per day
1986	Purchase of the industrial gas company Big Three, Houston, Texas
	Acquisition of the AGEFKO CO_2 Company, Germany
	Acquisition of Seppic, France
	Joint venture with AGA terminated
	Foundation of Air Liquide GmbH, Germany
1987	Expansion of liquid hydrogen production in Europe, North America, and Japan
1988	Founding of Medal in United States for the production of membranes
1989	First Chinese operation in Guangzhou
1990	First large carbon monoxide hydrogen unit, in Estarreja, Portugal
1991	Development of a very high purity analyzer, APIMS, for electronic gases

1992	Startup of the first compact oxygen VSA unit
	Startup of the Sarox process
1994	Air Liquide America founded
1997	Construction of the largest Oxyton in the world in Antwerp, Belgium, capacity 3000 tons per day

References

1. Dennery, F. M. (1992). L'Air Liquide and the development of cryogenics in France. In *History and Origins of Cryogenics* (R. G. Scurlock, ed.), Chapter 5 pp. 144–179. New York: Oxford University Press.
2. L'Air Liquide (1952). *Cinquantenaire de la Société L'Air Liquide Oct 1902-Oct 1952. L'Air Liquide 50 years*. Paris:
3. Deschars, C. (1979). L'histoire de l'Air Liquide. *IOMA Broadcaster* **1979** (July/August). Reprint.
4. L'Air Liquide. (1992). *Yesterday and today: A brief history of L'Air Liquide. L'Air Liquide 90 years*.
5. L'Air Liquide. (1992). The Air Liquide Saga. Alizé historical series. In *Alizé* 41–54.
6. Claude, G. (1926). *Air liquide, oxygène, azote, gaz rares* (2nd ed). Paris: Dunod.
7. Claude, G. M. (1919) *Communications sur l'ammoniaque synthétique*. Paris: Dunod.
8. Smith, H. K. (1981). The Big Three Story. *IOMA Broadcaster* **1981** (November–December) pp. 8–11.
9. Storck, W. J. (1980) Liquid Air: The firm that acquisitions built. *Chemical and Engineering News* **1980**(March 4):11–13.

5.5. AGA[1-4]

Background

The French company that purchased the rights to L. M. Bullier's acetylene patent was in represented in Sweden by Axel Nordvall. In 1898 Nordvall founded Svenska Carbid & Acetylen AB to supply rural areas with acetylene lighting. In 1901 the company acquired the Scandinavian patent rights for another French invention, the dissous gas of CFAD. Nordvall also began working with the inventor Gustaf Dalén. Dalén was at that time employed by de Laval Steam Turbines, but he also made inventions with acetylene for his company Dalén & Celsing, which he had formed with Henrik von Celsing. They designed gasworks for the production of acetylene from calcium carbide.

Nordvall received technical assistance from Dalén & Celsing and in 1901 moved his enterprise from Gothenburg to Stockholm; Gustaf Dalén became his chief engineer. The company built acetylene gasworks and installed town lighting in some Swedish towns. Dalén also demonstrated welding with oxyacetylene for the first time in Scandinavia, at the Finnboda shipyard in 1902. The experts considered this welding process to be interesting, but of no practical value.

In 1904 the financial situation of Svenska Carbid & Acetylen AB weakened and the company was reformed with new backing and renamed AB Gasaccumulator (later AGA), starting acetylene production in small premises in Järla near Stockholm with 15 employees (Fig. 5.5.1). One problem was that acetylene had acquired a bad reputation due to accidental explosions. Dalén left in early 1904 and returned to de Laval

Figure 5.5.1. First AGA plant, in Järla, Sweden. Courtesy of the AGA Historical Picture Archive.

Steam Turbines. Nordvall became the Director of Sales for Gasaccumulator, and after some persuasion Gustaf Dalén accepted a position as consulting engineer for this new company.

Early Development

Lighting

The first major task for the company was to supply railway lighting in Sweden and Denmark. The next challenge came from the National Swedish Administration of Pilotage, Lighthouses and Buoys. Because of its long coastline, Sweden had an ongoing interest in improving lighthouse design. Over the preceding 30 years development had been primarily in two areas, namely variable flashing lights to distinguish one lighthouse from another, and automation to reduce costs. Swedish engineers had developed ingenious designs for flashing lights. Lighthouses were few and very primitive at the time; strong enough light sources did not exist and transport of fuel to the remote locations where the lights were used was very difficult. Acetylene, with its bright light, proved to be an excellent fuel, but it was too expensive when burned continuously 24 hours a day.

The Lighthouse Administration tested acetylene gasworks in lighthouses and buoys, and in 1900 the Marstrand lighthouse was fitted with acetylene lighting. In 1902, Chief Lighthouse Engineer John Höjer, who had previously developed lamps and rotating mirror systems for lighthouses, fitted an experimental gas accumulator in a light buoy. He demonstrated to AB Gasaccumulator the tremendous potential market for acetylene in lighthouses, and pointed out the necessity of using flashing lights, partly to save gas and partly to provide varying lighting characteristics.

The AGA Lighthouse System

This was the beginning of a chain of inventions which, link by link, was to create the AGA lighthouse system (Fig. 5.5.2) and make the real breakthrough for the AGA company. Gustaf Dalén's development of an automatic lighthouse mechanism first included an intermittent light regulator, the AGA flasher, invented in 1905, and the sun valve, devised in 1906, which turned the light on automatically at dusk and turned it off at dawn. The introduction of the AGA flasher reduced fuel consumption up to 90 percent and the sun valve cut fuel consumption by another 4 percent. Lighthouses could now be operated at a low cost and left unattended for long periods.

There were more problems to solve. The acetylene gas accumulators using the containing mass in the original French design were not safe from explosion. In 1906 after a period of development (Fig. 5.5.3), AGA found that a mixture of charcoal, cement, asbestos, and water gave a very hard, porous, and homogeneous substance, the so-called AGA mass. Further ways of saving gas were achieved by creating an air/gas mixture. In large lighthouse beacons at sea it was necessary to obtain the greatest possible light intensity, and this was done by using incandescent mantles, so-called Dalén light.

The optimum gas mixture using mantles is 1 part acetylene to 10 parts air, a very explosive mixture. The Dalén mixer, developed in 1909, mixed the gas to the right proportions without the risk of explosion, and independent of external conditions. It also provided the power for the lens rotation system. There was, however, still the need for beacon maintenance to deal with mantles burning out. The mantle changer was developed in 1916, providing completely automatic replacement of burnt-out mantles. With a magazine of 24 mantles a beacon could operate for 1 year without service. These inventions achieved the final goal: "The AGA System" of 1916 provided fully automatic beacon service. Gustaf Dalen's inventions rapidly entered world markets thanks to the Pilotage Service contact network (Figs. 5.5.4 and 5.4.5).

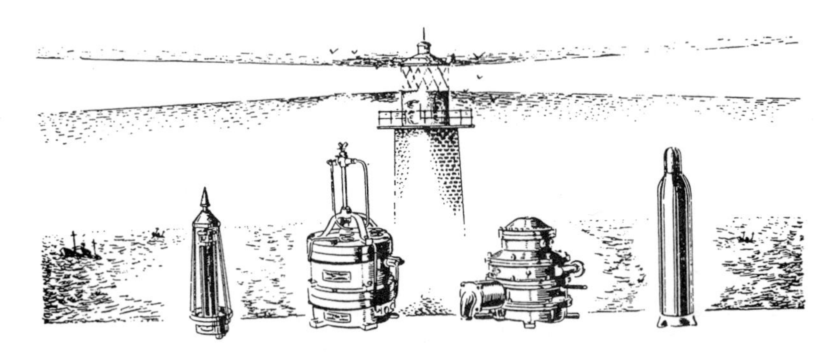

Figure 5.5.2. The AGA lighthouse system: sun valve, AGA-flasher, Dalén mixer, and AGA mass. Courtesy of the AGA Historical Picture Archive.

Figure 5.5.3. The AGA mass for acetylene cylinders was developed in this laboratory, 1905. Courtesy of the AGA Historical Picture Archive.

Market Development

Thomas Willson, the Canadian inventor of the industrial acetylene process, built an automatic lighting system for lighthouse beacons in 1904 for the International Marine Signal Co. in Ottawa, Canada. This system was introduced in 1905 as the standard for Canada, and by 1910, 250 units had been installed. Initially Willson used compressed acetylene in steel bottles without a porous mass, which often led to tragic accidents, and therefore he changed to a method of generating acetylene in the buoy itself, which was not reliable when the water froze.

World-Wide Operations

The AGA system rapidly gained ground in Sweden and neighboring countries, but orders for lighthouse installations and light-buoys also began to come in from other countries, including Great Britain, Holland, and Russia. In 1908 the first deliveries were made of light-buoys for Japan, Australia, and North and South America. The increasing exports soon reached a level where it was necessary to extend the business. The delivery of acetylene lights also required that dissous gas plants were built. By the start of World War I, AGA had delivered acetylene plants to Denmark, Italy, Java, Australia, Japan, Finland, Panama, Hong Kong, and China among other countries.

In 1910 AGA sold 40 lighthouse lighting systems and 76 light buoys to the United States and some to Canada. The success of the AGA system was crowned in

1912 by an order for the complete Panama Canal lighthouse beacon system. Because of the shortage of petroleum products, World War I brought about a quick changeover to the AGA system in several countries. The war also gave new openings for the use of acetylene, which could be produced locally from calcium carbide, and it took over much of the petroleum lighting market.

From 1910 onward gas welding and gas cutting with acetylene/oxygen were used increasingly by industry. The need for oxygen increased sharply, which led to AGA starting oxygen manufacture on Lidingö in 1914 (Fig. 5.5.6) and then in many other places in the world.

AGA's first investment outside Scandinavia was in American Gas Accumulator Corp., founded in New Jersey in 1911. By 1915 there were AGA companies in seven European countries as well as in the United States and Brazil. Building on the

Figure 5.5.4. Axel Nordvall, one of AGA's founders, on a light buoy. Courtesy of the AGA Historical Picture Archive.

Figure 5.5.5. Gustaf Dalén with a lighthouse system, ca. 1925. Courtesy of the AGA Historical Picture Archive.

Figure 5.5.6. Interior of AGA's first oxygen plant, in Lidingö, Sweden, 1914. Courtesy of the AGA Historical Picture Archive.

war-time demand for acetylene for beacons, lighting, and welding, by 1928 subsidiary companies had been formed in a further 17 European and Latin American countries for the manufacture and marketing of acetylene. AGA companies were founded in Mexico and in all countries on the South American coast. As a rule, foreign subsidiaries were granted considerable independence, due partly to the extremely long distances between subsidiaries and headquarters and partly to the fact that gas production is a local business. AGA's stock in American Gas Accumulator Co. was sold in 1949, but new operations had been established in other industries in the United States. AGA Corporation in New Jersey, selling electronic equipment, was part of the group until 1980, when AGA concentrated its world-wide activities on industrial gas.

AGA developed an important market in Eastern Europe, where industrial gas companies were established in Estonia, Latvia, Lithuania, Russia, Poland, Czechoslovakia, Hungary, and Yugoslavia. One-third of AGA's world-wide market, including 26 gas plants, were nationalized and lost behind the Iron Curtain after World War II.

The collapse of the Iron Curtain gave AGA an opportunity to return to several of its old Eastern European markets either by acquiring gas companies or establishing new ones. The first such developments were in Hungary and the former East Germany, followed by Estonia, Latvia, Lithuania, the Czech Republic, Slovakia, Poland, Russia, and Romania. By acquiring one-third of the shares in the Icelandic company ISAGA, AGA also returned to that country.

Welding Development

Welding using acetylene and oxygen achieved a rapid breakthrough in the engineering industry. It was included in the AGA program in 1906 and was used, among other things, in the manufacture of buoys. Cutting techniques were developed starting in 1910 (Fig. 5.5.7). After building its first oxygen plant at the AGA headquarters in 1914, a welding school was started in 1916 (Fig. 5.5.8). Since then the welding market has been a stable area. The next AGA oxygen plant was erected in Helsinki, Finland, in 1922.

A Diversified Company

Welding was to play a small role in the company's economy into the 1920s. Until then the emphasis had been on the manufacture and sale of acetylene equipment for lighting; lighthouse lighting was the company's pride and joy. In later years new applications of the AGA system included acetylene lighting along air routes. In the 1920s AGA air traffic signals were used on the transcontinental air-mail route from New York to San Francisco and on the London–Paris air route. New Dalén inventions were added to the product line and the company became very diversified:

- Lighthouse production expanded to include all types of communication aids on sea, on land, and in the air. Subsidiaries based on acetylene production were formed mainly in Europe and Latin America.
- Welding and cutting were developed and became indispensable in industry and crafts.

Figure 5.5.7. Ship shaft cutting, 1913. Courtesy of the AGA Historical Picture Archive.

- Oxygen production together with the development of the lighthouse flasher led to the development of apparatus for breathing and for treating narcosis.
- From 1930 radio and sound film equipment was developed by the subsidiary AGA-Baltic.

In its first 50 years the AGA group manufactured such varied products as cookers, cars, film projectors, medical equipment, radios, and television sets. The AGA cooker still has outstanding heat economy and is manufactured in the United Kingdom according to Dalén's early design.

Concentration on the Marketing of Gases

The depression of the 1930s affected AGA, among other things through reduced exports. The crisis had considerable negative impact on the sales of lighthouse products and other exported AGA equipment. A new market opened, however, as domestic hospitals began to demand oxygen and carbogen (oxygen with ca. 5 percent carbon dioxide) for respiratory treatment and nitrous oxide for anesthetics and pain relief. AGA's gas technology was used in new medical products for oxygen and nitrous oxide. The Spiropulsator, a combined anesthetic apparatus and respirator, was developed in 1934. AGA started its own manufacture of nitrous oxide and carbogen in 1936. The Sedator, an anesthetic apparatus for pain relief during childbirth, was launched in 1937.

During World War II argon began to be used as a shielding gas for arc welding. Until the early 1950s the main interest in gases by AGA management was acetylene for lighting and welding applications. Since the acetylene market became stagnant after the war, new market opportunities for gas had to be found. Swedish metallurgists went to the United States after the war to learn and benefit from experiences in using oxygen for combustion. Many Swedish steel works applied this new technology for use in annealing and blast furnace processes. Since handling of the large quantities of oxygen needed was expensive and troublesome, large consumers promoted oxygen deliveries in liquid form by tankers. In 1951 AGA built its first liquid oxygen plant at their headquarters on Lidingö (Fig. 5.5.9) based on a German installation used for V2 missile fuel. At about the same time oxygen deliveries in liquid form also began at AGA subsidiaries in Belgium and Finland.

Expansion within the industrial gas market followed increased demand for nitrogen and argon. AGA responded to the new demands, but the suppliers of large "liquid" plants, which at the same time were competitors refused to sell such plants to AGA. AGA started to manufacture its own "liquid" plants, and gradually became a recognized player in the industrial gas market. From the early 1970s, however, AGA's efforts

Figure 5.5.8. Teacher demonstrating welding procedure, ca. 1920. Courtesy of the AGA Historical Picture Archive.

Figure 5.5.9. First AGA liquid plant, built in a rock shelter, Lidingö, 1951.

to get away from its widely diversified engineering manufacturing interests became apparent.

Back to Basics

AGA intensified its research and development, focusing on new applications for gas. All available resources were invested in these efforts, with the emphasis on air gases. In order to get a strong position in continental Europe for both parties, a joint venture with Air Liquide in West Germany and the Benelux countries was formed in 1971. As an important step in the developing the market for gas operations, AGA also reentered the American industrial gas market through the acquisition in 1978 of Burdox Inc., a local U.S. gas company in Cleveland founded in 1923. Since 1980 AGA has made large investments in gas operations in the eastern United States.

Frigoscandia, a large Swedish manufacturer of mechanical deep-freezing equipment, also with interests in the storage and transport of frozen goods, was acquired in 1978.

In 1980 the Pharos Group was established, composed of AGA's small, high-technology electronics and hardware manufacturing units. In 1980 AGA sold its shares in Pharos, thus ending its involvement in the high-tech equipment and electronics industry.

Changes were now made with the primary objective of strengthening the foundations of AGA's future course. What appeared to be a deviation from the former strategy of concentration was an acquisition which gave AGA greater opportunities to pursue its strategy with increased resources and financial strength. In 1985 AGA acquired Uddeholm, a leading manufacturer of tool steel with three operating areas: tool steel, world-wide specialty steel trading, and electric power production. The acquisition of Uddeholm doubled total assets within 2 years and made a tripling of investments possible. During 1985–1989 very great investments were made by acquisitions in distribution networks and new production facilities. One of the leading European carbon dioxide companies, the Rommenhöller Group, was acquired in 1986 and the French industrial gas company Duffour et Igon was acquired in 1987. At the same time the joint venture with Air Liquide in West Germany and the Benelux countries was dissolved.

In 1988 AGA decided to refocus on its core business. Uddeholm's steel operations were sold, but the electric power part of Uddeholm was kept by AGA. A continued focus was made on electric power, which was considered to have favorable future prospects and market conditions similar to those of the to gas operations.

During the 1990s Frigoscandia and the energy operations were sold, and what remained of AGA was the gas operation. In 1999 AGA was acquired by Linde AG to form Linde Gas AG.

Short Biography

Gustaf Dalén (1869–1937) was born in 1869 on a small farm in Stenstorp, in the county of Västergötland, Sweden (Fig. 5.5.10). He was the third of five children, who all obtained high positions in their careers. His parents were poor, but capable farmers with wide interests: has father was very active in local politics and his mother had a studious nature. As a child Gustaf had a practical disposition and was rather lazy at school. Therefore, he was looked upon as most suitable of the brothers to take care of the farm.

As a young farmer and dairyman in 1891 he invented a milk-fat tester for checking the quality of milk deliveries. This invention was shown to Gustaf de Laval, founder of AB Separator (later Alfa Laval) in Stockholm. Laval had recently patented a similar piece of equipment and was much impressed by the creativity of this totally self-taught man. He advised Gustaf Dalén to take up formal study, which he did, taking his degree at Chalmers Technical University and spending a year in further study at the Technical University ETH in Zürich. Returning to Sweden, he was employed by AB de Lavals Ångturbin as turbine designer.

When the project failed Gustaf Dalén became interested in acetylene, which was another of de Laval's spheres of operation. Together with a fellow former student from Chalmers, Henrik von Celsing, he started the company of Dalen & Celsing to launch an acetylene works for lighting. Dalén was hired by AGA in 1904.

As chief engineer he invented the main part of his famous lighthouse lighting system during 1905–1909 and was elected president of the company in 1909. At an

Figure 5.5.10. Gustaf Dalén, Nobel Laureate in Physics, 1912.

early stage, Dalén had become interested in advanced technology in different areas of transport, including flying and railways. He was one of the guarantors for the balloon *Svenske* in 1902, and became a very active member of the board of the Swedish Aeronautical Society from that time up until 1920. He was also a part owner of Enoch Thulin's Aeroplane Factory in Landskrona, which manufactured around 100 airplanes during World War I.

Gustaf Dalén retained both his intense curiosity in new technology and his inventive ability throughout his life despite being blinded as a result of an accident when "bonfire-testing" an acetylene accumulator in 1912. In the same year he was awarded the Nobel Prize in Physics. Dalén continued to take a very active role in leading and developing the AGA company.

Milestones

1901	Svenska Carbid & Acetylen starts association with Gustaf Dalén
1902	Gustaf Dalén demonstrates oxyacetylene welding for the first time in Sweden

1904	Formation of AB Gasaccumulator (AGA) to exploit acetylene gas technology
1905–1906	Gustaf Dalén invents the AGA flasher, the AGA mass, and the sun valve
1908	First railway signal using AGA's flashing light goes into operation
1909	The Dalén mixer
	Economic crisis, restructuring with new owners; Gustaf Dalén becomes president
1910	AGA lighthouses established in Latin America
1911	AB Svenska Gasaccumulator, AGA, listed on the Stockholm Stock Exchange
1912	Historic lighthouse order for the Panama Canal
	Gustaf Dalén receives the Nobel Prize in Physics
1914	AGA starts its own oxygen production
1916	AGA buys AB Lux
	The first AGA welding school starts training gas welders
1919	Focus on radio technology together with ASEA and LM Ericsson
1929	Introduction of the AGA cooker and sound film equipment for movie theaters
1934	Launch of the Spiropulsator, the first significant medical AGA technology product
1936	Increased medical demand justifies AGA's own nitrous oxide production
1937	Death of Gustaf Dalén
1939–1945	Strong diversification due to war demands for high-tech products
1951	Construction of first AGA liquid oxygen plant at Lidingö
1970	New strategy with strong focus on gases
1971	Cooperation agreement with L'Air Liquide in the Benelux countries and West Germany
1975	Breakthrough for AGA's oxygen bleaching of wood pulp
1977	Significant penetration of U.S. market through acquisition of the gas company Burdox
	Launch of the welding gas Mison
1978	Acquisition of AB Frigoscandia, a leading company in deep-freezing equipment, cold storage, and transport of food
1979	Oxy-fuel technology established in cooperation with Smedjebackens Valsverk
1980	Creation of the Pharos Group
1981	Marketing of AGA's first gas freezer
1982	Sale of the Pharos Group
1983	Establishment of a strong presence in the Norwegian gas market by acquisition of Norgas
1986	Acquisition of the German Rommenhöller Group
1987	Purchase of the French gas company Duffour et Igon strengthens position in France
	Dissolution of joint ventures with Air Liquide in Germany and Benelux

1991	Return to Eastern Europe with gas companies in the Baltic and other countries

References

1. Almqvist, E. (1992).Technological changes in a company. AGA—The first 80 years. In *Deadalus, the Swedish Technical Museum Annual 1992*. pp. 78–109.
2. AGA. (1929). AGA 25 years. *AGA Journal* 1929 (June).
3. AGA. (1944). AGA 40 years. *AGA Journal* 1944 (June).
4. AGA. (1954). *AGA 50 år* [AGA 50 years].

5.6. BRITISH OXYGEN[1]

Background

Brin's Oxygen Company

Brin's Oxygen Company (BOC) was founded in England 1886 to produce oxygen for lighting and medical use. In the middle of the 19th century several scientists were experimenting with ways of producing oxygen on a large scale. The French chemist Jean Baptiste Boussingault (1802–1887) showed that barium monoxide when heated to 538°C would absorb oxygen from the atmosphere and when heated further to 871°C it would discharge this oxygen.

Two of Boussingault's pupils at the Conservatoire des Art et Métiers in Paris became interested enough to start development of this new process. They were the brothers Arthur and Leon Quentin Brin. A demonstration plant was erected at the Inventions Exhibition in South Kensington, London, in 1885. An English stoneware manufacturer, Henry Sharp (Fig. 5.6.1), saw the demonstration and found the concept commercially sound. He persuaded his family and friends to join in the formation of a company to produce oxygen. The company was registered in January 1886 and called Brin's Oxygen Company Limited. Henry Sharp was the first chairman, from 1886 to 1890. The difficulty with the barium oxide process was that it was a thermal cycling process. The ingenuity and skill of the Scottish engineer Kenneth S. Murray (Fig. 5.6.2) (later chairman of BOC, 1925–1933) converted this process into a successful one by using pressure changes at constant temperature. The oxygen purity was about 95 percent and with this improved process the U.K. output in 1889 was about 27,000 Nm^3 (normal cubic meter) of oxygen.

Early Markets for Oxygen

Selling oxygen in these early years was difficult. The chief market was limelight, at that time the best stage lighting used by theaters and music halls. Another developing market was medicine. Although the output of oxygen was increased through improved manufacturing methods, the new company found it more difficult to increase sales. An attempt to use oxygen for quickly maturing whisky failed. In spite of the claims, the process did not gain the universal approbation of the whisky distillers.

Figure 5.6.1. Henry Sharp, Brin Oxygen chairman, 1886–1890. From BOC (1986) with permission.

Figure 5.6.2. Kenneth Sunderland Murray, BOC pioneer and chairman, 1925–1930. From BOC (1986) with permission.

New Process for Separating Oxygen

At the turn of the century, a new process for separating oxygen from air was developed. At the same time new developments took place in metal-cutting and welding techniques. The use of an oxyacetylene flame, the hottest flame then known, for cutting and welding began to open up a large new market for oxygen. In the early 1900s the Brin Company recognized the potential importance of the air liquefaction process and entered into arrangements with one of the inventors, William Hampson, for the exclusive rights for his patents.

Even at this early stage, some attention was being directed to operation at lower temperatures. A hydrogen liquefier based on a design by Hampson was marketed by the Brin Company. A unit was exhibited at the World Fair in St. Louis in 1904 and was purchased by the National Bureau of Standards for use in their Washington laboratories. It was not, however, until Linde developed a process for rectifying liquid air using the single–column system in 1901–1903 and, in particular, after the double-column system was launched in 1910 that the cryogenic air separation industry really began to develop.

BOC Formed[2]

In 1906 Brin's Oxygen Company took over the rights to the Linde patents in Great Britain in exchange for shares in the company and the election of Carl von Linde to a seat on the board. A change in name was felt to be appropriate and it became the British Oxygen Company Ltd. (BOC). Linde participation in the British Oxygen Company ceased in 1914 with the start of the war. At this time the Société L'Air Liquide set up in Great Britian the British Liquid Air Company to operate the Claude process for liquid air separation. The British Oxygen Company claimed that this infringed the Hampson and Linde patents. BOC initially lost its case, but on appeal it was upheld in the House of Lords.

Market Development

Domestic Market

Air Liquide ceased operating in the United Kingdom and the British Oxygen Company acquired the rights to the Claude patents in the United Kingdom and British territories. BOC began the manufacture of oxygen-producing plants of improved design, using the best features of the Linde and Claude systems. Oxygen purity was raised to 99.5 percent or higher and by 1914 such plants were in operation in London, Birmingham, Manchester, Glasgow, Newcastle, Cardiff, Sheffield, and Greenwich (Fig. 5.6.3). During the war the demand for oxygen-cutting and oxyacetylene welding greatly increased and BOC built new works at Coventry and Wembley.

International Market

Earlier attempts to sell the licenses for the Brin oxygen process on the European continent and New York had been quite successful. BOC began to market its new oxygen process internationally and associated companies were set up in Australia in 1912, South Africa in 1927, and India in 1934 and later in many of the other colonies

Figure 5.6.3. Interior, Westminister works, 1911. From Anonymous (1911). Courtesy of BOC, Windlesham, England.

of the British Empire. By acquisitions BOC nearly reached a monopoly position for industrial gases in the Commonwealth countries. The main developments were in expanding business interests, rather than in new technology.

Product Development[3]

During the first part of the 20th century oxygen was distributed under pressure in steel cylinders and as the amounts required increased, this became increasingly

Figure 5.6.4. Early cylinder delivery van, ca. 1915. From BOC (1986) with permission.

Figure 5.6.5. One of the first BOC tankers. From BOC (1986) with permission.

unsatisfactory because of the high dead weight of metal that had to be transported. In the 1920s Heylandt in Germany developed a new process for large-scale liquid oxygen production. Oxygen then could be delivered in liquid form (Fig. 5.6.5).

The Metal Industries company acquired the U.K. rights to the Heylandt patents and started using the Heylandt process in a subsidiary called Oxygen Industries Ltd. They used oxygen in their ship-recycling and scrap yards, selling the surplus to other users. BOC acquired Oxygen Industries Ltd. in exchange for a major Metal Industries interest in BOC, which they sold in the late 1950s. BOC installed many Heylandt plants, initially with an approximate capacity of 11 tons per day, at sites in the United Kingdom (Fig. 5.6.6).

Figure 5.6.6. Interior of the BOC Heylandt plant. Courtesy of BOC, Windlesham, England.

Figure 5.6.7. Hydrogen-filled barrage balloon being prepared during World War II. From BOC (1986) with permission.

After World War II the Heylandt process was superseded by large tonnage plants. These produced part of the output as liquid, and contained large liquefiers with turbocompressors and expanders operating at lower pressures, significantly reducing the relative capital cost with improved reliability. Machine and cycle efficiency improvements also brought the power requirements down to comparable or lower levels. As economies began to revive after the war, a new era of oxygen usage began, with steel industry customers using oxygen in tonnage amounts (>100 tons per day). The use of nitrogen from air separation plants also began a significant increase. The emphasis was therefore increasingly on plants to meet these demands.

On-site plants or gas pipeline supplies were necessary because the shipment of liquid oxygen with subsequent evaporation was uneconomical on a large scale. Emphasis was therefore mainly on cycles for the large-scale production of gaseous oxygen.

The Rescol cycle was developed by Paul Schuftan in an attempt to reduce power costs by using a thermodynamically more reversible operation. In the 1950s two plants each producing 110 tons per day were supplied to British Gas for a coal gasification project at Westfield in Scotland, where the oxygen was required at a pressure of about 30 bars; the units used a cycle with internal liquid oxygen "compression" to this pressure and vaporization in the heat exchanger system. This cycle was developed for

Figure 5.6.8. BOC tank car, 1965. Courtesy of BOC, Winderlsham, England.

large-scale oxygen supply to steel works in the United Kingdom by pipeline at about 40 bars. It was termed the Tonnox IC process and the plant capacity was generally 230 tons per day of oxygen.

In 1970 BOC built in Widnes, Lancashire, the largest liquid gas plant then known, producing 1200 tons per day of liquid nitrogen, liquid oxygen, and liquid argon. The liquid products were distributed throughout the United Kingdom by high-speed dedicated rail trains.

Further development followed the availability of liquid helium made possible by the introduction of the Collins helium liquefier, initially by the A. D. Little Company in the United States. This greatly stimulated research into phenomena at liquid helium temperatures, and the discovery of type II superconductors in the early 1960s further increased the effort to develop large-scale cryogenic applications.

In the 1960s BOC became the sole provider of European helium by extracting and liquefying helium from the natural gas fields at Odalanov, Poland, for distribution throughout Europe.

Competition from Air Products

Up to the 1950s there was relatively little competition for BOC in the United Kingdom. In 1951, however, the Butterfly Co. entered into an agreement with Air Products to manufacture small, packaged oxygen/nitrogen generators in Great Britain. A number were supplied, mainly to the Air Ministry. In 1957 Air Products took a controlling interest in a joint venture with the Butterfly Co., and Air Products (Great Britain) Ltd. was founded. The first tonnage plants ordered from them in 1958 were for ICI at Billingham and for Stewart and Lloyds at Corby to supply oxygen to their steelworks. A plant was built at Fawley in the early 1960s to supply the Esso refinery there with nitrogen by pipeline and formed the basis of the entry of Air Products into the U.K. industrial gases market, using some surplus capacity.

Cryogenic Application Development by BOC[4]

One of the earliest applications of cryogenic cooling by BOC was the use in the 1930s of liquid oxygen, nitrogen, or air for shrink fittings, in which metal parts are cooled and thereby contracted prior to insertion into the appropriate cavity. On warming up and expanding, a firm fit is obtained.

In the late 1960s and early 1970s BOC entered joint ventures with Airco for air separation plant development, with Big Three for oil-well service activities, and with German Linde for food applications.

In the early 1960s liquid nitrogen came into use for cooling concrete and for freezing of biological material and food. In the 1970s other cryogenic applications such as rubber deflashing and freeze grinding were developed and marketed.

In the beginning of the 1980s Airco in the United States, part of the BOC Group from 1978, developed a solvent recovery technology (ASRS) for the drying of ovens used in the manufacture of coated tapes such as magnetic tapes and photographic film.

BOC's gas-bearing expansion turbines (see Fig. 4.3.10) were licensed to the Helix Corporation. Development and sale of containers, insulated pipelines and transfer systems for the cryogenic fluids liquid helium, liquid hydrogen, and liquid oxygen and nitrogen were concentrated in BOC Cryoproducts, incorporated into BOC's Edwards High Vacuum operation.

In the early 1970s BOC's superconducting magnet activity was incorporated into the Oxford Instrument Company under a mutual agreement by which BOC acquired shares in the Oxford company. In 1980 BOC sold its operations in the helium refrigerator/liquefier equipment area. These activities were transferred to Sulzer and formed the basis for Sulzer UK Cryogenics. BOC's interest in Oxford Instrument Company was sold in the mid-1980s.

Geographic Expansion

By the late 1960s the BOC Company had grown into a highly diversified group with world-wide operations. In 1978 BOC acquired Airco, originally called the Air Reduction Company, which was founded in 1915. During 1982, BOC increased its South American operations by acquiring 100 percent of Chem-Sub, Inc., which operated industrial gas and welding equipment businesses in Venezuela (GIV) and Colombia (Indugas). The same year BOC bought a 43 percent interest in Osaka Sanso Kogyo KK, the third largest industrial gas company in Japan. In the beginning of 1984 BOC sold its Brazilian company, Brasox, to White Martins (Union Carbide). The carbon dioxide part of the business was transferred to Liquid Carbonic.

BOC entered the gas business in Taiwan in 1984 by acquiring 50 percent of Lien Hwa, the largest gas company in Taiwan. In 1987 BOC entered the markets of Turkey and the Peoples Republic of China through 50:50 joint ventures.

Milestones

1850s	Discovery of a chemical barium oxide process for extracting and storing oxygen from the atmosphere
1880	Brin brothers' first patent
1886	Incorporation of Brin's Oxygen Company Ltd. in England; first plant in operation
1906	Change of company name to the British Oxygen Company Ltd.
1910	Founding of Ohio Chemical and Manufacturing Company in Cleveland, Ohio
1927	Formation of African Oxygen & Acetylene Pty (after Allen-Liversidge joined BOC in the United Kingdom)
1933	Introduction by Ohio Chemical and Manufacturing Company of cyclopropane, a pioneering anesthetic gas, to the U.S. market
1935	Formation of Commonwealth Industrial Gases Ltd. (CIG) by amalgamation of several complementary and competing units (including the Comox Co.) in Australia
	Foundation of the Indian Oxygen & Acetylene Co. Ltd.
1937	Acquires Odda Smelteverk A/S, a calcium carbide works in Norway
1940	Air Reduction Company (Airco) acquires Ohio Chemical and Manufacturing Company
1969	Founding of Irish Industrial Gases Limited and BOC Northern Ireland
1973	Formation of Bangladesh Oxygen Ltd. (BOC share 60 percent)
1974	Acquires 35 percent interest in Airco Inc. in the United States
1975	Change of company name to BOC International Ltd.
	Federal Trade Commission orders BOC to divest its holding in Airco
	Merger of Singapore Oxygen with Singapore interests of Far East Oxygen (L'Air Liquide group), giving BOC a 50 percent share in the new company, SOXAL
1977	Successful appeal against FTC's order to divest Airco holding
	Edwards acquires German freeze-drying business Kniese Apparatbau GmbH
1978	Acquires remaining shares of Airco, now 100 percent owned by BOC
	Share issue reduces the BOC holding in IGL (Nigeria) to 30 percent
	Reduction of share in Indian Oxygen Limited to 40 percent following government pressure
	Reduction of share in East African Oxygen Limited to 66 percent
	Reduction of share in African Oxygen Limited 60 percent
	Merger of Malaysian Oxygen Ltd. with L'Air Liquide local associate; BOC's share in the new company is 40 percent
1979	Formation of joint ventures formed in Taiwan and the Philippines
1981	BOC International Ltd. becomes BOC International plc (6 March)
	Formation of joint venture Daisan–BOC Ltd with Osaka Sanso (Japan)
1982	Change of company name from BOC International plc to the BOC Group plc
	Acquires 100 percent of Chem-Sub Inc., an industrial gas and welding equipment business in Venezuela, Colombia, and Aruba

	Acquires 43 percent interest in Osaka Sanso Kogyo KK, Japan, for an initial investment of approximately £30 million
1983	Sale of U.K. welding equipment manufacturing operations to ESAB AB of Sweden
1984	Sale of interests in Brazilian industrial gases business
	Several U.S. welding businesses shut down or sold
1985	Acquires 50 percent stake in Lien Hwa Industrial Corp., Taiwan
1986	Sells North American welding electrodes business
1987	Forms BOS, a 50:50 industrial gases joint venture in Turkey
	Forms 50:50 industrial gases joint venture in China with Wusong Chemical Works
	Acquires remaining 39 percent stake in New Zealand Industrial Gases
	Sells remaining northern hemisphere welding equipment manfacturing operations
1988	Holding in CIG increases to 87 percent
	Sells 49 percent investment in Zambia Oxygen Ltd.
1989	Acquires industrial gas and CO_2 divisions of AmeriGas Inc. for \$143.5 million, and makes 12 other small retail gas acquisitions in the United States
	Sells a gas plant and 10 retail gas outlets in Texas to Tri-Gas Inc. for \$40 million
	Forms Generon joint venture with Dow Chemical Company to exploit noncryogenic nitrogen production technology
	Acquires 50 percent interest in Haedong Industrial Co. Ltd., renamed Korea BOC Gases Ltd.
	Sells CIGWELD, Australian welding equipment manufacturer
1990	Acquires remaining minority shares in CIG
1991	Doubles equity stake in Industrial Gases Lagos (IGL), Nigeria, to 60 percent
1992	BOC Cryoplants changes name to BOC Process Plants
	Forms gas joint venture in Tianjin, northern China, with Hua Bei Oxygen
1993	Acquires German hydrogen business from Hüls AG
	Purchases 70 percent interest in part of Poland's state industrial gases sector based at Siewierz, Wroclaw, and Poznan
	Increases holding in IOL Ltd. (now BOC India Ltd.) to 51 percent by subscribing for new shares
1994	Afrox acquires LPG business from Engen
	Establishes new joint venture with Fushun Iron & Steel Co. at Fushun, northern China
	Changes to a BOC trading identity for gas companies in the United States, Canada, New Zealand, Taiwan, Zimbabwe, Ireland, Australia, and Pakistan
1995	Changes to a BOC trading identity for gas companies in India, Bangladesh, Venezuela, the Pacific Islands, Kenya, Curacao, Aruba, South Africa, Papua New Guinea, and Korea
	Invests \$50 million in Indura SA Industria y Commercio, the leading industrial gases company in Chile, to acquire a 41 percent holding.

References

1. BOC. (1986). *Around the Group in 100 years*. Windelsham, England:
2. Anonymous. (1911). The manufacture of oxygen. A visit to British Oxygen. *Acetylene* **1911** (November):252–257.
3. Aitken, B. (1998). Industrial oxygen. *Chemical Engineer* **1998**(June 11):16–18.
4. Gardner, J. B. (1992). History of cryogenics in BOC. In *History and Origins of Cryogenics* (R. G. Scurlock, ed.), Chapter 6, pp. 180–215. New York: Oxford University Press.

5.7. PRAXAIR[1]

Background

When Carl von Linde successfully separated oxygen, he realized the potential this process would have in the United States, where the iron and steel industry was highly developed and industry was growing at a rapid rate. To protect his process, he applied for and was granted two U.S. patents in 1903.

The Dream of America

Von Linde needed someone to sell his concept to American investors, survey the existing and the potential demand for oxygen, and determine the best location for a Linde oxygen plant. In 1906 he turned to Cecil Lightfoot (Fig. 5.7.1), son of Timothy Bell Lightfoot, Managing Director of the Linde British Refrigeration Company, Linde's agent in Great Britain.

Von Linde referred Cecil Lightfoot to Charles F. Brush, part owner of the U.S. Linde process patents, who headed the Brush Electric Company. Another contact was Adolphus Busch, head of the Anheuser-Busch brewery in St. Louis. A third was Fred Wolfe of Chicago, the U.S. agent for Linde's ice machine business. Lightfoot was

Figure 5.7.1. Cecil Lightfoot. Courtesy of Praxair, Inc., Danbury, Connecticut.

unable to obtain the necessary backing to get a company underway, however, because of sceptisism about the future of the oxygen industry in America. Tripler's failures some years earlier had killed off interest in investing money in liquid air, and large-scale applications had not yet developed.

At the end of 1906 von Linde travelled to the United States, so convinced of the potential success of a commercial oxygen plant there that he was willing to finance it himself if American businessmen were not ready to get involved. After initial negative results with Brush and Wolfe, von Linde got Brush and three Cleveland businessmen to agree to support the new venture. Lightfoot found that Buffalo would be the ideal place for a Linde oxygen plant, near the hydroelectric power provided by Niagara Falls and in the middle of the major industrial centers at New York, Cleveland, and Chicago.

Linde Air Products Established

On January 24, 1907 the Linde Air Products Company was incorporated in Cleveland, Ohio, and became America's first large-scale commercial oxygen producer (Fig. 5.7.2). Cecil Lightfoot was named Managing Director of the company with responsibility for supervising the construction and startup of the Chandler Street facility in Buffalo (Fig. 5.7.3). The challenge of getting this operation off the ground was difficult and there were problems from the start. The backers of the venture were unsure of the future and somewhat reluctant to advance funds.

Additionally, once the plant was running, there were problems with the lack of demand for the product. The oxyacetylene welding process using oxygen and acetylene was at that time popular in Europe, but it was virtually unknown in the United States. Hence, the new company had to create a demand for its product by introducing welding to America. Before the arrival of the Linde system in the United States, oxygen gas was produced by chemical processes or by electrolysis of water. It was sold in large, riveted cylinders at low pressure.

Figure 5.7.2. Early transport truck, 1911. Courtesy of Praxair, Inc., Danbury, Connecticut.

Figure 5.7.3. Horse transport of gas from the first Linde plant, 169 Chandler Street in Buffalo, in 1910. Courtesy of Praxair, Inc., Danbury, Connecticut.

Early Development

Linde developed new, light-weight cylinders made of high-alloy steel drawn in a single piece. By compressing the gas, a considerable quantity of oxygen could be transported in a comparatively small package. Linde Air Products Company started with 5000 German oxygen cylinders to meet early needs. The first high-pressure cylinders to be produced in the United States were made especially for Linde Air Products by the National Tube Company, McKeesport, Pennsylvania. Soon manufacture of cutting and welding torches and gas cylinder valves began at the Buffalo factory.

The company's second oxygen plant was built in East Chicago, Indiana, in 1910 and the following year two other plants were built, in Elizabeth, New Jersey, and Trafford, Pennsylvania. The company also sold a complete line of acetylene generators as well as regulators, torches, and related equipment.

The Union Carbide Company

Many of Linde Air Products's customers made their own acetylene by purchasing calcium carbide from the Union Carbide Company (Fig. 5.7.4), a leader in the field. Union Carbide kept an eye on the commercial oxygen industry and the oxyacetylene process, which created so much demand for their product. Before cheap oxygen came on the market, carbide had been used primarily to generate acetylene gas for lighting purposes (Figs. 5.7.5 and 5.7.6).

Union Carbide's interest in the oxygen business was so great that it acquired a part of the Linde Air Products Company in 1911, thus assuring the young company capital funds necessary for expansion. During 1913, air separation plants were built

Figure 5.7.4. The Acetylene Light, Heat and Power plant in Niagara Falls, 1897, forerunner of the Union Carbide Company, founded in 1898. Courtesy of Praxair, Inc., Danbury, Connecticut.

in Cleveland, Ohio, Oakland, California, Morristown, Pennsylvania, and Detroit, Michigan, followed in the next 3 years by 11 more plants spread over the country.

Linde Air Products Becomes Part of Union Carbide Corporation

Business was going very well. In 1917 a partnership was formed between the Linde Air Products Company, the Prest-O-Lite Company (dissous gas, founded in 1906), the National Carbon Company (electrodes and batteries, founded as Carbon Company in 1876), the Union Carbide Company (calcium carbide and acetylene, founded in 1898), and the Electro Metallurgical Company (metal refining) to form the Union Carbide and Carbon Corporation. This was later to become the Union Carbide Corporation. In 1919 German assets in the United States were confiscated and Linde Air Products was bought by Union Carbide; in 1963 it became the Linde division of Union Carbide.

Because the fast and efficient distribution of gas products was of growing importance, the group signed a contract with the American Teaming Company to haul cylinders by motorized truck rather than horse-drawn dray. In 1920 Linde Air Products had oxygen plants scattered throughout the country as well as 25 acetylene plants. In 1921 there was a severe economic crunch. As part of Linde Air Products's cutback program, the East Chicago plant was shut down. When it opened again in the end of 1922 the demand had come back to normal levels.

Research and Development

From the beginning Linde Air Products had a cooperation agreement with the U.S. Government. First it dealt with welding and cutting research and development

Figure 5.7.5. Old advertisement, acetylene lighting. Courtesy of Praxair, Inc., Danbury, Connecticut.

of corrosion-resistant ferroalloys, among others. During World War II Linde Air Products scientists made contributions to nuclear development by perfecting a uranium-refining process.

In 1920 research on synthetic organic chemicals began. The antifreeze agent ethylene glycol is among the products introduced by Union Carbide. Things developed very well until the stock market crash of 1929 and the depression of the 1930s.

Linde Air Products pioneered and developed technology for storing and transporting cryogenic fluids. This meant that more product could be transported in much less space with less weight. Linde Air Products changed over from producing and shipping oxygen in a compressed gaseous state only to producing and distributing both gaseous and liquid oxygen. In order to convert the entire production system from gaseous to liquid, Linde Air Products purchased the rights to the liquid oxygen production and distribution system developed by the Heylandt company of Germany in 1927. The production equipment was quite similar to that already in use, but involved

Figure 5.7.6. Old advertisement, acetylene heating. Courtesy of Praxair, Inc., Danbury, Connecticut.

more refrigeration. The transportation and storage systems consisted of insulated tanks and evaporators. The insulation quality was, however, not so good at that time.

Production and Distribution of Liquefied Gas

About 1930, Leo Dana, since 1923 in charge of research and development at the Linde Air Products Buffalo laboratory, started practical cryogenic development by improving the vacuum insulation for use in all types of cryogenic storage vessels. Dana received a patent on the design of a vacuum-insulated rail car. At the same time, work had begun to develop a pump capable of handling cryogenic liquids (Fig. 5.7.9).

Because it was costly to install the necessary storage and conversion equipment on a customer's site, early liquid distribution was limited to very large customers. The customer's on-site liquid oxygen equipment was called the Driox oxygen system. The

Figure 5.7.7. Liquid oxygen transport, 1931. Courtesy of Praxair, Inc., Danbury, Connecticut.

first customer installation was put on-line in 1931 at Merchant's Dispatch, a manufacturing company in Rochester, New York.

In 1935, Linde Air Products introduced the Cascade oxygen system, to allow smaller users of oxygen to benefit from the economy of using liquid oxygen. An evaporator unit carried on the delivery truck converted liquid oxygen to high-pressure gas immediately upon arriving at the customer's site. The gas was then stored in conventional cylinders or tubes permanently kept by the user. Later, smaller Driox units were developed so that customers with less demand could store oxygen in its liquid state on their premises. In the 1950s Linde Air Products introduced liquid cylinders for up to 100 liters of oxygen, movable on a hand truck (Fig. 5.7.10).

Figure 5.7.8. Linde air separation plant in Tulsa, 1926. Courtesy of Praxair, Inc., Danbury, Connecticut.

Figure 5.7.9. Tonawanda research and development center, 1940. Courtesy of Praxair, Inc., Danbury, Connecticut.

Figure 5.7.10. Linde small liquid oxygen containers LC-3 in the mid 1950s. Courtesy of Praxair, Inc., Danbury, Connecticut.

The development of liquid oxygen production and distribution technology was a long, difficult process with improvements coming bit by bit. During this period, Linde Air Products engineers continued to design and build gas plants, which also became much improved. In the 1930s, Linde began to build smaller gas plants at strategic locations in order to reduce the cost of gaseous oxygen distribution to customers who would previously have been served by larger, more remotely located facilities.

Success and Catastrophe

At the end of World War II, the demand for liquid oxygen by large customers, especially steel mills, was becoming so great that even the trucking in of cryogenic product could not keep up with demand. The response was to develop the on-site plant, a low-pressure facility built either on or adjacent to a customer's property. The on-site plant produced gaseous oxygen on the spot and fed it to the customer via a pipeline. The first such facility supplied Union Carbide's Texas City chemical plant. The second unit supplied a Du Pont facility at Belle West, Virginia. These plants produced 360 tons per day. The XA plants at the end of the 1940s produced from 7 to 20 million cubic feet per month (700–2000 tons per day). By the 1980s about 100 such plants were in operation. In the 1950s and 1960s the Union Carbide Linde Division kept its position as America's number one producer of industrial gases. In the 1960s the growth of oxygen-fired furnaces for steel production and the use of increased nitrogen in cryogenic cooling expanded the Linde Division's markets.

In the early 1980s Union Carbide was a very diversified company. About 60 percent of its business was chemicals, 70 percent of which was in the United States, the rest mainly in Europe. Linde Division accounted for about 11 percent of the business. Because of the terrible Bophal catastrophe in India, when poisonous gas was released to the environment by leakage in the methyl isocyanate manufacturing process, Union Carbide was forced to reorganize. In 1992 the Linde Division was spun off to Union Carbide shareholders as Praxair.

Applications Development

New central research and development laboratories were built in Tonawanda, about 10 miles from Buffalo, New York in 1937 (Fig. 5.7.9). The Tonawanda laboratories became very important for Linde's gas application developments.

The mid and late 1940s saw a number of significant changes at the Linde Air Products Company. In 1942, Oxweld became part of Linde, as did Prest-O-Lite in 1947. The "Air Products" part of the name was dropped in 1957, and in 1963 the company became known as the Linde Division of Union Carbide Corporation.

Noble Gas Recovery

Linde Air Products scientists also began researching ways of recovering the rare or noble gases argon, krypton, xenon, neon, and helium. Historically, a major market for such gases was in laboratories and in the manufacturing of incandescent and tungsten filament lamps. Argon is now used widely in welding and in steelmaking in the argon oxygen decarburization (AOD) process for decarburizing high-chromium raw steel when making stainless steel. The process was invented by W. A. Krivsky of Linde Air Products in 1954. Approximately 75 percent of the world's stainless steel is produced by AOD.

PSA

Synthetic molecular sieves were invented in 1949 by Robert M. Milton, a Linde researcher at the Tonawanda Labs, who was looking for new methods to separate atmospheric gases. Molecular sieves are crystalline products used in many industrial processes to dry, purify, and separate a wide variety of gases and liquids.

In the early 1980s the Linde Division made technological breakthroughs by introducing the noncryogenic vacuum-pressure swing adsorption process (VPSA) as an economical alternative to cryogenic air separation.

Railroad Business

Railroads have played an important role in the history of Linde Air Products. There was, in fact, a time when sales to railroads constituted a larger portion of its business than did sales to the steel industry. This was due to the efforts of the Oxweld Railroad Service (ORS) Company in Chicago, founded about 1912. Chicago was at that time the center of the U.S. railroad industry. A controlling interest in ORS was held by the welding equipment manufacturer Oxweld Acetylene Company, a subsidiary of the Union Carbide Company.

ORS made a deal with the Oxweld Acetylene Company to use the Oxweld name in contracting with railroads for full service relating to the oxyacetylene process. In return ORS agreed to purchase all necessary equipment from the Oxweld Acetylene Company. ORS entered into similar agreements with the Linde Air Products Company and the Commercial Acetylene Company for the purchase of oxygen and acetylene for resale to the railroads. ORS supplied all of the necessary welding equipment, including cylinders, torches, regulators, and cutting machines, as well as acetylene generators, annealing furnaces, and Driox and Cascade bulk oxygen systems to the railroads free of charge.

Ribbonail

Ribbonail was a Linde Air Products welding product sold to the railroads through ORS. It involved factory-manufactured, all-welded rails in a single, continuous strand, up to 1/4 mile in length. These rails were then placed on a string of flat cars and transported to where they were laid. The welding of rail segments involved a Linde Air Products-patented process known as pressure butt welding. It became popular around 1940, but in spite of its success, it partly caused the demise of ORS later.

A Swiss company developed a flash electric welding process for joining rail, and the welding repair of locomotives decreased considerably with the introduction of diesel locomotives. The ORS was disbanded in 1953 and partly absorbed by the Linde Air Products Railroad Department.

Steel Gouging and Deseaming

Around 1915 the oxygen lance came into use in steel plants for tapping furnaces and open hearths and for lancing holes in skull and other heavy scrap. Gouging and deseaming seem to first have been employed in the 1920s. About 1929, gouging was attempted using a separate jet of oxygen, which produced a contour satisfactory for rolling. This was the first practical step in using the oxyacetylene flame for conditioning steel.

The new process of deseaming in steel mills used oxygen so rapidly that it threatened to produce a shortage of cylinders. To avoid this, Linde Air Products introduced the Driox liquid oxygen system of distribution in 1932. One tank car of liquid oxygen delivered to the customer's Driox installation could be converted to 25,000 liters of gaseous oxygen, which is equivalent to 11 freight cars of cylinders.

The deseaming process also continued to receive intensive development. In 1934 the first Linde Air Products Desurfacer was demonstrated for a steel company and was promptly put to work.

Arc Welding and Cutting

A difficult metal engineering problem was to find procedures for cutting high-alloy steel. In 1945 that problem was solved by the addition of an iron powder.

Unionmelt

Other new metal-working processes were developed by Linde Air Products laboratories during the same period. In 1938, an entirely new welding method was introduced, named Unionmelt. High currents, high penetration, and high speeds are characteristics of this process.

Tungsten Inert Gas (TIG), Heliarc, Sigma

The need for larger quantities of argon was generated by Linde Air Products's development of the Heliarc welding process, which used inert gas to shield a tungsten arc. It was developed during World War II for the welding of aluminum and magnesium. Heliarc welding was a simple and easy method for welding the difficult-to-weld metals important to the war industry. It gave rise to another welding process in which the use of a consumable electrode replaced the nonconsumable tungsten electrode. This was called sigma welding. Sigma apparatus was developed for manual and fully automatic applications, the process being used on almost all commercially available metals.

Unionarc

Another welding first was Unionarc welding, which used a continuously fed electrode for the welding of mild steel and operated in all positions.

Market Developments and Acquisitions

1994 Praxair set up China's first helium transfill plants for magnetic resonance imaging. In 1995 the company made a hostile bid to buy CBI, owner of the

Figure 5.7.11. A modern semitrailer in front of an air separation unit. Courtesy of Praxair, Inc., Danbury, Connecticut.

industrial gas company Liquid Carbonic. Success of this bid made Praxair the world leader for CO_2, ahead of Air Liquide and BOC. In 1995 Praxair began operations in Peru and India. In 1996 Praxair acquired Maxima Air Separation Center Ltd., Israel's second largest producer of industrial and specialty gases. In 1996 the Linde trademark and name in North America were sold back to Linde AG. More than 50 percent of the revenues of Praxair are from operations outside North America.

Milestones

- 1907 Organization of Linde Air Products in Cleveland, Ohio, with a plant at Chandler Street, Buffalo, New York
- 1910 Opening of the second oxygen plant at East Chicago, Indiana
- 1912 Construction of the Newark factory as part of Oxweld Acetylene Company
- 1914 Establishment of the Buffalo laboratory for testing and experimental work
- 1916 Linde Air Products becomes the first company in the United States to produce argon for commercial use (Cleveland plant)
- 1917 Formation of Union Carbide and Carbon Corporation
- 1918 Operation of Linde Air Products in Canada under the name Dominion Oxygen Company
  Oxweld Acetylene Company joins Union Carbide Corporation
- 1919 Speedway Laboratory, originally known as Prest- O-Lite Acetylene Research Division, is set up at Speedway
  Purchase of Oxweld Railroad Service Company by Union Carbide Corporation

1927	Buffalo Laboratory begins experimental work to develop the Driox system of liquid oxygen production, storage, and distribution
1928	To cope with rising costs of distribution, smaller acetylene and oxygen plants are constructed to cut cylinder transportation costs
	First bulk liquid oxygen is made at East Buffalo plant
1929	Purox Company joins Union Carbide Corporation as part of Linde Air Products
1931	Linde Air Products becomes first to make USP oxygen available to hospitals in large cylinders and at lower cost
1932	Supply by Linde Air Products of large quantities of oxygen, via the Driox oxygen system, to some of the country's largest steel producers and other large industrial accounts
	Opening of first commercial liquid oxygen plants at Trafford, Pennsylvania, and East Chicago, Indiana, to supply oxygen to steel centers
1935	Development of the Cascade oxygen system to make available to smaller customers the benefits of low-cost bulk liquid production and distribution
1937	Construction of factory and laboratory buildings at Tonawanda, New York
1947	General acceptance of Heliarc welding gives impetus to demand for Linde Air Products argon
1954	Introduction by Linde Air Products of new type of adsorbents called molecular sieves
1956	Distribution of liquid oxygen in LC-3 cylinders as an alternative for distribution in high-pressure cylinders
	Construction of Linde Air Products's first large-volume acetylene plants at Moundsville and Montague, Michigan
1957	"Air Products" dropped from the company name
1963	Company becomes known as the Linde Division of Union Carbide Corp. (UC Linde Division)
1992	UC Linde Division splits off from Union Carbide and becomes Praxair
1994	China's first helium transfill plants set up by Praxair
1995	Liquid Carbonic becomes part of Praxair
1996	Praxair acquires Maxima Air Separation Center Ltd. in Israel
	Linde trademark and name in North America sold back to Linde AG

Reference

1. The Linde Story 1907–1982. History articles in *Focus* Issues 2–5 (Summer 1982-Winter 1984).

5.8. NIPPON SANSO[1,2]

Background

Nippon Sanso was founded in 1910, and was until 1960 mostly a supplier of industrial oxygen. The company began manufacturing of oxygen plants in 1917 and developed its first air separation plant in 1934. The first "home-made" liquid oxygen production unit was manufactured in 1954 and the first on-site plant was built in 1964. After that Nippon Sanso promoted the idea of selling plants to on-site customers or operating them in joint ventures with customers. The production of specialty gases for electronics was started in 1969 with a new plant, and the company thereby acquired knowledge of electronics gases in the early 1970s.

Origins

The origins of Nippon Sanso date to 1910, when Takehiko Yamaguchi (Fig. 5.8.1), an importer of tools and machinery and the founder of Yamatake Honeywell, set up a limited partnership with the assistance of Korekiyo Takahashi (later Prime Minister). At that time Japanese industry was still in its infancy and the manufacture of oxygen was just beginning, even in Germany and France.

Early Market Development

In 1911, the first oxygen manufacturing plant was imported from Firma Hildebrand of Germany and started in Osaki in Tokyo (Figs. 5.8.2 and 5.8.3).

Figure 5.8.1. Takehiko Yamaguchi, founder of Nippon Sanso. Courtesy of Nippon Sanso Corporation, Tokyo, Japan.

Figure 5.8.2. First oxygen column. Courtesy of Nippon Sanso Corporation, Tokyo, Japan.

During the first decade the new company found increased demand for oxygen for welding and medical use. In 1918 the company was formally established as a stock corporation and became Nippon Sanso KK.

Because of the great need for air gases in Japanese industry during the first decades of Nippon Sanso's existence, 16 oxygen plants and 33 nitrogen plants were imported from German and French manufacturers between 1910 and 1935.

Production Plant Development

During 1917–1930 Nippon Sanso built 17 oxygen separation units (single column) mainly for its own use. In the early 1930s Nippon Sanso decided to begin

Figure 5.8.3. Inside the first Nippon Sanso gas plant (the Osaki plant). Courtesy of Nippon Sanso Corporation, Tokyo, Japan.

building air separation plants with double rectifying columns, and the first one was built in Kamata in 1934. In 1954, Japan's first plant for producing liquid oxygen was completed in Kawasaki (Figs 5.8.4 and 5.8.5).

After World War II Nippon Sanso made the strategic choice to manufacture and sell plants to the steel and chemical industries and sell the gases through a chain of distributors. Some years later the distributors grew to the extent that Nippon Sanso lost control over most of their air separation capacity. In 1964 the first on-site plant was erected in Shunan for supplying gases by pipeline to the nearby large steel mills and petrochemical complexes.

Market Development

In 1969 Nippon Sanso built the first plant in Japan for manufacturing specialty gases especially for the semiconductor industry. In the same year a joint venture company was set up to produce liquid oxygen and nitrogen utilizing cold from the nation's first liquid natural gas facility. In 1970 a joint venture with American firms, Japan Helium Center, was set up for the purpose of importing and refilling liquid helium.

The know-how acquired from gas technology was applied to two new consumer areas at the beginning of the 1970s. Vacuum and insulation technology led to the development of the vacuum flask division and the manufacture of thermos bottles, whereas cryogenic technology knowledge led to entry into the frozen food business.

Figure 5.8.4. Interior of the first Nippon Sanso LOX plant in Kawasaki, 1954. Courtesy of Nippon Sanso Corporation, Tokyo, Japan.

Figure 5.8.5. Interior of the first LOX plant, LOX compressor. Courtesy of Nippon Sanso Corporation, Tokyo, Japan.

Figure 5.8.6. Nippon Sanso's largest helium liquefier. Courtesy of Nippon Sanso Corporation, Tokyo, Japan.

Figure 5.8.7. Tukuba research and development facility. Courtesy of Nippon Sanso Corporation, Tokyo, Japan.

International Expansion

Overseas activities resulted in the establishment of the following operations in the United States: Japan Oxygen, Inc. (marketing of Nippon Sanso products) in 1980, Ansutech, Inc. (manufacture of air separation units) in 1982, Matheson Gas Products, Inc. (specialty gases) in 1983, and KN Aronson, Inc. (gas cutting equipment) in 1985. In 1982 National Oxygen Pte. Ltd., Singapore, was established and offices were set up in Beijing, China, in 1986 and in Seoul, Korea, in 1989. In 1988 the Malaysian company Industrial Oxygen Inc. (IOI) became part of the Nippon Sanso group.

With the large growth of semiconductor manufacturing in Japan in the early 1980s, the demand for nitrogen exceeded that of oxygen for the first time. To keep up

Figure 5.8.8. The Nippon Sanso head office in Tokyo. Courtesy of Nippon Sanso Corporation, Tokyo, Japan.

Figure 5.8.9. A modern Nippon Sanso cylinder-filling facility. Courtesy of Nippon Sanso Corporation, Tokyo, Japan.

with these new demands new PSA and membrane separation technologies began to be used. Since that time Japan has maintained a leading position in the world in these new areas of air separation technology.

Milestones

1910	Nippon Sanso established
1911	Construction of the first air separation plant, in Osaki, Tokyo, by Firma Hildebrand of Germany
1934	Construction of the first air separation plant by Nippon Sanso
1954	Construction of the first liquid oxygen plant in Japan in Kawasaki
1962	Supply of the first high-purity nitrogen gas by a 5-kilometer pipeline to a major semiconductor manufacturer
1964	Construction of the first on-site plant in Japan in Kure, to supply oxygen and nitrogen to a steel mill
1969	Construction of the first air separation plant in the world using cold liquid natural gas, in Yokohama
1971	Construction of the first specialty gas facility, in Oyama
1972	KK Freck starts to manufacture liquid-nitrogen-frozen food
1978	Starts manufacture of vacuum bottles
1980	Establishment of Japan Oxygen, Inc., in United States
	Office opens in Singapore

1982	Establishment of National Oxygen Pte. Ltd. in Singapore
	Establishment of Ansutec Inc. in the United States with AmeriGas
1983	Acquisition of Matheson Gas Product Inc. (United States) by Nippon Sanso and AmeriGas
1984	First "total gas center" for a major semiconductor manufacturer put on-stream in April
1985	Acquisition of KN Aronson Inc. (United States) by Nippon Sanso and Koike Sanso Kogyo Co., Ltd.
1986	Opens office in Beijing, China
	Matheson Gas Product Inc. acquires Isotech Inc.
1988	IOI, Malaysia, joins Nippon Sanso group and is renamed Nissan Industrial Oxygen Inc.
1989	Acquires Thermos Group (United Kingdom, Canada, and United States)
1990	Establishment of Tanaka Thailand Co. Ltd. in Thailand for manufacturing of welding/cutting machines
1990	Establishment of Top Thermo Manufacturing in Malaysia for manufacturing of vacuum flasks
1990	Establishment of AVANCETECH GmbH in Frankfurt, Germany, as a jointly owned company with AGA AB for supplying semiconductor equipment to the European market

References

1. Anonymous. (1990). A slice of postwar history. 35 years of industrial gas history in Japan. *Gas Review Nippon* **1990** (Winter):7–8.
2. Nippon Sanso. (1985). *Nippon Sanso Annual Report. 75th Anniversary*. Tokyo.

5.9. AIR PRODUCTS[1]

Background

In the late 1930s an oxygen salesman in the American Mid West, Leonard Pool (1906–1975), realized that the transporting of heavy steel cylinders filled with compressed oxygen from a central plant to customers scattered over the vast distances of the United States was uneconomical. For many orders the cost of transport exceeded the cost of the gas. Instead of delivering oxygen in cylinders, Pool proposed to build and lease oxygen generators, which would be placed at the customer's premises. Although it was already possible to purchase generators in the United States and Europe, Pool's business concept was something new.

With the aid of a young engineering student at the University of Michigan, Frank Pavlis, Pool built a small oxygen generator that could produce 10 cubic meters per hour (350 cubic feet per hour). They assembled the unit in the corner of a building owned by a Detroit transport company and set out to build a business based on the idea of providing on-site oxygen generators and offering the gas under long-term contracts.

Early Development

Pool founded his company, Air Products, in 1940, and in addition to Pavlis he hired as engineer Carl Anderson, who had extensive experience with oxygen generator manufacturing with the Gas Industries Company (Fig. 5.9.1).

The new company did not at first find a civilian market for oxygen generators and Air Products got its first order to build gas generators for the Army Air Force. Pool changed course when he saw the chance to sell generators to the U.S. Government during World War II. The idea of leasing contracts had to wait until after the war.

When the United States entered the war the demand for mobile oxygen generators increased so greatly and the company grew so rapidly that it had to move in 1943 (Fig. 5.9.2). The Government offered new premises in Chattanooga, Tennessee, and gave Air Products financial help to move. About 1942 Air Products entered the helium and argon markets, which were developing due to new arc welding methods.

Competing companies like the Union Carbide Linde Division (before 1963 Linde Air Products) and the Air Reduction Company (from 1971 named Airco) were reluctant to sell gas-making equipment to the Government and the armed forces. They were afraid of leaving their gas-making experience in other hands and preferred delivering centrally made gas in cylinders all over the world. In 1943 Leonard Pool hired his brother George (1910–1973) to handle Government contracts. George had experience in the industrial gas business as a District Manager of the National Cylinder Gas Company. At Air Products he later became Vice President for Sales, responsible among other things for the successful entry into the cylinder gas market. In 1945 Air Products moved from Chattanooga to Emmaus near Allentown in Leigh Valley,

Figure 5.9.1. Carl Anderson and Leonard Pool with the company's first equipment order. From Butrica (1990) with permission.

Figure 5.9.2. Mobile oxygen generators being used in Alaska by 1944. From Butrica (1990) with permission.

Pennsylvania, an area where the flourishing steel and mining industries were growing gas markets.

Building a Civilian Market

After the war the military contracts were terminated and Air Products introduced their on-site concept to the commercial market under the motto "Keep the cow—sell the milk." The company managed to successfully enter one of the largest U.S. markets: the steel industry. Among the first lease customers were Weirton Steel Company and Ford Motor Company.

Leonard Pool, with the help of his brother George, decided to move the company in a second new direction—the cylinder gas business—in a further effort to compete with Linde Air Products and Air Reduction. Air Products entered the oxyacetylene business in 1944 by buying an interest in Wolverine Products, a cylinder oxygen and acetylene gas distributor in Michigan. In 1946 Air Products started their own cylinder business in Indiana, just at the time oxygen sales were beginning to recover after the postwar decline. Studebaker Corporation was the first main customer.

Leonard Pool now entered the welding equipment business with the acquisition of K-G Welding and Cutting Company, Inc., of New York in 1946. Air Products also acquired Parco Compressed Gas Co. and Harris Calorific Sales. By 1950 the cylinder gas and leased plant operations were the major revenue sources for Air Products.

In May 1946 Air Products went public. From that time on the stockbroker and investment specialist Reynolds & Co. in New York had an influence on the financing of Air Products developments. The concept of leasing on-site oxygen generators attracted Wall Street because this was a business method in vogue, as applied by IBM among others. During the years after the war a large part of the Air Products generator market was abroad, in South America, but also Europe, the Middle East, India, China, and South Africa.

Demand for Liquid Oxygen Equipment

In 1948 aircraft carriers were equipped with jet planes, which flew at much higher altitudes during part of their flight than did the old propeller planes. Therefore more oxygen was needed on board. Linde Air Products and Air Products received contracts to develop liquid oxygen generators and other equipment for distribution, storage, and handling of gases in liquid form. At the same time the Navy needed liquid oxygen apparatus for breathing in submarines. This marked the start of Air Products's entry into cryogenic equipment development.

U.S. participation in the Korean war during the 1950s led to new contracts for the Air Products mobile generator business, now for delivering liquid oxygen generators. In the first part of the 1950s the military and the largest commercial producers began to emphasize the advantages of liquid oxygen. The first application for oxygen in its liquid form came in 1949 when Linde Air Products was asked to build equipment for rocket fuel applications. Neither Linde Air Products nor Air Reduction, however, was interested in building small, mobile liquid oxygen generators, so the military went to Air Products. In 1952 the U.S. Air Force and the U.S. Navy gave Air Products a contract for generators delivering 50 tons per day of liquid oxygen.

Plant Development

Tonnage Oxygen and Nitrogen

The tonnage oxygen market was Pool's next endeavor (Fig. 5.9.3). The main industrial gas producers had to change their business to produce cheap oxygen in tonnage quantities. Leonard Pool's creative financing allowed Air Products to outbid its big competitors (Linde Air Products and Air Reduction), and he was able to move his small, capital-poor firm into this important field without spending large sums of money. In 1953 Air Products sold tonnage plants for liquid oxygen to its competitors Air Reduction and National Cylinder Gas Company. In 1955–1956 Air Products built a 600-ton per day plant for rocket fuelling in Santa Susana, California, and a 300-ton per day plant for the Wright-Patterson air field. Some years later Air Products began to operate and compete in the merchant market and acquired a number of small U.S. producers.

The trend to use basic oxygen furnace (BOF) technology in the steel industry in the 1960s led to increased oxygen demand. At the same time there was a growing of use of nitrogen in metallurgy for heat treatment and annealing. Nitrogen found large new markets in the new float-gas process used by the glass industry and for the protection of silicon and germanium in the electronics production industry.

In 1971 Air Products built a 50-mile pipeline to deliver high-purity nitrogen to Silicon Valley (Fig. 5.9.4).

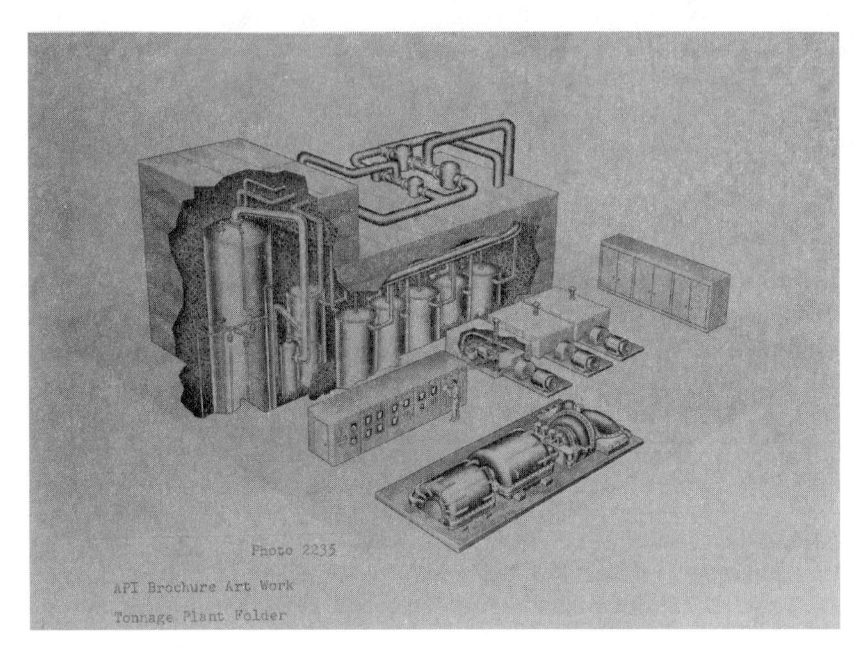

Figure 5.9.3. Drawing of the completed Weirton tonnage plant, 1951. From Butrica (1990) with permission.

Liquid Natural Gas: Helium Recovery

Shortly after World War II Air Products began purchasing helium for resale from the U.S. Bureau of Mines, the country's only helium producer. Air Products soon became involved in research and development work on liquid natural gas (LNG) for the Bureau, developing equipment for helium extraction and purification in the 1950s. In 1946 Air Products received a preliminary contract for building a nitrogen removal plant for natural gas. In 1959 it got the contract to build a large (300 million cubic feet per year = 10 million cubic meters per year) helium recovery plant at Keyes, Oklahoma.

In 1965 Air Products received a contract to build a huge LNG plant for Esso in Libya, to be used for shipping natural gas in liquid form to other continents. Next it built the world's largest LNG plant for Shell in Brunei, Borneo, in 1972.

Liquid Hydrogen

In the 1950s the space race began between the Soviet Union and the United States, and in 1957 the Soviet Sputnik was the first satellite to be launched. When in the mid-1950s the U.S. military wanted to develop and use hydrogen as a rocket fuel Air Products moved into the liquid hydrogen tonnage business. They designed, built, and operated the world's first large-scale liquid hydrogen plants for the U.S. Air Force in 1957, located in Painesville, Ohio, and West Palm Beach, Florida. Air Products established itself as the nation's premier supplier of liquid hydrogen in tonnage quantities. It thus helped to put a man on the Moon and became the world's largest supplier of hydrogen.

Figure 5.9.4. The air separation plant at Santa Clara, pipeline serving Silicon Valley with high-purity gas. From Butrica (1990) with permission.

Market Development

New Subsidiaries

When Air Products was still fighting for survival, Pool had the vision to expand internationally, first to the United Kingdom, then to the European continent. In 1947 Sperry Gyroscope Co. Ltd. in England obtained the right to manufacture and sell Air

Figure 5.9.5. Air Products liquid helium truck. From Butrica (1990) with permission.

Products non-tonnage generators in the British Empire except Canada. In 1950 the Butterfly Company acquired the contract from Sperry.

In 1957, Air Products entered the overseas market for industrial gases through a joint venture in the United Kingdom with the Butterfly Company, to which Air Products licensed its processes and designs for the manufacture of cryogenic equipment for industrial gas production. The new venture was called Air Products (Great Britain) Ltd. (APL). Air Products later acquired a 100 percent interest in this business and expanded its operation into the supply of industrial gases as well as cryogenic equipment.

Markets in the Benelux countries, Germany, and South Africa were penetrated in the 1960s. During the 1970s Air Products's operations expanded to France (Prodair) and Brazil. Diversification via acquisitions occurred during the 1960s into chemicals and chemical process engineering and welding and medical equipment.

Air Products and the Merchant Market

The company's production strategies were further developed during the 1950s. Air Products had mastered the small-quantity cylinder market and the large-quantity tonnage market quite well, but until 1957 the middle range, known as "the merchant market," was left mainly to competitors. These had long delivered gas in cylinder trailers and tankers. Some regional gas distributors were now acquired and the marketing concept known as "piggybacking" was introduced, by which extra gas liquefaction capacity was added to on-site plants. The company now built plants for tonnage quantity customers, such as steelmakers, and delivered additional liquefied gases economically to "merchant" gas customers.

The next step was to build large stand-alone liquid oxygen plants for the merchant market with a production of about 75 tons per day. These cost-effective methods of production combined with a strengthening of the distribution network soon gave Air Products a profitable position against the competition even in the merchant market.

Air Products Unlimited

Air Products used APL as a base for penetrating merchant gas markets in the Benelux countries, France, Germany, and South Africa. The plant built in Belgium was especially important for the firm's start on the European continent. Sidmar, a large steel company, was planning to build a large basic oxygen operation. The plant needed 400 tons per day of oxygen, and Air Products won the contract in 1964. It also decided to open sales offices in Paris, Brussels, and Düsseldorf. In Germany the Rheinstahl Hüttenwerke planned to expand its basic oxygen capacity and needed 400 tons per day. Air Products built the plant and from 1964 they operated the plant and formed a subsidiary, Air Products GmbH.

In late 1962, the Iron and Steel Corporation of South Africa (ISCOR) asked for bids for a low-pressure oxygen generator of capacity 200 tons per day for its steel mill near Johannesburg. In 1969, ISCOR, Air Products, and Babcock & Wilcox joined to create Air Products South Africa (APSAP), with headquarters in Pretoria. This joint venture operated ISCOR's tonnage plants and sold industrial gases. In 1968, Air Products incorporated Air Products of Puerto Rico Inc. in the state of Delaware. The Asian market was more difficult to penetrate, although Air Products had acquired a minority equity position in Asiatic Oxygen Limited of India as early as 1963. The most attractive market was Japan, which had a growing steel industry. In 1970, Air Products started a wholly owned subsidiary, Air Products Pacific Inc., in Tokyo.

In the 1980s Air Products acquired minor positions in Korea, Malaysia, Hong Kong, China, Thailand, and Taiwan and started joint ventures in Mexico (Infra), Canada (Inter City Gas Co.), Italy (Sapiro), and Spain (Carburos Metalicos; Fig. 5.9.6). In the early 1990s Air Products started a joint venture in Czechoslovakia (Cryotech) and new companies in Poland, Singapore, and Indonesia. In Japan Air Products started joint ventures with Daido Hoxan and the semiconductor gas specialist Showa Denko. Air Products also acquired the manufacturing companies Separex (PSA production) and Permea (membranes).

Applications Development

Cryogenics and Advanced Product Development

The first commercial cryogenic process was for freezing food with liquid nitrogen. Today, nitrogen, once an unwanted by-product of air separation, is the leading industrial gas in annual sales volume mainly due to its use in cryogenic applications.

One of the first applications by Air Products was Cryo-Guard, a liquid nitrogen system for truck refrigeration, which during the early 1960s competed with the

Figure 5.9.6. Air Products acquired Carburos Metalicos in 1996. Courtesy of Air Products and Chemicals Inc., Allentown, Pennsylvania.

UC Linde Polarstream system, which was marketed from 1961. Other early cryogenic equipment were the Cryo-Quick system for freezing food (1965), Cryo-Trim, a process for deflashing of molded rubber parts (1967), and the Cryo-Grind, a process used for grinding virgin material and recovering rubber and plastic scrap.

At the Advanced Products Department, formed in 1959, laboratory research was started in advanced chemistry, physics, astronomy, and related fields. In the early 1960s Air Products developed infrared sensors for military applications as night-vision devices or missile-guidance systems used in the Sidewinder, Maverick, and Chaparral missiles. A new type of helium refrigerator was also developed that could be used in vacuum pumping.

Specialty Gases

In 1963 Air Products organized a Specialty Gases Department. Specialty gases and gas mixtures of ultrahigh purity or special composition were sold in cylinder quantities. Like the Advanced Products Department, the Specialty Gases Department grew out of Government-funded research, by combining Government-funded fluorine research with the company's small trade in xenon, krypton, and gas mixtures.

Going into Chemicals

In 1952, Leonard Pool led his company into the chemical industry market, with the contract of a 200-ton per day oxygen and nitrogen plant for the Spencer Chemical Company. When oxygen, hydrogen, and other industrial gases became important as building blocks for chemicals and polymers in the early 1960s, Air Products diversified into chemical development, from that time one of their main business areas.

In 1961 Air Products started a joint venture with the Tidewater Oil Company to produce oxyalcohols. They then acquired the Houndry Process Corp., which specialized in process catalysts for refining petroleum. The next large chemical acquisition occurred in 1969, when the Escambia Chemical Corp., active in industrial and agriculture chemicals such as synthetic ammonia and fertilizers, was added.

In 1953 Air Reduction started its Chemical and Plastics Division, building plants in Calvert City, Kentucky (Fig. 5.9.7), where they used coal and electricity for acetylene production with organic chemicals as the end product. Because large-scale ethylene production from natural gas made acetylene obsolete as a raw material for

Figure 5.9.7. The Calvert City petrochemical plant. From Butrica (1990) with permission.

organic chemicals in the 1960s, Air Products was able to purchase Airco's chemical plants at a low price in 1971. Air Products's chemical business, especially in petrochemical intermediates, has since reached a leading position in the world.

Short Biography[2]

Leonard Parker Pool (1906–1975) was employed in the early 1930s by the Burdett Oxygen Company to sell welding and cutting equipment and industrial gases (Fig. 5.9.8). Burdett was acquired by Compressed Industrial Gases (CIG). In 1936, Pool incorporated his own company, the Acetylene Gas and Supply Company, which bought and resold gases from the National Cylinder Gas Company, founded in 1933. Some years later the company was close to bankruptcy, but was attractive enough to be acquired by CIG in 1938. As part of the deal Pool became manager of railroad sales for CIG in Chicago.

He found that the chief cost of oxygen to the customer was the cost of shipping it in heavy containers. Pool's idea was to generate oxygen next to the customer's plant. This plan would only work if an inexpensive oxygen generator could be designed. Pool suggested that customers would lease the on-site plants and pay for

Figure 5.9.8. Leonard Parker Pool. Courtesy of Air Products and Chemicals Inc., Allentown, Pennsylvania.

oxygen they consumed, thus avoiding transportation costs. During World War II, however, the main business of the company was selling mobile oxygen generators to military customers.

After the war ended, Pool's company struggled for many years to convert to a peace-time economy. The facility was moved to Emmaus, Pennsylvania, in order to get closer to the major markets. In order to get working capital, Air Products went public in 1946, with Leonard and his wife, Dorothy, as the largest stockholders. Now Pool tried his idea of leasing generators to the civilian market. A big break came with a lease agreement with Ford Motor Co., one of the largest users of oxygen in the United States. Over a 5-year period after the war, lease revenues gradually increased and by 1950 they accounted for 38 percent of Air Products's net sales.

In the early 1960s Air Products added chemicals to their business. Now the diversified company had to change its organizational pattern from a family enterprise to a professionally managed corporation. The name was changed to Air Products and Chemicals and the main products and business changed to consumable products such as gases and chemicals instead of hardware. Pool's original strategy, "keep the cow, sell the milk," by leasing generators to the customer was now changed. Air Products owned and operated tonnage plants connected to main customers in order to sell huge quantities of gas. The customer had a "take-or-pay" contract and also paid a base charge regardless of consumption.

After the death of his wife in 1967 Pool left the presidency of the company, but remained as CEO and Chairman until 1973. Leonard's brother George had been in the company since 1943, and was Vice President. Among other things he was responsible for Air Products's successful entry into the cylinder gas market. He died in 1973 and Leonard Pool passed away in 1975.

Milestones

1940	Air Products is founded in Detroit, Michigan
1941	First sales of mobile oxygen generators to the U.S. Government
1942	Enters the helium and argon markets
1945	Moves from Chattanooga, Tennessee, to Emmaus in Leigh Valley, Pennsylvania
	The first lease customer in the commercial market
1946	Starts its own cylinder business in Indiana
	Company goes public
1957	Builds the world's first large-scale liquid hydrogen plants for the U.S. Air Force in Painsville, Ohio, and West Palm Beach, Florida
	Forms Air Products (Great Britain) Ltd. (APL), a joint venture with Butterfly Corp.
1961	Diversifies into chemical development by joint venture with the Tidewater Oil Co.
1962	Acquires Houdry Process Co.
1963	Organizes a Specialty Gases Department
1964	Forms 60 percent-owned industrial gas subsidiary in Belgium for Benelux market

1965	Introduces Cryo-Quick system for freezing food
1967	Introduces Cryo-Trim process for rubber deflashing
1969	Forms Air Products South Africa (APSAP), with headquarters in Pretoria, together with ISCOR and Babcock & Wilcox to operate ISCOR's tonnage plants and sell industrial gases
1971	Purchases Airco's chemical plants in Calvert City, Kentucky
1975	Death of Leonard Pool
1980s	Acquires minority positions in industrial gas companies in Korea, Malaysia, Hong Kong, China, Thailand, and Taiwan
	Forms joint ventures in Mexico (INFRA), Canada (Inter City Gas Co.), Italy (Sapiro), and Spain (Carburos Metalicos)
1990s	Starts joint ventures in Japan with Daido Hoxan and the semiconductor gas specialist Showa Denko and in Indonesia
	Establishes a new company in Singapore
	Starts a joint venture in Czechoslovakia (Cryotech) and a new company in Poland
	Acquires manufacturing companies Separex (for PSA) and Permea (membranes) for noncryogenic gas separation
1996	Completes acquisition of Carburos Metalicos

References

1. Butrica, A. J. (1990). *Out of Thin Air: A History of Air Products and Chemicals, Inc. 1940–1990*. New York: Praeger.
2. Lopez, S. (1996). Leonard Pool—Visionary, entrepreneur, philanthropist. In *Pioneers in the Cryogenics and Industrial Gases fields*. Allentown, PA: Air Products.

6

EXPANSION OF THE INDUSTRIAL GAS BUSINESS

6.1. INTRODUCTION: ORIGIN OF THE INDUSTRIAL GAS BUSINESS IN WESTERN EUROPE (FIG. 6.1.1)

Leak-prone metal vessels as well as leather or textile bags had been used for transporting hydrogen, oxygen, coal gas, carbon dioxide, and nitrous oxide from about 1850. In 1886 the first patents were issued to Howard Lane and Richard Taunton in England for manufacture of the first industrial production methods for high-pressure gas cylinders. In Germany in 1886 Max and Reinhard Mannesmann patented a new, revolutionary production method for manufacturing seamless steel tubes by extrusion, a method which was soon adapted to the manufacture of gas cylinders. These innovations marked a major step toward the birth of the industrial gases business. Gases could now be transported in a safe way at pressures as high as 100 bars. The first international safety regulation for the transport of high-pressure gas was established in 1890. The new cylinders were first used for the transportation of oxygen from Brin Oxygen in London, which pioneered commercial gas production by chemical processes. In the United States carbon dioxide transport in cylinders was pioneered by Jacob Baur, who established the Liquid Carbonic Company in 1888.

Although the transportation problem was solved, the market for commercial oxygen was not there. Limelight for theater and music halls represented by far the greatest demand for oxygen, and attempts to develop sales to other applications such as oxygenated water failed. Things began to change early in the 20th century. The rapidly increasing use of oxygen and acetylene in industry led to the establishment of companies all over the world for the production and distribution of compressed gases. Of the seven current leading industrial gas companies, six were founded before World War I. At the turn of the century Linde AG in Germany started to produce oxygen based on its expertise in cryogenics. After forming in 1903 a domestic joint marketing organization for oxygen with two existing sellers of chemically produced oxygen gas, manufacturing companies were also established outside Germany by Linde or with Linde's participation. For the international business Internationale Sauerstoff Gesellschaft AG (ISG) in Berlin was formed jointly with Societé d'Applications de Acétylène in Paris in 1906. ISG formed companies to manufacture gases in several other European countries and in the United States.

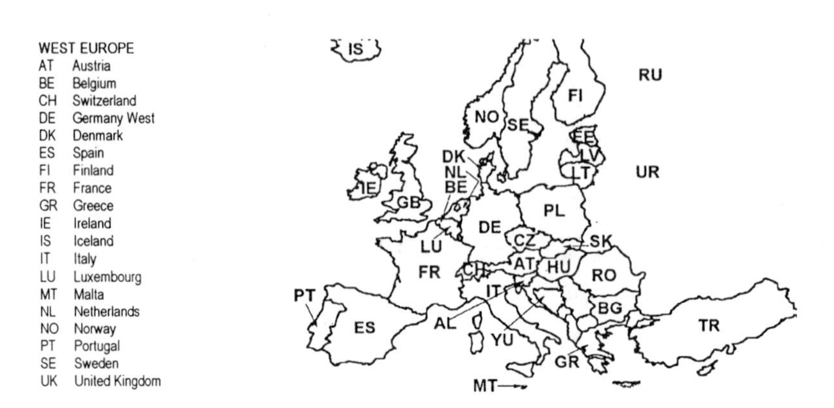

a= affiliated

WEST EUROPE	AT	BE	CH	DE	DK	ES	FI	FR	GR	IE	IS	IT	LU	NL	NO	PT	SE	UK	MT
Air Liquide	1963	1906	1957	1986	1972	1911	1993	1902	1920			1907	1931	1987		1923	1913		
BOC British Oxygen										1969								1886 1906	
Air Products		1964		1964		1982 (m)		1994		1984		1987		1967				1957	
Linde	1906 (a)	1905 (a) 1978	1906 (a)	1902		1906 (a)		1905 (a) 1973	1963			1909 (a)		1973		1987	1906 (a)		
AGA	1916	1928	1917	1906?	1915	1926	1919	1913	1951		1919 1990			1916	1911		1904	1913	
Messer	1972	1978	X	1896		1970	1997	1973	1998			1980		1973			1998	1913 (1918) 1987	
PRAXAIR Praxair		1979		1977		1955		1980				1984		1984				1985	
Liquid Carbonic						1965													
Early local industrial gas companies	Hansmann & Co 1905	Oxyhydrique 1896	Carbagas 1893	Hanseatische Acet.-Gasindustrie 1898	Dansk Ilt & Brint 1913	Carburos Metalicos 1899		Duffour et Igon 1899		Irish Oxygen 1948		Sapio 1923		Hoek Loos 1916	Norsk Surstoff & Vandstoff 1912		Nordiska Syrgas-verken 1907		Multigas 1926

Figure 6.1.1. Origins and early expansion of the industrial gas business in Western Europe.

L'Air Liquide in France, founded in 1902, pioneered the production of dissous gas (dissolved acetylene) as well as cryogenic oxygen production and took a very active part in the development of welding and cutting methods using these gases.

For the international market Air Liquide preferred selling gas from its own subsidiaries, whereas Linde chose to sell plants to customers together with a license to make and market gas for a defined geographic area. After some disputes Linde and Air Liquide made an international marketing agreement in 1908 according to which some markets in Europe and South America were divided. In the same year Air Liquide transferred its patent rights in Germany to Griesheim Elektron. Another plant manufacturer, Messer & Co., preferred to build and sell air gas plants and not sell the gases themselves on the international market. The strategy was to build air separation plants directly on the customer's premises.

British Oxygen (BOC), formed in 1906, the successor to Brin Oxygen in England, undertook international expansion to Australia in 1912 and to India and South Africa in the 1930s. By acquisitions BOC then nearly reached a monopoly position for industrial gases in the Commonwealth countries.

The simple method of producing acetylene in large quantities by using calcium carbide and water was discovered simultaneously in the United States by Thomas Willson and in France by Henri Moissan. The Gasaccumulator Company of Sweden (AGA), formed in 1904, was a pioneer in using dissolved acetylene for lighting. The AGA lighthouse system rapidly gained ground all over the world before World War I. Because acetylene could be produced locally from calcium carbide, it took over much of the petroleum lighting market during the war. The delivery of acetylene lights also led to the construction of dissous gas plants for customers. By 1928 AGA subsidiary companies had been formed in 24 European and Latin American countries for the manufacture and marketing of first acetylene and later air gases. In Latin America AGA subsidiaries were founded in Mexico and in all countries on the South American coast, which all used AGA lighthouse equipment.

6.2. EASTERN AND CENTRAL EUROPE (FIG. 6.2.1)

Background: Prehistory of Industrial Gases

Acetylene

In Eastern Europe the use of acetylene as a light source began early. The news of Willson's discoveries was brought to Europe by Armin Tenner in 1894 and acetylene manufacturing by hydrolysis of calcium carbide was started in many countries. The use of acetylene for town lighting began in 1897–1898 in France, Germany, and Austria-Hungary, but serious acetylene explosions in Paris and Berlin in 1896 delayed the introduction until the supposedly safer method of acetylene storage using "acetylene dissous" appeared some years later.

In Budapest and Vienna the Acetylen-Gas-Actien-Gesellschaft started in January 1897 to instal acetylene lights. An order from the Emperor for the illumination of a large park (at least 250 lamps) got the Acetylen Gas company off to a fast start. The first European towns illuminated by acetylene were Vezpren, Tata, and Tovaros (Totis) in Hungary starting from May 1897. In Austria (Vienna), Hungary (Budapest), Romania (Bucharest), Russia (St. Petersburg), and Poland (Warsaw) as well as in Slovenia and Croatia acetylene was much used in headlights and for railway lighting before 1910.

Industrial Gases

Soon after the introduction of the new autogenous welding and cutting method the first oxygen plants were built in the region, mainly by the Linde cartel Vereinigte

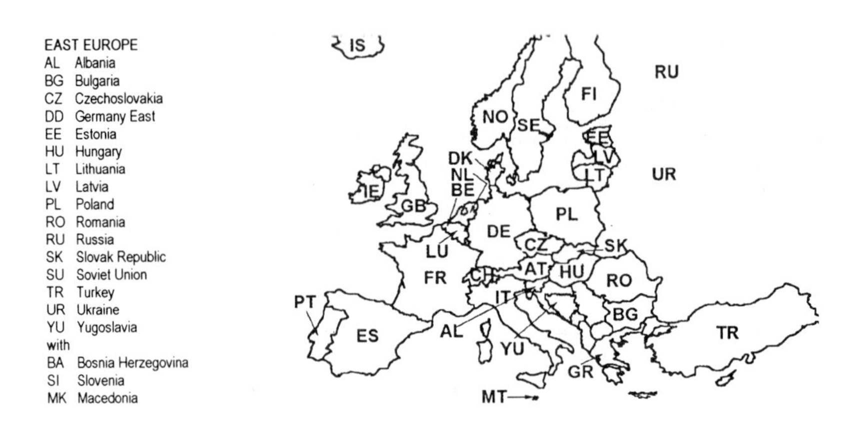

EAST EUROPE	AL	BG	CZ	DD	EE	HU	LT	LV	PL	RO	RU	SK	UR	YU
Air Liquide			1928			1929 1990			1910		1910			1929
BOC British Oxygen									1977 1991		1996			
Air Products			1992						1991			1998		
Linde			1991	1991		1992								
AGA			1927 1991	1991	1921 1993	1916 1990	1936 1993	1926 1993	1927 1991		1913 1993		1996	1927
Messer		1995	1991	1990	1998	1991		1999	1991			1991		1992
PRAXAIR Praxair														
Liquid Carbonic									1993					
Early local industrial gas companies			Hubočepy plant 1899			Franz Krükl & Co 1899			ZZGT-Polgas 1911		Volgograd 1934			Tovarna duskia Ruse 1915

Figure 6.2.1. Origins and early expansion of the industrial gas business in Eastern and Central Europe.

Sauerstoff Werke (VSW). At the turn of the century Hungary, Czechoslovakia, and Yugoslavia were part of the Austro-Hungarian Empire, and prior to its dissolution in 1918, Austrian companies could operate in the whole monarchy. With the loss of this privilege after the World War I, many new gas companies started in these countries during the 1920s. In Russia and the Baltic states all foreign companies were nationalized after the Russian Revolution in 1918. The same fate befell all private industrial gas companies in Eastern and Central Europe after World War II.

Development in Different Countries[1,2]

Hungary

Starting in 1899 the Austrian Franz Krükl (Fig. 6.2.2) pioneered the manufacture of industrial acetylene. He purchased the patent rights to produce dissolved acetylene from the French CFAD for the Austro-Hungarian Empire and founded Allgemeine Carbid- und Acetylen-Gesellschaft Franz Krükl & Co. (Fig 6.2.3). In 1904 he started the first dissous gas plant in Central Europe, at Kelenföld near Budapest. In 1908 the first Austrian dissous gas plant was started in Möllersdorf, near Vienna, and a network of filling stations was built up.

In Hungary the acetylene/dissous gas applications were developed by Firma Krükl's affiliate from 1906, which in 1908 became the representative for AGA. The Swedish AGA group then took over Magyar Dissous-Gáz and set up a subsidiary, Magyar Autogen Gázaccumulator RT, in 1916.

At the World Exhibition in Paris in 1900, Franz Krükl met another acetylene pioneer, the Swede Gustaf Dalén, who later laid the foundation for AGA. They kept contact and during the war in 1916 Dalén initiated a merger between Franz Krükl & Co. and the industrial gas manufacturer Sauerstoff- und Stickstoffindustrie Hausmann & Co., founded in 1905, to form Autogen-Gasaccumulator Krükl & Hausmann GmbH.

The German Linde took part in the foundation of the Österreichisch-Ungarische Sauerstoffwerke GesmbH in Vienna, which was started in 1906 by Medinger & Söhne of Gumpoldskirchen. In 1908 they set up a Linde air separation

Figure 6.2.2. Franz Krükl, industrial gas pioneer in Austria-Hungary.

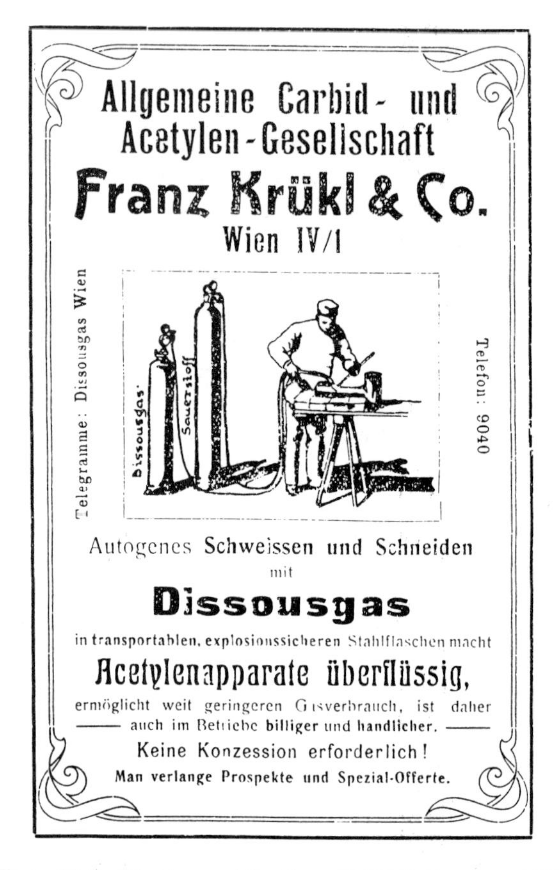

Figure 6.2.3. Advertisement for Franz Krükl & Co., about 1906.

in plant Pensing. This led to new companies in Budapest, Hungary; Budweis, Czechoslovakia; and Barcola, Trieste. In Hungary, Magyar Oxigéngyár RT (1909) and Dr. Wagner Szénsav és Oxigényár RT (1912) were formed. In 1931 Air Liquide built an oxygen and acetylene plant in Miskolcz in northern Hungary.

Czechoslovakia

In 1899 Mazin and Martinek industrially manufactured the first calcium carbide in Czechoslovakia. They used a water works at Lobkovice as a source of electrical energy for a calcium carbide factory. The carbide was first sold predominantly for use in carbide lamps and later for gas production in acetylene gas generators used for metal brazing and welding.

The need for dissolved acetylene for welding purposes was closely dependent on the development of oxygen manufacturing. Until 1910 compressed oxygen in pressure vessels was not produced in Czechoslovakia, but was imported from Austria and Germany.

Oxygen

The first oxygen produced in Czechoslovakia by rectification of liquefied air was manufactured in 1910 at the Hlubo£epy plant near Prague, where CO_2 production by chemical absorption had started in 1898. The Hlubo£epy plant management saw the rise in oxygen consumption for welding purposes abroad and started negotiations with the Austrian company Sauerstoff und Stickstoffindustrie Hausmann in Vienna to purchase a Linde apparatus with an output of 10 cubic meters per hour of oxygen. The production capacity was doubled in 1914.

In 1910 problems of producing oxygen pure enough for autogenous iron and steel welding and cutting were solved at the Tkoda works in Pilsen. Industrial production of oxygen and hydrogen by the electrolysis method was started, but when the demand increased, a Linde oxygen separation apparatus with an output of 45 cubic meters per hour of oxygen was also installed. There was also keen interest in oxygen in Slovakia before World War I. In 1910 the first oxygen plant in Slovakia, located in the engineering works at Piesok, was put into operation. Linde equipment for production of liquid air was also built at Dynamit Nobel, Bratislava, in 1913. Another oxygen manufacturer was Hydroxygen A.G., whose plant at Isti was the first oxygen factory in the northern Bohemia region. In 1912 a Linde air separation unit for 15 cubic meters per hour of oxygen was installed. Oxygen production and sales within Austria-Hungary around 1914 were dominated by the Vereinigte Sauerstoffwerke (VSW) manufacturing cartel in which the following three Czech companies participated: Hlubo£epy, Prague; Vitkovice coal mines, Ostrava; Egerer Sauerstoffwerke, Cheb. In 1916 the chemical company Hochstatter and Schickhardt in Brno built an oxygen plant in their factory. The plant was equipped with a Linde apparatus yielding about 30 cubic meters per hour of oxygen, which was filled into steel cylinders.

Dissolved Acetylene

The first company to manufacture dissolved acetylene in Austria-Hungary was Krükl and Co., Vienna, who collaborated closely with the Swedish company Gasaccumulator AB, Stockholm. In 1912 a business agreement was negotiated with the VSW cartel under which Czech oxygen manufacturers who were VSW members were supplied with dissolved acetylene from the factory at Möllersdorf near Vienna. Thus the first dissolved acetylene in steel cylinders used in Bohemia and Moravia was imported from Austria.

When World War I started in 1914 Karel Schulz, who manufactured welded pipes, purchased and installed a Messer oxygen plant in his factory at Komoňany near Prague. He also invested in an oxygen compressor for filling compressed oxygen into steel cylinders and a Griesheim acetylene plant. In 1920 this was replaced by a dissous gas plant. Oxygen manufacturers in the VSW cartel, observing Schulz's method of getting around their cartel rules, saw him as competition in the acetylene market.

At the end of World War I Krükl and Co. did not want to lose their contact with the Czech region and sought to maintain their market through links with Eisenbau Gesellschaft, Brünn (Brno). In 1918 Krükl opened distribution offices in

Figure 6.2.4. Autogen Gasaccumulator, Brno plant (Czech Republic), 1927.

Brno, Ostrava, and Prague in order to maintain its monopoly in dissolved acetylene supplies and further its markets in Czechoslovakia. In 1919 it built its first dissolved acetylene gas plant in Czechoslovakia, near Prague. When the new plant burnt down completely because of an explosion, Schulz's filling plant at Komoňany filled their cylinders with dissolved acetylene. After that Krükl and Co. cooperated with Schultz by expanding his gas plant.

Krükl and Co. launched further development in Czechoslovakia by building dissolved acetylene gas plants in Frodek (1920), Brno (1921), and Ceska Lipa (1922), from 1926 on operated by Krükl and Hausmann, Ltd., Brno that in 1927 became Autogen Gasacumulator a.s. Brno (Fig. 6.2.4).

Russia

The need for railway acetylene lighting initiated local calcium carbide and acetylene manufacture in the early 20th century. Trial production of dissolved acetylene began in St. Petersburg in 1902 for acetylene lighting on the St. Petersburg–Warsaw rail line.

Air gas production in Russia began with a privately operated Linde plant in St. Petersburg about 1910. Perun, a subsidiary of Air Liquide formed in 1910, started oxygen production in St. Petersburg in 1913 and acetylene production in Dnepropetrovsk, Ukraine, and Baku, Azerbaijan. Swedish AGA, which had a representative in Russia for selling lighthouse and railway equipment from 1908, started a subsidiary in St. Petersburg in 1913. Foreign-owned oxygen and acetylene in St. Petersburg and Moscow were nationalized after the Russian Revolution in 1918.

Baltic States

In the Baltic states, Estonia, Latvia, and Lithuania, industrial gas production started in the 1920s. Surrounding countries first exported gas cylinders to these countries. AGA, as the only major gas company, started subsidiaries to sell acetylene for lighthouses and oxygen for cutting. In Reval, Estonia AGA started acetylene and oxygen production in 1921–1922 after forming the subsidiary Eesti-AGA. Eesti-Aga was nationalized in 1940, merged with the oxygen producer Kemikaal, and renamed Autogen. AGA at first exported acetylene to the Latvian market from Sweden. Acetylene production started in 1922 in Riga, where an oxygen plant was also built in 1926. The subsidiary Lett-AGA was formed in 1926. In Lithuania AGA started oxygen production in Kaunas by acquisition in 1934. The AGA subsidiary Liet-AGA was formed in 1936 and an acetylene plant was built. A cartel agreement was signed with a local oxygen competitor, Technideg, which had been in the market since 1930. In 1938 Liet–AGA was forced by the Lithuanian government to change its name to Akcine Bendrove. As in Russia, all foreign properties were nationalized around 1940.

Poland, Danzig

In Poland Air Liquide began business in Warsaw linked to their Russian subsidiary Perun in 1910. The Acetylen- och Sauerstoffwerke started an industrial gas business in the harbor city of Danzig (presently Gdansk) in 1915. At first gas had probably been supplied from neighboring countries. Gdansk became a free state in 1919 after having been the capital of West Prussia. AGA offered to purchase Acetylen- och Sauerstoffwerke in 1923. When AGA took over in 1925 it built new acetylene and oxygen plants and also sold its products in Poland, where strict cartel rules existed in particular among Perun (Air Liquide), Linde, and AGA.

Gasaccumulator Spolka Akcyjna (SPAGA) was formed 1928 in Katowice. Its main competitor was Air Liquide's subsidiary Perun, which had started in Poland some years before AGA. A special niche for SPAGA was the provision of gas and equipment for railways. Besides Perun and AGA, also active in the Polish market in 1929 was Gaz Trzebina, with an acetylene plant and small oxygen plant in Trzebina near Katowice. It owned 49 percent of Slaski Gaz, with the rest owned by Schlesi. In 1929 Slaski built a dissous plant in Mata Dabrowka. Later both Gaz Trzebina and Slaski were taken over by Perun.

Romania

From 1908 AGA was represented by Societea Anon Romana Pentru Industria Oxigenului, Acitilinei si alte gaze in Bukarest.

Bulgaria

In the 1930s AL built oxygen and acetylene works in Sofia. AGA was at that time represented in Bulgaria by AGA-Ruse of Yugoslavia.

Yugoslavia

In 1912 Montkemija Tehnicki Plinori and Dalmacija Tehnicki Plinori was formed in the Split region for carbide production. In 1920 a Linde air separation plant was installed. Montkemija originated with the SODOAD company of Zagreb, which had been formed by French interests in 1930 for the production of oxygen and acetylene.

Tovarna dusika Ruse was started at Maribor (Marburg) in 1915 due to a government decision to build three nitrogen plants for making gunpowder. Österreichische Stickstoff-Werke AG of Vienna was founded to finance, build and operate the factory. A water power company was linked to the project and in 1918 the nitrogen plant was connected to a calcium carbide and lime nitrogen factory. In 1921 a filling station for surplus oxygen was built, and in 1928 the largest (100 tons per day) carbide furnace in Europe was installed. In 1929 AGA acquired 50 percent of the shares and the company was renamed AGA Ruse. AGA started selling industrial gases and welding and cutting equipment. The industrial gas part of the company expanded until it was nationalized in 1946. AGA Ruse started acetylene production in 1929 in Bosnisch-Brod and 1934 in Belgrad-Rakovica. In 1930 AGA-Ruse purchased an oxygen plant in Belgrad–Rakovica.

References

1. Unpublished articles and private communication Narcis Larger, AGA Gas, Vienna, Austria, and Josef Kaplickò and Petr Aunický, AGA GAS, Prague, Czech Republic.
2. Air Liquide. (1996). *PERUN 1910–1996*. Unpublished article

6.3. NORTH AMERICA (FIG. 6.3.1)

United States[1]

Linde Air Products

Late in 1907 in Buffalo, New York, Cecil Lightfoot began operating the first plant to produce oxygen from the atmosphere. His company the Linde Air Products Co., which had been formed to exploit the American rights to the liquid air process developed by Carl von Linde. Although industry had developed a pressing need for the oxyacetylene process and torches had been produced, there was a lack of low-cost oxygen. Much of the limited amount available was being produced by chemical processes or electrolytically from water at high cost.

In 1912 the Union Carbide Co. secured a major interest in the Linde Air Products Co., which turned over the manufacture and development of oxyacetylene apparatus to a newly formed organization, Oxweld Acetylene Co.

American Oxygen

In 1908 the American Oxygen Co. was producing oxygen from potassium chlorate and manganese dioxide. The Albany banker Robert C. Pruyn sought a cheaper production method for pure oxygen for oxyacetylene welding. He found a partner in Percy A. Rockefeller, nephew of John D. Rockefeller, and they hired a salesman with welding experience, Hermann van Fleet. Initial experiments were carried out at a small plant in Camden, New Jersey. Several chemical methods were tried, of which the barium oxide process proved the most economical for volume production. Plant

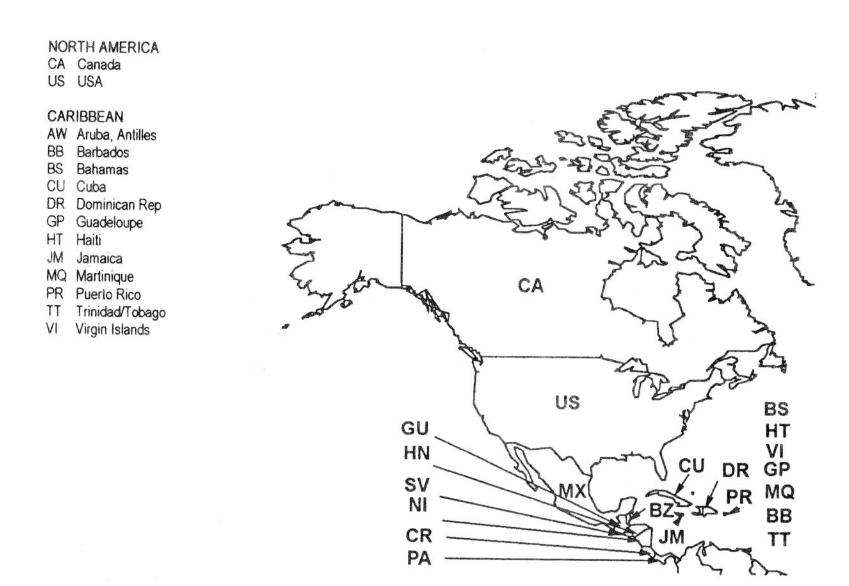

	NORTH		CARIBBEAN											
NORTH AMERICA	CA	US	BB	BS	CU	DR	GP	HT	JM	MQ	NA	PR	TT	VI
Air Liquide	1911	1915					1963					1980		
BOC British Oxygen	1951	1978												
Air Products	1976	1940										1968		
Linde		1907												
AGA		1911 1978												
Messer	1990	1912 (1918) 1974											1997	
Nippon Sanso		1983												
PRAXAIR Praxair	1918	1898 1917										1993		
Liquid Carbonic	1945	1888?						1974	1958				1952	
Early local industrial gas companies		Burdett 1923 Big Three 1923							Jamaica Ox. & Ac. 1952				Ind. Gases 1935	Island Gases 1958

Figure 6.3.1. Origins and early development of the industrial gas business in North America.

costs were too high, however, for large-scale use of oxygen. About 1911, upon Van Fleet's recommendation, Pruyn and Rockefeller discontinued operations at the Camden plant and began to experiment with a method for producing oxygen from air. In 1912, construction of a small oxygen unit was started. By 1914 the unit had been put in operation and was producing about 300 cubic feet (10 cubic meters) of oxygen per day.

In the meantime F. W. Clifford, a chemist with wide experience in oxygen production, began manufacture of oxygen by an electrolytic process in 1908 at Suspension Bridge, New York, as the United Oxygen & Chemical Co. By 1911 seven such plants were installed in the United States: two in Chicago, one in Cleveland, one in Pittsburgh, two in Elizabeth, New Jersey, and one in Jersey City, New Jersey.

Air Reduction

Air Liquide sent W. T. P. Hollingsworth to the United States to explore the possibility of exploiting the air liquefaction process developed by Georges Claude. Hollingsworth opened negotiations with Rockefeller and Pruyn of American Oxygen Co., which resulted in an exchange of patent rights and the formation of Air Reduction Co. Inc. in 1916. Air Liquide was a major stockholder together with U.S. financiers. With the acquisition of U.S. and Mexican patent rights to the Air Liquide process, Air Reduction began using the Claude process with equipment purchased from Air Liquide. Air Reduction Co. soon became a leading producer of oxygen and other atmospheric gases; it became Airco in 1968 and was taken over by the BOC Group in 1978.

Union Carbide

The formation of the Union Carbide and Carbon Corporation (Union Carbide Corporation) in 1917 brought together the Linde Air Products Company, the Prest-O-Lite Company, the National Carbon Company, the Union Carbide Company, and the Electro Metallurgical Company.

By 1916, 25 U.S. companies were producing 358 million cubic feet of oxygen per year (12 million cubic meters per year). Of these companies, only the three largest (Linde Air Products Company, Air Reduction Company, and Superior Oxygen Company) were producing oxygen by the liquefaction of air. They accounted for approximately 91 percent of total U.S. oxygen production. The remaining 9 percent was produced by the 22 companies who were producing oxygen and hydrogen by the electrolytic process.

Matheson

Matheson Products was founded in 1926 as the first commercial producer and supplier of specialty gases.

Prest-O-Lite

The history of Prest-O-Lite goes back to 1904 when P. C. Avery interested two Indiana businessmen, James Allison and Carl Fisher, in investing in his acetylene

Figure 6.3.2. Prest-O-Lite employees at the Indianapolis plant. Courtesy of Praxair, Danbury, Connecticut.

container development. Acetylene was most used for lighting purposes, but until then could only be piped from the generator to the point of consumption. Allison and Fisher immediately recognized the potential of the portable container. Their first thought was to make it possible for automobile, bicycle, and motorcycle riders to drive at night. They started the Concentrated Acetylene Company and set up a small factory in a bicycle shop. In 1906 Fisher and Allison organized the Prest-O-Lite Company and went on to a larger plant. The huge growth of the business led them to move again in 1908 and seek premises for future expansion. They bought farmland outside of Indianapolis, built a factory, and moved into it in 1912 (Fig. 6.3.2).

The Prest-O-Lite Company promoted the development of transportable containers, which were essential for the growth of welding and cutting applications of acetylene. Their work on acetylene also played an important role in the development of chemistry. They sponsored a fellowship at the Mellon Institute in order to develop a new source for producing acetylene because of their problems with the heavy carbide industry. The Mellon Institute found that acetylene is only one of the building blocks in the part of organic chemical industry then called the petrochemical industry.

By 1914 the electric automobile starter was introduced by Ford, and along with it the electric storage battery. This was clearly the beginning of the end of automotive acetylene lighting. Allison and Fisher decided to join the new lighting generation. They acquired the Pumpelly Battery Company and began production of the Prest-O-Lite storage battery.

Oxweld

Two competing companies, the J. B. Colt Co. of New York and the Acetylene Apparatus Co. of Chicago, were manufacturing acetylene generators and lighting fixtures for home and farm use. In 1911, the Acetylene Apparatus Co., backed by interests of the Peoples Gas Light and Coke Co., took over Colt. The following year a new Chicago company, the Oxweld Acetylene Co., formed by Union Carbide, took over the Acetylene Apparatus Co. and moved to Newark, New Jersey, where it opened a factory to produce acetylene generators.

To this point, Oxweld was not involved in the welding business. In 1914, the Oxweld group took control of Linde Air Products, which at that time manufactured oxyacetylene welding and cutting equipment. Soon after, the manufacture of Linde Air Products's welding equipment was transferred to the Oxweld factory in Newark. During this period, Union Carbide also purchased an interest in Prest-O-Lite.

Linde Air Products–Prest-O-Lite Cooperation

In 1917, the federal government asked Linde Air Products to undertake the manufacture of helium for World War I observation balloons. Helium is separated from natural gas by high-pressure distillation. The natural gas in wells in the Fort Worth, Texas, area contained 1 percent helium. Together with Prest-O-Lite, Linde built a plant which produced the required helium. Linde Air Products (now Praxair) has been in the helium business ever since, now operating a large helium facility at Bushton, Kansas. Even though Linde Air Products and Prest-O-Lite worked together through these early years, it was not until 1947 that Prest-O-Lite became part of the Linde Air Products family.

Canada[2]

In 1911 Air Liquide sent an agent, R. J. Levy (Fig. 6.3.3), to Canada to start a new subsidiary named Liquid Air. In Montreal he built Canada's first plant for the

Figure 6.3.3. R. J. Levy was sent to Canada in 1911 by Air Liquide. Courtesy of Air Liquide, Paris.

commercial production of oxygen by the Claude liquefaction process. Then he concentrated his efforts on introducing oxyacetylene welding to Canada's industries. He set up and operated a welding shop to which broken machinery could be brought to be welded by French specialists, who had accompanied him to Canada. During World War I new industrial production demands caused a great need for the new oxyacetylene welding and cutting processes.

Having thus laid a good foundation for the company's future in Canada, Levy, returning from a visit to France, tragically vanished in the sinking of the *Titanic* in April 1912. The young company continued on, and a second oxygen plant was built and put into operation in Toronto in 1913 to serve Ontario industries. In that year Liquid Air also moved west and opened an office and warehouse in Winnipeg, Manitoba.

Oxyacetylene welding and cutting had become accepted for maintenance and repairs throughout Canada's relatively small, but rapidly growing industries when World War I broke out in 1914. During the war, demands for Liquid Air's products skyrocketed and necessitated building of new gas production facilities. The first plant to manufacture compressed acetylene was built in 1915, and in quick succession during the war years additional plants to produce oxygen, nitrogen, and acetylene were built across the country in Winnipeg, Halifax, Sudbury, Vancouver, and London. From 1915 the British Admiralty needed helium to fill observation balloons, and to meet this demand a plant for the separation of helium from natural gas by the liquefaction process was built in Hamilton, Ontario. This marked the first use in the world of this gas for lighter-than-air aviation applications.

Between the two world wars, the company continually developed new uses for its industrial gases and processes. Liquid Air conducted welding schools for the free training of its customers' employees. In 1938 at Sydney, Liquid Air pioneered a new concept in gas distribution by supplying a steel mill with large volumes of oxygen by a direct pipeline from the oxygen plant. Because of the high consumption of oxygen in the Hamilton steel mills, Liquid Air in 1941 built a large-capacity oxygen plant directly connected to the major users with pipelines.

Figure 6.3.4. One of the first Air Liquide delivery trucks in Canada. Courtesy of Air Liquide, Paris.

Figure 6.3.5. Liquid Air's fleet of delivery trucks in Montreal 1937. Courtesy of Air Liquide, Paris.

A similar plant was built in North Vancouver for West Coast shipyards. Due to the increased industrial gas demand after the war, gas-producing plants were built and older plants in Montreal, Hamilton, and Vancouver were replaced by large-production liquid oxygen and liquid nitrogen units. In addition, hydrogen plants were built in Montreal and Edmonton, argon plants in Montreal and Hamilton, and medical gas plants in Montreal and Vancouver.

Liquid Air's long experience enabled the company to pioneer in the building of the first Oxytons. From the early 1950s Liquid Air became a world leader in this new field. Since these plants became available, the demand for tonnage oxygen plants of ever larger capacity has become greater and greater each year, especially in steel mills, whose refining processes are being completely revolutionized by tonnage oxygen.

The next major industrial gas company on the Canadian market was Union Carbide Linde, with its subsidiary the Dominion Oxygen Company, which started in 1918.

References

1. Baker, C. R., and Fisher, T. F. (1992). Industrial cryogenic engineering in the USA. In *History and Origins of Cryogenics* (R. G. Scurlock, ed.), pp. 217–254. New York Oxford University Press.
2. Canadian Liquid Air. (1961). *Liquid Air Review. 50th Anniversary Edition.*

6.4. SOUTH AND CENTRAL AMERICA (FIG. 6.4.1)

Background

In Brazil, Argentina, Chile, and Mexico oxygen companies were formed locally mainly for welding and cutting applications in the 1910s. For the first 80 or so years of the 20th century, however, South and Central America was poorly developed

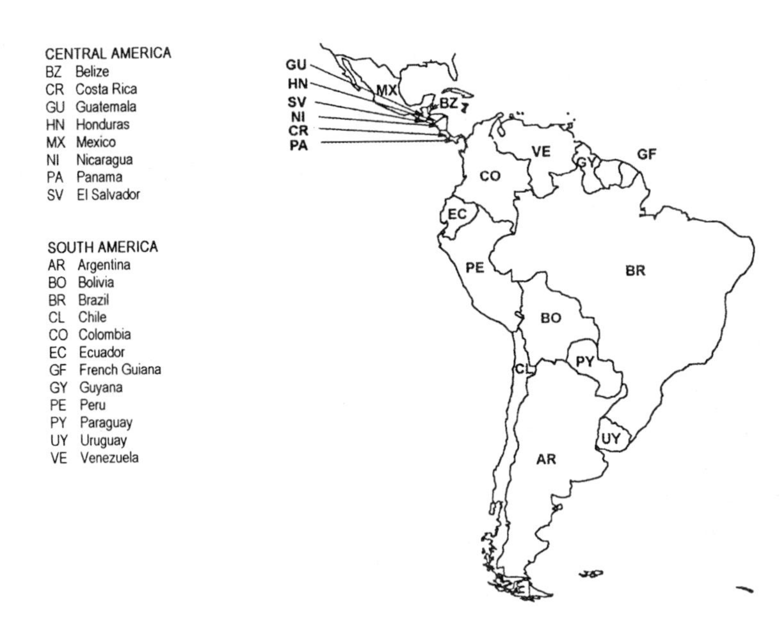

LATIN AMERICA	SOUTH AR	BO	BR	CL	CO	EC	GF	GY	PE	PY	UY	VE	CENTRAL BZ	CR	GU	HN	MX	NI	PA	SV
Air Liquide	1938	1967	1946				1990	1963		1939	1991									
BOC British Oxygen			19xx 1999		1995	1982						1982					1998			
Air Products			1973														1986 (m)			
Linde			1974																	
AGA	1920	199x	1915	1920	1931	1956			1951		1947	1948					1930			
Messer	1998		1995						1997						1997	1998	1991			
PRAXAIR Praxair			1944						1995								1960			
Liquid Carbonic	1962	1967	1944	1962	1965	1974			1958		1953	1944					1943			
Early local industrial gas companies	Oxigena 1914		White Martins 1912	Indura 1947 1913?										Fabrica de Ox. Miller 1930	Fabrigas 1925	Gases Industriales 1956	Infra 1919			Oxigeno y Gas 1959

Figure 6.4.1. Origins and early development of the industrial gas business in South and Central America.

because of a lack of basic infrastructure; what existed was controlled by state-owned monopolies and under heavily bureaucratic management. More recently Latin America has been transforming into a modernized, profitable region where foreign companies wish to invest.

The market for industrial gases has grown during recent decades and attracted most of the international players in the industrial gas business. The main part of the market is in Brazil, followed by Mexico and Argentina, whereas the other countries are still very small in terms of international comparison.

AGA along the Coast

Among the main international players, AGA started in 1906 and has had since the 1920s a well-established position in these countries. It sold acetylene-fueled lighthouse equipment in Latin America from 1906. From 1906 to 1910 its representatives concentrated on Argentina, Chile, and Uruguay. From 1910 extensive sales of beacons were made also in Brazil, where AGA opened its first subsidiary in the region in 1915. From that time acetylene gas plants were also sold and controlled by AGA in the countries around the Latin American coastline.

Subsidiaries were formed in Chile and Argentine in 1920, in Mexico in 1930, in Colombia in 1931, and in Uruguay, Venezuela, Peru, and Ecuador during the years after the World War II. Oxygen was also produced and sold from the 1920s.

Other Early Players

In Brazil, White Martins was founded in 1912 and acquired by Union Carbide in 1944. In Argentina, Oxigena was formed in 1914 and acquired by Air Liquide in 1938. In Valpariso, Chile, a small Linde oxygen plant was built by German interests at the start of World War I.

6.5. ASIA

Gas Markets in Asia

The Far East has attracted international players in the industrial gas market, whereas the slower development in Central Asia and the Middle East has left these regions waiting for more international investment (Figs. 6.5.1 and 6.5.2). The gas market in the Far East has been developing very fast, and some of the gas companies were established there as late as the 1980s. The yearly growth rate has been very high (over 10 percent) during recent years in many Far East countries.

In the Asian Far East region most of the local companies tend to be involved in joint ventures with the leading international gas companies. BOC, Air Liquide, Air Products, Iwatani, Linde, Messer, Nippon Sanso, and Praxair are all active in joint ventures in Asia. BOC and Air Liquide had an early start in Asia because Great Britain and France had colonies there and business relations already were established. Air Products and Messer were the last major companies to enter the local markets in Asia.

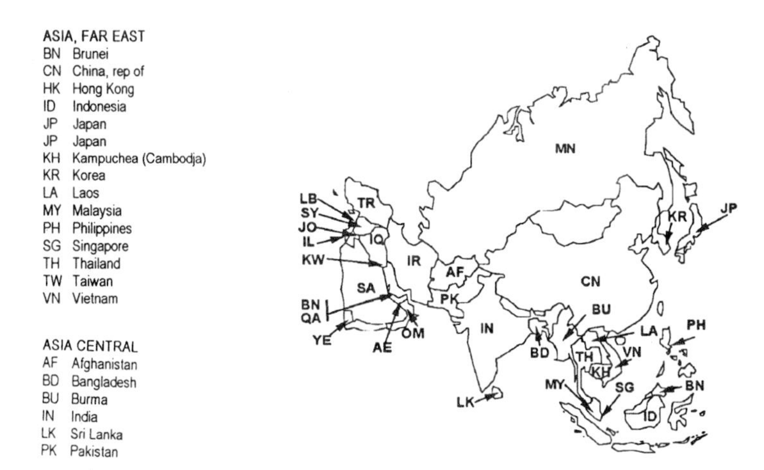

	ASIA, FAR EAST													CENTRAL ASIA			(m= minority owned)		
FAR EAST C. ASIA	BN	CN	HK	ID	JP	KH	KR	LA	MY	PH	SG	TH	TW	VN	AF	BD	BU	IN	LK	PK
Air Liquide		1929	1959	1998	1907		1979		1927		1927	1928	1987	1909				1991		
BOC British Oxygen	X	1987	1959	1971	1980		1989		1960	1979	1946	1970	1985			1973		1935		<1970
Air Products		1986 (jv)	1987 (m)	1990	1973 (jv)		1992		1988 (m)		1989	1986 (jv)	1988 (m)					1963 (m)		
Linde		1995																		
Messer		1995		1998?			1995		1998		1994	1997 (m)	1998	1998				1995	1999	
Nippon Sanso		1993			1910				1988		1982									
PRAXAIR Praxair		1988			1984		1975					1985						1995		
Liquid Carbonic					1971		1971					1972								
Early local industrial gas companies				Aneka 1971	Iwatani 1930 Osaka Sanso 1934		Korea Ind Gas 1973 Union Gas 1975		Malaysian Oxygen 1960	Ingasco 1951	SOEAO 1916	Thai Ind Gases 1970	Lien Hwa 1955					Asiatic Ox. & Ac. 1942		

Figure 6.5.1. Origins and early development of the industrial gas business in the Far East and Central Asia.

Japan and China

Japan[1–4]

Early in 1902, Hantaro Nagaoka (1865–1950) purchased an air liquefier from the Linde Co. of Germany and demonstrated air liquefaction at Tokyo University for the first time in Japan. He was much interested in low temperatures and wrote

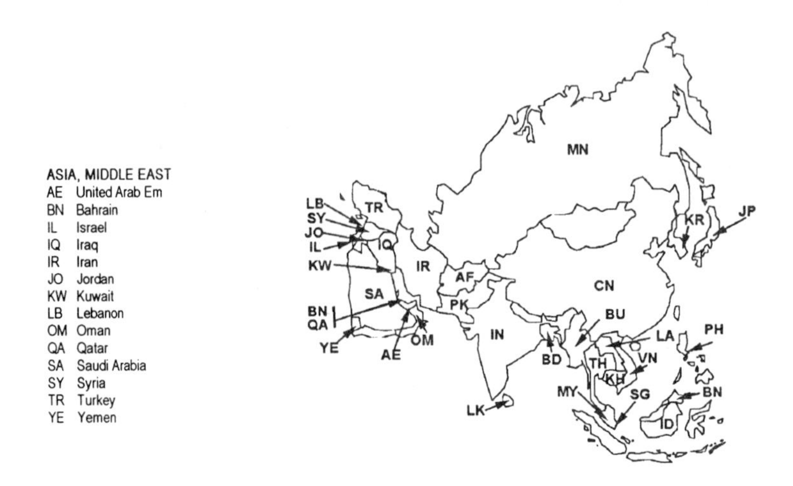

W. ASIA MIDDLE EAST	AE	BN	IL	IQ	IR	JO	KW	LB	OM	QA	S-A	SY	TR	YE
Air Liquide	1979				1933			1928			1928		1988	
BOC British Oxygen													1987	
Air Products	1990?													
Linde														
Messer													1998	
Nippon Sanso														
PRAXAIR Praxair			1996											
Liquid Carbonic													1985	
Early local industrial gas companies			Oxygen & Argon Works 1935				Kuwait Ox. & Ac. 1955				Saudia Ind. Gases 1954			

Figure 6.5.2. Origins and early development of the industrial gas business in the Middle East.

introductory review articles in Japanese scientific magazines not only on the physics of the processes, but also on applications.

Importation of oxygen gas from Germany commenced in 1906 for the welding and cutting of steel. Because the transportation of gas cylinders via the Siberian Railway was very costly, plans were made for domestic gas supply.

Teisan

Air Liquide started business in Japan in 1907. The Swiss merchant Charles Fawre-Brandt approached Air Liquide's subsidiary in Indochina and proposed

Figure 6.5.3. Early cylinder gas transport by Air Liquide in Japan. Courtesy of Air Liquide, Paris.

starting a gases business in Japan. Air Liquide rented him a plant producing 20 cubic meters per hour, which was set up at Osaka Steel Works, the predecessor of Hitachi Works. At the same time Air Liquide introduced the oxyacetylene process in Japan. In 1910 Air Liquide established the Société d'Oxygène et d'Acétylène du Japon (SOAJ), which marked Air Liquide's start in producing gas in Japan (Fig. 6.5.3). After the outbreak of World War I SOAJ changed its name to Teikoku Sanso Acetylene. In 1930 Teikoku Sanso (Oxygene Imperial) was established as a Japanese–French joint venture, in which Sumitomo Goshi Kaisha became part owner. Sumitomo obtained 15 percent of Teikoku Sanso's shares, with Air Liquide retaining the other 85 percent.

When in 1941 Japan declared war on the United States and England, it seized their oxygen plants in Japan and militarized them. A Japanese vice-admiral was appointed to manage Teikoku Sanso. After the war, Air Liquide's initial stake in the company was restored, but almost 10 years was required to rebuild Teikoku Sanso's industrial capacity.

Nippon Sanso

In 1910 the Japanese Finance Minister, M. E. Takahashi (later Prime Minister), saw the importance of having a domestic oxygen supply. He created and invested in what is now called Nippon Sanso Corporation. Nippon Sanso imported an oxygen plant from Hildebrand in Germany, which, without rights or permission, sold copies of Linde's plants. The first Nippon Sanso plant in Osaki, outside Tokyo, started producing in 1911. The plant was located in the countryside, where the air was thought to be better than in the city.

There was intense competition between Nippon Sanso and Teikoku Sanso. In the first decades Japan was geographically divided between the two: Nippon Sanso, with German systems, in the eastern part and Teikoku Sanso, with French systems,

Figure 6.5.4. Air Liquid's Nagoya plant. Courtesy of Air Liquide, Paris.

in the western part. First the main customers were the railroads, the army, and the navy. The government favored the Japanese company, and Air Liquide, as a foreign-owned company, had a very hard time.

Development of the Oxygen Industry

Between 1910 and the 1930s, other oxygen companies were established, which supplied oxygen gas produced by imported air separation plants. The main additional companies included Toyo Sanso KK (1918), Daido Sanso KK (1933), and Osaka Sanso Kogyo Ltd. (1934). From 1910 to 1935 Japanese oxygen companies imported 16 oxygen separation plants from Europe. Japanese fertilizer companies imported 19 nitrogen separation plants from Air Liquide and 14 from Linde up to 1935. While oxygen companies were selling gas produced by the imported air separation plants, they used their experience from the installation and operation of this equipment to develop their own techniques, and started manufacturing air separators and related plant components based on the designs of the overseas companies.

Between 1917 and 1930 Nippon Sanso made 17 oxygen separation units with single rectifying columns, mainly for its own use. Oxygen plants with double rectifying columns were used from 1932. The development of domestic air separation plants was also initiated in the 1930s by Kobe Steel Ltd., Mitsubishi Kakohki Kaisha Ltd., and others.

After conflict between Japan and China broke out in the 1930s, Japan became gradually isolated from the international community and the importation of raw materials and technologies became more difficult. The need for oxygen gas grew very rapidly with the development of Japanese industries. The Japanese development of gas technology was far behind the Western countries until the end of World War II. The oxygen plants of the 1930s were usually of small and medium size intended for mobile military use. Nippon Sanso delivered about 230 air separation units to the navy from 1934 to 1945, mostly for the famous secret weapon, the oxygen-fuelled torpedo. Damage to the oxygen industry during the World War II was comparatively slight and its productive capacity was rapidly recovered.

Oxygen for Steelmaking

After the war, tonnage oxygen was needed in the steel industry. A research team from the Institute of Physical Industries Chemical Research (IPCR) led by Professor Y. Ohyama started development of tonnage oxygen production in 1946. After 4 years work in collaboration with Nippon Sanso they had a working pilot plant for 100 Nm^3 per hour of O_2.

In 1953 a prototype plant of capacity 500 Nm^3 per hour of O_2 was delivered by IPCR to the steel industry. By 1967 Hitachi, Nippon Sanso, and Kobe Steel had manufactured more than 200 oxygen tonnage plants of this type.

In 1948 Teisan KK experimented at the Amagasaki ironworks with a method already introduced in France of using an open-hearth furnace into which oxygen gas was blown from cylinders. This method was adopted by Japanese steel companies, but was generally replaced by the Austrian LD oxygen process from 1956. In order to keep up with international levels of technology, the Japanese steel industry decided to import further technology; for example, Nippon Sanso contracted with Linde, Germany, and Process Engineering, United States, and Teisan contracted with Air Liquide, France, for the latest air separation technology.

***China*[5,6]**

Until recently the main industrial gas production in China came primarily from very small air separation units, which locally manufactured and produced just gaseous oxygen. Besides these small plants, there were bigger tonnage plants supplying gaseous oxygen mainly for steel works. It is still rather common in China for the plants to belong to the gas user. With the entrance of international players this picture is now changing, but the old infrastructure and poor road conditions are slowing development. Because acetylene often is produced in small generators at the user's site, there are only a few acetylene plants and acetylene has not been connected to other industrial gases.

Background

Prerevolutionary China had no manufacturers of air gases and gas separation plants. In 1933 there was a Chinese gas factory in Qingdao which could only produce oxygen from barium peroxide and in 1934 this factory imported a plant of capacity 15 Nm^3 per hour of oxygen from Japan. By 1947 there were 91 oxygen plants in the coastal cities and regions, among which the biggest (in Shanghai) was a Canadian made plant of capacity 92 Nm^3 per hour of oxygen.

The demand for oxygen, nitrogen, hydrogen, rare gases, and gaseous hydrocarbons increased with the development of industries such as metallurgy, chemical engineering, petroleum, machine manufacturing, and electronics. Domestic production companies manufacturing gas separation plants and liquefiers were set up. Since the first successful trial production of air separation plants in 1951, China has produced a large number of different types of air separation and liquefaction units. Before international players arrived, there were about 1000 air separation plants around the country, primarily producing gaseous products. Most of these were manufactured in China and the domestic industrial gases companies in China were state-owned. Different ministries were involved in different parts of the business, such as joint ventures and investments with foreign industrial gas companies, for medical gases, electronics, space fuels, and transportation.

Gas production facilities range from small plants, 50–150 Nm^3 per hour of oxygen, to large tonnage production units. Because there is little coordination between the local production and distribution enterprises, there has been a tendency for over-establishment of small companies engaged in the same activity within a given area. There are vast distances between major population and industrial centers with regard to industrial gases production. Railways provide the basis for land transportation of cylinder gas. Hundreds of small-capacity oxygen plants were installed throughout the country even during the 1990s because liquid transport by rail has not been implemented.

New Trends

The major new development is for international gas companies to build plants with significant liquid capacity. This has resulted in the gradual conversion from gaseous to liquid distribution of gas. However, liquid distribution beyond a 200-kilometer radius of a production facility is currently considered uneconomical. In the 1990s the international players or their joint ventures introduced pipeline gas supplies from central production facilities to multiple customers. This was first successfully done in the Beijing, Tianjin, Pudong (Shanghai), and Suzhou areas. The international companies are also introducing PSA and membrane technology for smaller on-site gas requirements. On-site supply schemes have been introduced for very high purity nitrogen products for the electronics and semiconductor industries.

Acetylene production and distribution is usually kept apart from the other industrial gases. The situation for acetylene is similar to other Far East industrial gas markets, where the multinational companies have left fuel gas activities to local enterprises.

Air Separation Technology in China

After the People's Republic of China was founded in 1949 the First Machinery Factory in Harbin started production of small-scale air separation units. These were based on the flow diagrams and technical specifications of a small Russian air separation unit with a capacity of 30 Nm^3 per hour of oxygen.

From 1955, the General Machine Manufactory in Hangzhou (in 1958 renamed Hangzhou Oxygen Plant Manufactory) began trial production of air separation plants for the government by imitating liquid plants using the Heylandt cycle and gaseous plants using the Claude cycle. In 1958, the Hangzhou Air Separation Plant Research Institute was established in Hangzhou and began to study and design equipment for tonnage gas production. From 1958, when steel production in China increased and large-scale gas separation and liquefaction techniques were needed, the manufacture of Chinese-designed air separation plants was done with the help of Soviet specialists. A medium-sized plant manufacturing industry started.

At about the same time noble gas components were being recovered from air in air separation plants. Chinese engineers improved two such plants from East Germany to yield ultrahigh-purity liquid and gaseous argon. Since 1961, recovery plants for helium from natural gas have been in operation.

During the Chinese cultural revolution in the late 1960s steel production played a key role in the national economy. The technical standard of gas separation units and liquefiers was raised partly by use of imports, and the domestic development of gas separation techniques was promoted. Since that time air separation technology in China has kept to a high standard.

The Entry of Foreign Industrial Gas Companies into China

Air Liquide entered Japan in 1907 and arrived in China very early; before World War II it operated in Hong Kong, Canton, Shanghai, and Tientsin. The Air Liquide subsidiary Hong Kong Oxygen was established in 1959 as a 50:50 joint venture with the BOC Group.

The BOC Group first came to China in late 1984, but has owned 50 percent of Hong Kong Oxygen since 1959. After exploring various places, it selected Shanghai as the entry point. The first joint venture company partially owned by BOC was Shanghai BOC Industrial Gases, established with Wusong Chemical Group in 1987.

Linde, through its Process Engineering and Contracting Division, has participated in the Chinese air separation plant market for many years. Within the last 15 years Linde has built more air separation plants in China than any other foreign air separation plant manufacturer and a significant number are tonnage plants. In 1997, Linde's industrial gases section took a major step forward by establishing the 100 percent Linde-owned company Linde Gas Xiamen Ltd., which started with local production and distribution of CO_2. Since then an air separation unit with argon recovery and a H_2 production unit were also built.

In 1965 Iwatani International first entered into China by selling equipment. In 1989 the first joint venture, Dalian Iwatani Gas Machinery, was established with the Dalian provincial government.

Nippon Sanso started in China during the 1960s with sales of four air separation plants to the Chinese government. In 1977, a plant with high capacity was sold to Anshan Iron and Steel. After 1977 several air separation plants with related equipment were exported. The Nippon Sanso Beijing Office opened in 1986.

Air Products first did business in China in 1977, when CNTIC in Anshan purchased a large cryogenic air separation plant, which became operational in 1979. Several other plants have since been sold. In 1986, Air Products formed a joint venture company, Chun Wang Industrial Gases (CWIG), in Shekou to provide industrial gas products and services to China's Guangdong Province.

Praxair entered into its first joint venture in China in 1992, when, together with Beijing Oxygen Company, it formed a 50:50 joint venture company named Beijing Praxair, Inc., located in Beijing, where it supplies gaseous oxygen and nitrogen to the Dong Fong Ethylene Oxide facility at the Beijing Eastern Chemical Works complex northeast of Beijing. Product is supplied through 25 kilometers of pipelines, the first long-distance industrial gases pipeline distribution network in China.

Since Messer Griesheim started its subsidiary Messer China in 1997, the company has been awarded several large on-site supply contracts for N_2 PSA.

Main Industrial Gas Companies in Other Far East Countries ca. 1990[7]

Hong Kong

The market in Hong Kong has been dominated by Hong Kong Oxygen & Acetylene Ltd., a joint venture from 1959 between Air Liquide and BOC. New Sino Gases was established in 1979 by a former production manager for Hong Kong Oxygen & Acetylene. The company is a wholesaler of gas and gas application equipment. Chung Wang Oxygen, situated in Shekou, started business in 1981. Since 1986 Air Products, China Merchant, and Wang Tak Engineering & Shipbuilding have been its owners.

South Korea

The South Korean gas market is dominated by three companies: Union Gas Co., Korea Industrial Gases with Taisei Sanso, and KBOC. Union Gas, founded in 1975, is owned 86 percent by Praxair. Korea Industrial Gases, formed 1980, is partly owned by Air Products. KBOC was established 1972.

Taiwan

The dominating companies in the market are Lien Hwa Industries, started in 1955, of which BOC has owned 50 percent since 1984; San Fu Chemical, of which Air Products acquired 45 percent of the shares in 1988; and Liquid Air Far East, which was established in 1987.

Thailand

The market consist, of many small gas companies and many distributors. Air Products Industries (not Air Products of the United States) was established in 1977. Thai Industrial Gases was formed in 1970; BOC has a 45 percent interest. Bangkok Industrial Gases, founded in 1987, is 49 percent owned by Air Products (U.S.). Praxair Thailand is 74 percent owned by Praxair. Also in this market is Air Liquide Thailand.

Singapore

SOAEO (Far East Oxygen and Acetylene) was formed by Air Liquide in 1916. The current dominating company is SOXAL, established in 1961, and owned by BOC and Air Liquide. National Oxygen was formed in 1982; Nippon Sanso holds 53 percent of the shares. AIG is partly owned by Lien Hwa Industries of Taiwan (joint venture with BOC).

Malaysia

Three companies dominate the market: Malaysian Oxygen, established in 1960, owned 55 percent by BOC and Air Liquide; Industrial Oxygen Incorporated, established 1969, from 1988 a subsidiary of Nippon Sanso; and Super Oxygen, started in 1974, a subsidiary of Sitt Tatt and partly owned by Air Products.

Indonesia

Two large (Aneka and IGI) and many small gas firms exist in Indonesia. Aneka Gas Industries is state-owned. Industrial Gas Indonesia (IGI), formed in 1971, is 51 percent owned by BOC. Other firms include United Air Products (Air Products, 55 percent; Astra Group), Praxair Indonesia (Aneka–Iwatani–IGI), and Air Liquide Indonesia (Air Liquide, 51 percent; Indonesian government, 49 percent).

Philippines

Firms include Consolidated Industrial Gases Inc. (CIGI), established in 1979 and partly owned by BOC; Ingasco Inc., started in 1951 and partly owned by Nippon Sanso; and Superior & Gas Equipment.

Central Asia

India

The largest company in India is Indian Oxygen Ltd., followed by Industrial Oxygen and a group of companies owned by B. K. Jalan. The rest of the market (over 50 percent of the market share) is mainly controlled by many small individual companies. Indian Oxygen was established in 1935, with BOC now having 40 percent of the shares (Fig. 6.5.5). Since the late 1987s Industrial Oxygen has had a technical agreement with Air Products. Asiatic Oxygen is the largest company in the Jalan gas group.

Pakistan

Pakistan Oxygen Ltd., established before 1970, is a subsidiary of BOC (60 percent owned).

Middle East[8]

The industrial gas business in the Arabian Peninsular has mainly grown on the basis of the development of the oil and petrochemical industry. Several companies

Figure 6.5.5. BOC delivery of cylinders in India. Courtesy of BOC Group, Windlesham, England.

established their industrial gas business in the 1950s supplying oxygen and acetylene to oil companies. Some of the major international industrial gas companies are now present as plant suppliers, but only BOC (50 percent in BOS, Turkey) and Air Liquide (99 percent in Aligaz, Turkey) have major ownership in the region. The largest market for merchant oxygen and nitrogen in the region is in Turkey and Saudi Arabia.

Gulf Region (Saudi Arabia, UAE, Kuwait, Bahrain, Quatar, Oman)

In Quatar the National Industrial Gas Plant (NIGP) built the first air separation plant in 1954. Oman Industrial Gases first imported gas cylinders from the United Arab Emirates (UAE). In 1975 they started the first domestic gas production in Oman. In Saudi Arabia, Saudi Industrial Gases (SIGAS) is the oldest industrial gas firm in the kingdom. It was founded by Abdullah Hashim. In Kuwait, Kuwait Oxygen (KOAC) was also started by Abdullah Hashim in 1955. In 1961 the Refrigeration and Oxygen Co. (ROC) started business in Kuwait.

In UAE, Emirates Industrial Gases is the largest producer, and in Yemen the Yemen Oxygen Factory is the only industrial gas company. Messer Primeco in the United Arab Emirates. Started nitrogen service for enhanced oil recovery in Dubai.

Eastern Mediterranean

In Israel the Oxygen & Argon Works Ltd. was formed in 1935 with an oxygen manufacturing capacity of 17 cubic meters per hour. This company is still a major industrial gas company, with a present capacity of 130 tons per day of air gases. They also deliver liquid helium for research and development, industrial, and medical applications.

References

1. Oshima, K., and Aiyama, Y. (1992). The development of cryogenics in Japan. In *History and Origins of Cryogenics* (R.G. Scurlock, ed.), (Chapter 13) pp. 520–546. New York: Oxford University Press.

2. Anonymous. (1990). A slice of postwar history. 35 years of industrial gas history in Japan. *Gas Review Nippon* **1990**(Winter):7–8.
3. Komatsubara, M. (1987). How we succeeded in Japan: XIV Teisan, K. K. *Keidanren Review* **105**:554–556.
4. Anonymous. (1990). Running through the Japanese market as a market leader. History of Teisan celebrating its 90th anniversary. *Gas Review Nippon* **1997**(Summer):16–17.
5. Hong-zhang, Q. (1992). The cryogenic air separation and liquefaction industry in China. In *History and Origins of Cryogenics* (R. G. Scurlock, ed.), Chapter 15, pp. 559–575. New York: Oxford University Press.
6. Anonymous. (1993). China Fever. Gas majors entering the market. *Gas Review Nippon* **1993** (Winter):4–10.
7. Careless, H. P. (1996). The industrial gases market in Asia. *IOMA Broadcaster* **1996**(January–February):15–19.
8. Cryo Gas Consulting Ltd. (1996). *An overview of the industrial gases business in the Middle East.*

6.6. AUSTRALIA[1,2]

The Start of the Industrial Gases Business in Australia (Fig. 6.6.1)

F. S. Grimwade and Alfred Felton started a business as wholesale druggists in Melbourne in 1867, Felton, Grimwade and Company, from which a major fertilizer company was later developed. Russell Grimwade, one of F. S. Grimwade's four sons, graduated with a degree in science from the University of Melbourne in 1901. He visited England in 1902 and learned about Dewar's experiments with the storage of liquids at low temperature in vacuum flasks and saw Hampson's apparatus for liquefying air. In 1903 he joined his father's company and began a career which contributed to the development of welding and industrial gas production in Australia and later to the establishment of the Commonwealth Industrial Gases Ltd.

Young Grimwade had also some companions who contributed to the early development of the Australian industrial gas business. Among them were Philip Schemnitz, a German-born amateur wrestler; J. D. Wearn, the Swedish Consul in Melbourne; and Corrie Gardner, who was Australia's entire team at the 1904 Olympic Games in St. Louis.

Oxygen

Felton, Grimwade and Company were producing small quantities of oxygen for some years before the turn of the century by heating potassium chlorate and manganese oxide. Doctors used oxygen for resuscitation and theater owners needed it for their limelight projection, but the market for oxygen in Australia was meager. In 1909, Grimwade demonstrated to an Institute of Engineers meeting in Melbourne a new and cheap way of producing oxygen. He used a Hampson laboratory unit made by British Oxygen to distil about 1 quart of liquid air and he showed how Dewar flasks conserved the liquid. Later in 1909 Russell Grimwade and two of his brothers ordered a German liquid air plant capable of producing 175 cubic feet of oxygen per hour (5 cubic meters per hour). The plant was a Volant made by Maschinenfabrik Sürth,

AUSTRALIA PACIFIC	AU	FJ	NC	NZ	PF	PG	SB	VA	WS
Air Liquide	1956		1973	1983		1969		1977	
BOC British Oxygen	(1912) 1935	X		X		X	X		1974
Linde	1975								
Early local industrial gas companies	Australian Oxygen Company 1910								

Figure 6.6.1. Origin and early development of the industrial gas business in Australia and the Pacific.

imported by Krupp's agent in Australia. Krupp recommended the Hungarian Phillip Schemniz to install the plant. In 1910 the Australian Oxygen Company was established in Melbourne and after a steady growth moved in 1914 to a new site. By 1917 the company was also producing hydrogen and carbon dioxide.

In 1910, William Fyvie and Alexander Stewart started an agency for the British Oxygen Company (BOC) in Sydney and in the following year BOC decided to build an oxygen plant, using a Linde plant of capacity 750 cubic feet per hour (20 cubic meters per hour). In early years the name of the company was the Commonwealth Oxygen Company, incorporated in July 1912 (Fig. 6.6.3). Lever Bros., based in England, was operating electrolytic hydrogen plants in various parts of the world, and in 1915 BOC arranged with it to buy the oxygen by-product from its plants. In Sydney, the surplus oxygen from the Lever Bros. plant at Balmain exceeded Commonwealth Oxygen's total sales of oxygen at that time and from 1916 Lever Bros. and BOC became joint shareholders in Commonwealth Oxygen. In 1920, a plant of capacity 50 cubic meters per hour of oxygen was installed at Balmain.

Figure 6.6.2. The Acetylene Co. in Brisbane, Queensland, started in ca. 1899 to deliver calcium carbide, acetylene, and acetylene equipment.

Acetylene

Dissolved acetylene became available in Australia soon after 1904, when Melbourne athlete Corrie Gardner visited Stockholm after the St. Louis Olympic Games and returned to Melbourne with an agency for AGA dissolved acetylene. In Stockholm, he had arranged for AGA cylinders of acetylene to be exported to Australia and returned to Sweden for recharging. AGA's speciality was later lighthouse equipment, and Gardner, Waern & Company, owned by Corrie's father and the Swedish Consul in Melbourne, J. D. Waern, entered the market in Australia. In 1908, Gardner, Wearn ordered an AGA plant for manufacturing acetylene in Australia and they were the sole supplier of high-purity dissolved acetylene until 1919.

Figure 6.6.3. Commonwealth Oxygen Company plant in 1912. Capacity 70 oxygen cylinders per day.

Figure 6.6.4. Western Oxygen Ltd., Subiaco, Perth, 1929.

After the end of World War I two Australians, Tom Millner and Harold Morgan, were offered an agency by Allen-Liversidge Ltd., which had a close relationship with BOC, to manufacture dissolved acetylene and sell it in Allen-Liversidge cylinders, which contained a patented kapok mass. An acetylene plant was shipped to Australia and built at Annandale in 1919. In 1920 Allen-Liversidge (Australia) began manufacturing and marketing dissolved acetylene in cylinders. The Allen-Liversidge cylinders were much easier to handle than the heavy AGA cylinders, which could not be stood upright because of their convex bottoms. In 1920 Harold Morgan and Grimwade discussed the establishment of an Allen-Liversidge factory in Melbourne. Australian Oxygen bought shares in the company and obtained a contract to market its acetylene.

Calcium carbide, from which acetylene gas is produced, was not manufactured in Australia until 1917. Because the cost of overseas carbide was very high during the war, Hydro Electric Power and Metallurgical Company Ltd. in Tasmania began to produce carbide in 1916–1917 with electrodes, furnace, and a works manager obtained

Figure 6.6.5. COMOX medical gas delivery car, 1930s.

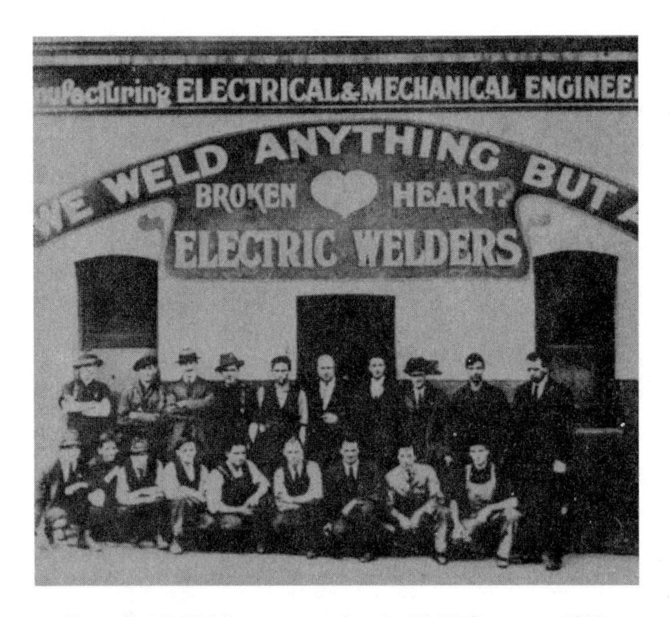

Figure 6.6.6. Welding personnel at the EMF Company, 1922.

from Sweden. It took, however, 2–3 years to achieve a satisfactory product and even then it was not as good as the imported material.

Welding

After leaving Grimwade in 1911, Schemnitz went to Germany to learn oxy-acetylene welding and cutting. He returned to Melbourne after 9 months to sell

Figure 6.6.7. Oxyacetylene welding on a leaking pipeline.

Messer acetylene generators and torches. In 1914 he joined the Newport workshops of the Victorian Railways, where he demonstrated the efficiency of the new welding and cutting methods. In Sydney before World War I oxyacetylene welding received a good start with the arrival of the brothers George and Alfred Kennedy, who formed Autogenous Welders of Australia Ltd., with boiler repairs as a specialty.

Not until after World War II did the other main industrial gas companies enter the Australian market: Air Liquide in 1957 and Linde in 1975.

References

1. BOC (1994). *1935–1995. Serving Australia for 60 Years (CIG)*. BOC Gases, Australia.
2. CIG. (1974). *The History of Welding in Australia*. CIG.

6.7. AFRICA

Northern and Central Africa[1]

In spite of its size and great natural resources, Africa has not developed as rapidly as other regions of the world (Fig. 6.7.1). The industrial gas business was started by the colonial powers France and England, which were interested in taking advantage of the possible development of the continent's vast natural resources. The gas companies Air Liquide and British Oxygen started activities in Africa in the 1910s. Because the possessions of the two countries were in different parts of Africa, Air Liquide is present in mainly the west and north and BOC in the east and south. The industrial gas market is still small (in 1997 about 575 million dollars), of which South Africa represents about 70 percent.

Many of the early subsidiaries have been nationalized: In 1956 the Egyptian government took over the properties of Air Liquide and British Oxygen. The situation became so difficult that it was impossible to keep friendly business relationships. As of 1999 the state-owned Industrial Gases Company (IGC) was the market leader with about 45 percent market share. From 1977 the government again allowed private domestic gas companies in the Egyptian merchant market. In the 1990s a number of family companies started operation. About 90 percent of Egyptian air separation plant capacity is presently captive (owned by its only user). BOC also had to surrender its properties in Zambia, Zimbabwe, Tanzania, and Uganda.

Air Liquide and Air Products are part owners of Sonatrach, which operates large liquid helium plants in Algeria.

South Africa[2]

In South Africa four of the main industrial gas companies are present, namely BOC, Air Products, Messer, and Air Liquide.

BOC's subsidiary Afrox was formed in 1935 when BOC took over a company owned by their partner in Australia, Allen-Liversidge. Afrox acquired Zamox (Zambian Oxygen) and can therefore reach into Zimbabwe and Malawi. In 1998 Air

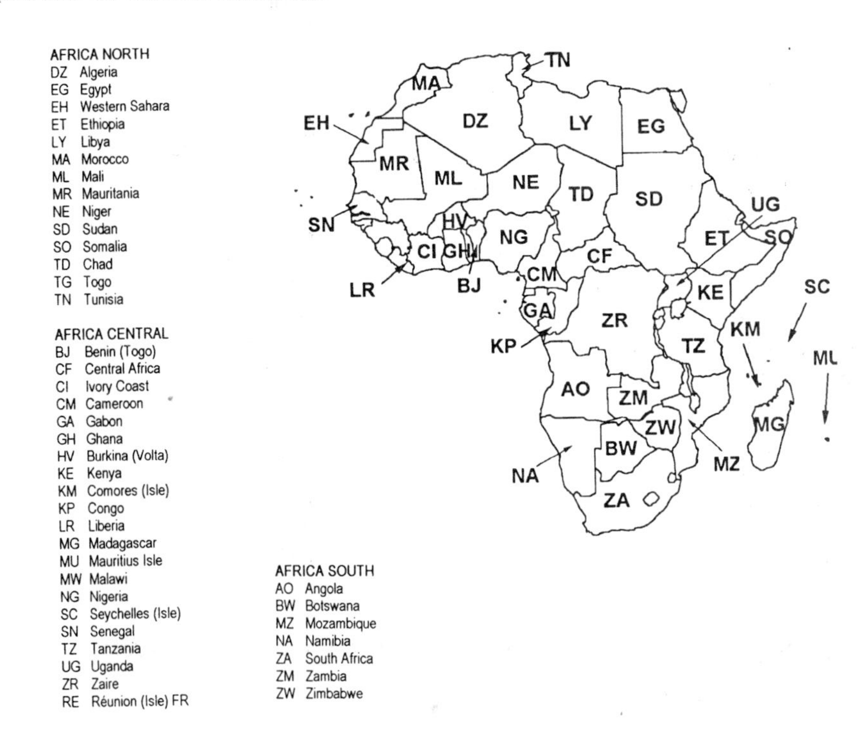

AFRICA	TD	DZ	EG	ET	LY	MA	ML	MR	NE	SD	SO	TN	EH	BJ	CF	CI	CM	GA	GH	HV
	NORTH													CENTRAL						
Air Liquide		1913	1920		1930	1919	1956	1966	1971	1949		1917		1954	1965	1945	1951	1958	1964	1927
BOC British Oxygen		X																		
Messer		1998 (jv)	1996									1998 (jv)								
Early local industrial gas companies																				

	KE	KM	KP	LR	MG	MU	MW	NG	SC	SN	TZ	UG	ZR	AO	BW	MZ	NA	ZA	ZM	ZW
														SOUTH						
Air Liquide		1983	1951		1948	1980		X	1980	1929					1977			1948		
BOC British Oxygen						X	X	1978?			X	X					X	1931	1988	X
Messer																		1973 (m)		1998
Early local industrial gas companies																		Alan Liversidge 1927		

Figure 6.7.1. Origins and early development of the industrial gas business in Africa.

Products received a 15-year contract supplying the world's largest coal gasification plant, Sasol, with 3000 tons per day of oxygen. A new on-site plant produces 2100 tons per day of oxygen, 1440 tons per day of nitrogen, and 50 tons per day of argon. The plant is owned by Sasol, but operated and managed by Air Products, which also delivers CO_2 and hydrogen in South Africa.

Air Liquide in South Africa has a contract for 2450 tons per day of oxygen to steel works, but otherwise its gas business is still quite small.

Messer started a new operation in Zimbabwe and operates an air separation plant in Durban for the merchant market. They also have a joint venture with the National Iron and Steel Company in Alexandria, Egypt. Messer is considering new business in Tanzania, Algeria, and Egypt.

References

1. Anonymous. (1993). News from Sub-Saharan Africa. *Alizé* **1993**(July).
2. Raquet, J. (1996). The industrial gases business in South Africa—An opportunity or need for rationalization? *CryoGas International* **1996**(Aug/Sept):25–27.

6.8. THE DEVELOPMENT OF THE MODERN INDUSTRIAL GAS BUSINESS

The Industrial Gas Market: Background[1]

The Demand for Industrial Gases

Air gas separation technology was developed in 1902 and became efficient from about 1910. Because of the rapidly increasing use of oxygen and acetylene by industry during the first decade of the 20th century, companies to produce and distribute compressed air gases and acetylene. Of the seven now leading industrial gas companies, all but one were founded before World War I. The decade after 1910 saw an increasing demand for the other main components of air. Nitrogen was used in ammonia production according to Haber's principle from 1913. Other chemical industries needed a new source of nitrogen because the importation of Chilean potassium nitrate was stopped during World War I. The newly discovered rare gases argon (Ramsay and Rayleigh in 1895) and neon, krypton, and xenon (Ramsay and Travers in 1894–1998) found technical uses for, among other things, filling electric light bulbs from about 1915.

Helium, which had previously been produced by heating minerals, became interesting as a balloon gas for military airships during World War I. Three experimental plants for production of helium from liquefied natural gas were started in 1917 in Clay County, Texas.

The growth in size of a typical oxygen production plant can be seen as follows:

Year	1915	1925	1950	1960	1970	1980	1989	1997
Production (tons/day)	10	35	200	500	1350	2250	2500	3000

The industrial use of air gases, hydrogen, helium, and natural gas increased sharply during the years after World War II. These gases are largely handled in the liquefied state because of advantages of volume and weight. The availability of liquefied gases at cryotemperatures (<120 K) increased greatly and led to a continuous development of applications and industrial handling techniques. This section summarizes the most important stages of development during the last four decades of the 20th century, when the demand in this industry really took off.

Business Overview since the 1960s

The World of Industrial Gases

The world-wide industrial gas business is oligopolistic. At the turn of 2000 the seven top players constituted at least 90 percent of the estimated world market. Within this oligopoly, some of the companies dominate a given geographic region. Joint ventures among the major participants are not unusual and tend to encourage cooperation in place of competition. Although production of industrial gases began in the early 1900s, the industry has thus far avoided stagnation by continual creation of new uses for its products.

Development of Production and Application (Fig. 6.8.1)

In the 1960s, 10 major players handled the industrial gas business in about 10 Western countries. As the steel industry converted from open-hearth furnaces to basic oxygen furnaces through that decade, the growth product was oxygen and the volumes involved were multiplied by a factor of 10. The industrial gas companies built very large oxygen plants and pipeline systems from the mid-1960s to supply the steel industry. In the 1970s the liquid, bulk market for nitrogen developed for food industry applications and in heat treatment of metals, and oxygen began to be used for environmental purposes. Another new, large gas application was the development of argon processes for the steel industry. Large-size liquid plants were built around the world, a number of them piggybacking on very large oxygen tonnage plants. In the United States very large liquid hydrogen plants were built because of the space program, and could also supply industrial customers.

The oil crisis in 1973 offered a new incentive for energy saving, although in the 1970s there was no dramatic change in production technologies.

In the 1980s the semiconductor industry became a rapidly expanding market for industrial gases, first in the United States and Asia, then much more slowly in Europe. The production of microcircuits requires high-purity bulk gases delivered to the points of use and new exotic specialty gases. At the same time the industry at large sought improved efficiency, purity, and quality due to competition and sometimes forced by regulatory authorities.

A strong growth in chemical demand for bulk industrial gases also arose in the 1980s. Environmental-related applications became a very significant market when

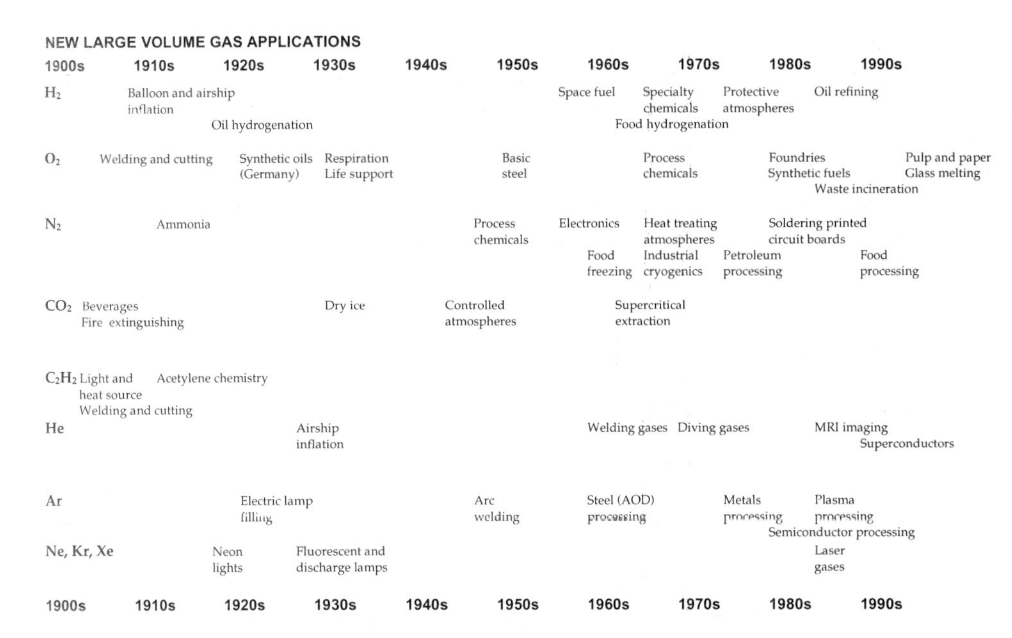

Figure 6.8.1. The development of new large-volume gas applications.

customers were forced by law to clean up effluents that were creating problems. Not only were gas products used for environmental purposes to clean up polluting effluents, but there was also an increased use of gases for cleaner production in among others the pulp and paper industry and ceramics and metal manufacturing. The very rapid increase in steel demand slowed down during the 1980s, but at the same time the search for higher steel quality led to considerable growth in argon and nitrogen demand.

As a result of the oil crisis in the 1970s, when the price of oil shot through the roof, there were great hopes in the early 1980s for applications of industrial gases in oil recovery and in coal gasification, and many huge plants were built, mainly in parts of the world lacking natural gas. This was not a very successful investment because oil prices started coming down in the middle of the 1980s.

Production

North America in the 1980s saw a major drive toward building very large liquid product plants with the hope of acquiring a very large market share. This resulted in significant overcapacity, which took years to resolve and burdened the industrial gas industry for a number of years.

Since the beginning of the 20th century air gas production has relied on cryogenics. In the early 1980s in Japan oxygen adsorption began to be a competing method. This production method was mostly driven by equipment manufacturers in the electric industry. Some years later PSA nitrogen separation, membrane technologies, and oxygen VSA were developed by chemical companies. Union Carbide, which had a chemical business and an industrial gas business, took a leading position in PSA development. This was a significant turning point and the major gas companies had to choose a partner in the chemical industry. BOC lined up with Dow, Air Liquide made a joint venture with Du Pont and took control of Medal, and Air Products acquired Permea from Monsanto.

Twenty years later, noncryogenic production has cut the growth rate of cryogenically produced liquid product, although the latter is still increasing. Cryogenic air separation will always have natural market niches where its product is used for its cryogenic value and extremely high purity or very large volumes are required. The market trend from the 1990s was to go to on-site technologies. Some companies started to move very actively into converting liquid customers to on-site production and promoting the use of on-site production to open new markets because of its lower costs.

Internationalization[2]

The industrial gas world in the 1960s was essentially made up a few industrialized nations of the Western world. Japan was a leading gas nation, but this was not recognized by the West. Eastern Europe and China were hidden behind the Iron Curtain and the Bamboo Curtain, but the Far East was just seen as a coming market. The existing Latin America market was moving very slowly. The 1970s saw the

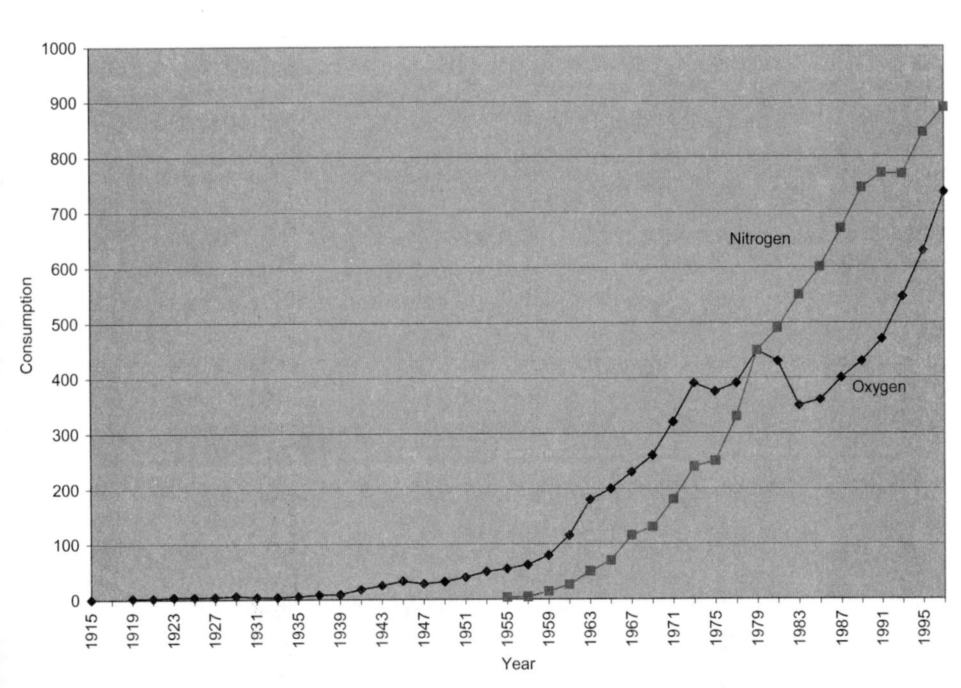

Figure 6.8.2. The oxygen and nitrogen demand in the United States.

beginning of a trend of internationalization across the Atlantic, with most of the major industrial gas companies starting or expanding their operations. In the middle of the 1980s Asian nations like Korea, Taiwan, Singapore, and Malaysia increasingly became partners of Western industrial gas companies. At the end of the decade, new international markets were opened by the collapse of the Soviet Union, the liberation of Eastern Europe, and the gradual opening up of China.

Many industrial gas mergers and acquisitions in Western countries started during that decade as well as some in Asia. In terms of internationalization the "gas world" became the real world in the 1990s.

References

1. Joly, A. (1999). *EIGA 75 year anniversary conference*, Brussels (January). Invited paper.
1. Lovett, J. R. (1991). *Global opportunities for industrial gases*. Allentown, PA: Air Products.

7

HOW NEW GAS APPLICATIONS WERE DEVELOPED

7.1. BIOLOGY AND MEDICINE

> *Oh, Tom! Such a gas has Davy discovered! Oh, Tom! I have had some. It made me laugh and tingle in every toe and finger tip. It makes one strong. And so happy! So gloriously happy! Oh excellent gas bag! Tom, I am sure the air in heaven must be this wonder working gas of delight.*
>
> Robert Southey (a friend of Humphry Davy)

Background to the Medical Use of Gases

Pneumatic Medicine

Pneumatic medicine, the use of gas in medical therapy, became popular in the early 1770s when doctors in England and France began introducing various gases into the body. The first gas used was carbon dioxide, which was used to treat diarrhea. It was also inhaled for diseases such as sore throat or tuberculosis and was applied directly on the skin as a treatment for breast cancer. Joseph Priestley tried breathing his newly discovered gas, oxygen, and found that "his breast felt peculiarly light and easy for some time afterwards." He was one of the first to suggest the therapeutic use of artificially obtained gases. He persuaded the British Lords of the Admiralty in 1772 to equip Captain James Cook's ships on his second voyage to the South Seas with devices to produce carbonated water in an attempt to cure or prevent scurvy.

Pierre Joseph Macquer (1718–1784) in 1777 proposed oxygen therapy for the treatment of asphyxia caused by suffocation. Jan Ingenhousz (1730–1799), who also was the first to observe that oxygen is liberated by plants exposed to sunlight, recommended oxygen therapy "to cure many diseases" and Erasmus Darwin advised the inhalation of oxygen together with ether for asthma and other pulmonary disorders, whereas hydrogen could be used to treat the "convulsive" type of asthma.

Pneumatic Institution

The leader of the physicians experimenting with pneumatic medicine was Thomas Beddoes (1760–1808), scientist and poet of Welsh descent. In 1793 he proposed to the Lunar Society (see Chapter 3.2) that a Pneumatic Medical Institution should be established. He was sponsored by some of the society members and a small hospital for the development of pneumatic medicine was established in Bristol in 1799.

His assistant, Humphry Davy (1778–1829), who later became more famous than his employer both as a chemist and a poet, soon experimented with nitrous oxide (factitious air) inhalation. He conducted many of his experiments on himself and realized the usefulness of nitrous oxide for alleviating toothache. His experiments and the conclusions on the associated pleasurable and mood-raising effects of this gas were described in a 580-page work published in 1800. Many famous people in Davy's social sphere, including the poets Robert Southey and Samuel Taylor Coleridge, the industrialist James Watt, the potter Wedgwood, and other dignitaries came to inhale Davy's new discovery primarily for recreational purposes.

However, the Pneumatic Institution had a short existence. It became apparent that inhalation of "factitious air" did not provide miraculous cures. Accidents occurred in the use of nitrous oxide and staid and respectable ladies were occasionally found to behave with impropriety. It was also realized that none of the other gases and gas mixtures tried seemed to have an effect on curing diseases. Furthermore, the early methods of administration and inhalation of gases were annoying to patients and equipment was poor. Davy's interest in nitrous oxide declined. In 1801 he left Bristol to take up a post at the Royal Institution in London. Even Beddoes became disillusioned with pneumatic medicine. Nonetheless Robert John Thornton (1768–1837) and Daniel Hill became strong advocates for oxygen therapy, which they developed until 1820.

N_2O: "Laughing Gas"[1]

The isolation of nitrous air was probably first described by John Mayow (1640–1679), a medical practitioner in Bath, best known for his description of the mechanism of respiration. His work on nitrous air, however, was forgotten for over 100 years. The actual discoverer of nitrous oxide was Joseph Priestley, who in 1772 reported production of nitrous oxide upon exposing nitric oxide gas to a mixture of iron filings, sulfur, and water. His experiments were based on studies done by Stephen Hales around 1733.

Forty years after the pioneering practical work by Beddoes and Davy, other people started to provide painless surgery as well as entertainment by the inhaling of nitrous oxide. In 1844 an American dentist, Horace Wells (Fig. 7.1.2), attended an exhibition of the effects of nitrous oxide. Wells had long been interested in finding an analgesic to use when repairing the teeth of his patients. He began manufacturing nitrous oxide, testing it on himself and some of his patients, with some success. Together with a partner, William Morton, he arranged an experiment before the students and faculty at Harvard University and in 1847 he documented his achievements. The documentation was sent to the Royal Academy to claim credit for the discovery of anesthesia, but he was discredited. In 1848, under the influence of chloroform, he sprinkled sulfuric acid on two girls in New York and was sent to prison, where he committed suicide. Soon after his death Wells's place in the history anesthesia was

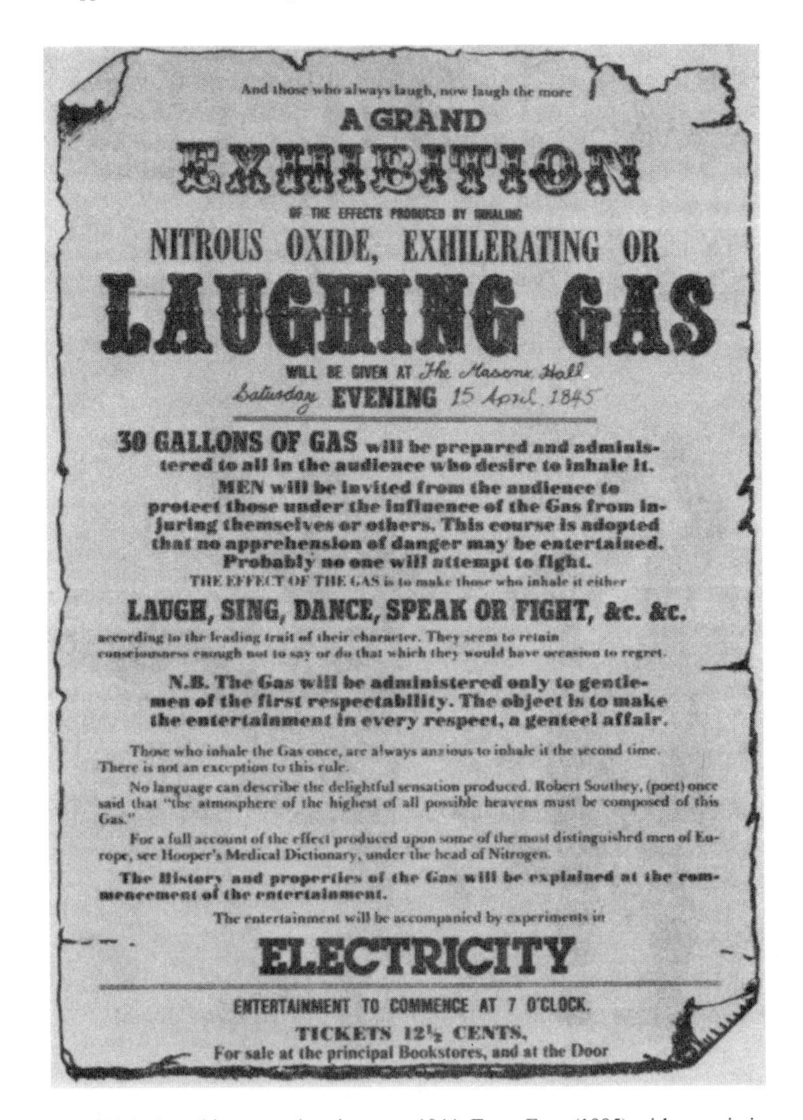

Figure 7.1.1. Laughing gas advertisement, 1844. From Eger (1985) with permission.

recognized. In 1848 he was elected posthumously to honorary membership at the Paris Medical Society.

Morton became the first to use ether as an anesthetic, at the Massachusetts General Hospital in Boston. Because ether was more potent as an analgesic, it was preferred to nitrous oxide as the main substance for pain relief. Chloroform and other volatile anesthetics soon followed.

Nitrous oxide had almost been abandoned when two dentists, G. Q. Colton and J. H. Smith, established the Colton Dental Institute in New York in 1864. By the turn

Figure 7.1.2. Horace Wells. From Eger (1985) with permission.

of the century they had treated nearly 200,000 patients, without problems. In France and England the method was established around 1870. Edmund Andrews, Professor of Surgery in Chicago, in 1868 suggested that oxygen should be added to nitrous oxide to make a more pleasant and safer anesthetic. Introduction of compressed oxygen in cylinders facilitated longer administration.

When inhaled in pure form, nitrous oxide is asphyxiating, but in high concentrations (80 percent or more) with sufficient oxygen present to avoid hypoxia, it induces rapid, but rather shallow anesthesia. The earliest gas anesthesia apparatus (about 1861) comprised a gasometer, a reservoir bag, and a valved face mask. The patient breathed in from the bag and out through the face mask. Since the nitrous oxide was undiluted, administration was asphyxial and very brief (Figs. 7.1.3 and 7.1.4).

Paul Bert of Paris realized that pure nitrous oxide could not be administered for more than 2 minutes without causing asphyxia. He speculated that a prolonged, yet safe state of anesthesia might be obtained if enough oxygen to sustain life and enough nitrous oxide to produce anesthesia could be introduced into the blood. In 1878, at the Academy of Science, he described animal experiments conducted in a hyperbaric chamber maintained at 2 bars. However, the awkwardness and expense of Bert's method precluded its common use.

In 1887, Sir Frederick Hewitt devised a machine that used two separate rubber bags to administer nitrous oxide and oxygen, although the delivered concentration of nitrous oxide could not be very well controlled. Between 1910 and 1925 tremendous advances were made in anesthesia machines.

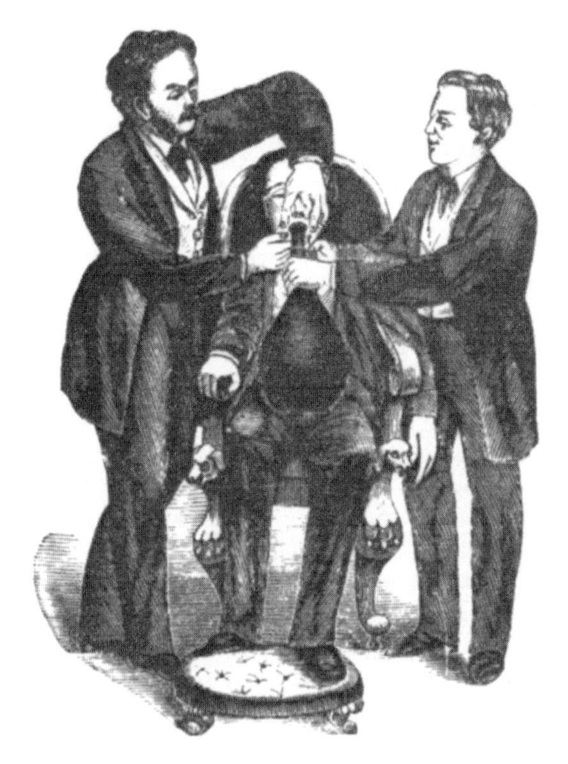

Figure 7.1.3. Administration of laughing gas, 1881. From Eger (1985) with permission.

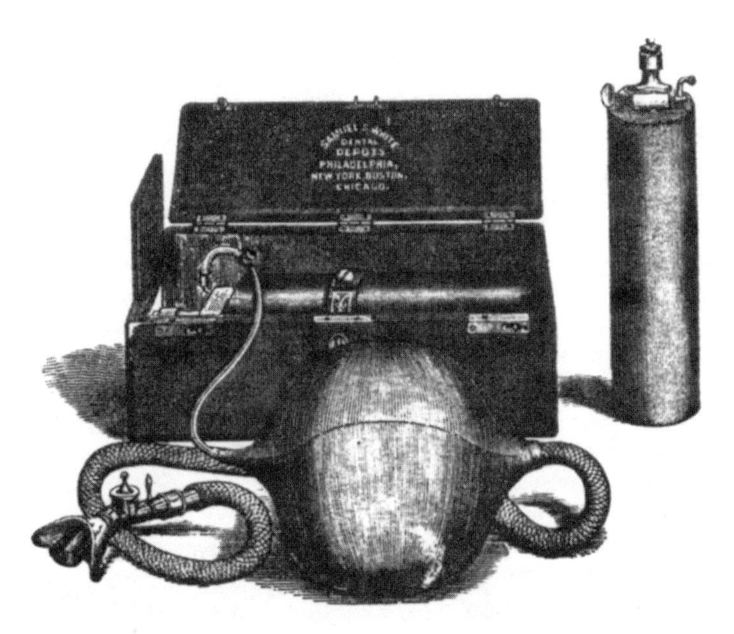

Figure 7.1.4. Portable nitrous oxide equipment, ca. 1880. From Eger (1985) with permission.

Breathing Aids and Treatment of Narcosis[2,3]

In 1913, James Tayloe Gwathmey devised the first water flowmeter. Nitrous oxide and oxygen passed through in separate tubes into a common water container. By controlling the number of bubbles from each gas, a rough estimate of the proportions of the gas mixture could be made. Between 1918 and 1937 several improved types of flowmeters were developed. In 1937 the industrial rotameter was introduced as a standard flowmeter for clinical use.

Partly as an economy measure to save gas, early anesthetists used the process of rebreathing. The patient rebreathed the contents of the reservoir bag for some time before being given a fresh gas mixture. It was thought, too, that the normal depth of respiration could not be preserved during anesthesia unless stimulated by a percentage of carbon dioxide. In 1923, John S. Lundy designed the Seattle gas machine, the first four-gas portable apparatus for the application of oxygen, nitrous oxide, carbon dioxide, and ethylene. The following year, the Lundy–Heidbrink model, which consisted of two yokes for each of the four gases plus an ether bottle, was introduced. It had an adjustable automatic shutoff valve that operated by pressure. This was used for a long time, the only major improvement to this equipment being the addition of an absorber system in 1932.

Already in 1912 McKesson had introduced the "intermittent-flow" apparatus, a demand regulator with controls to set any desired mixture of nitrous oxide and oxygen. The gases could be delivered to the patient under positive pressure without disturbing the percentage admixture. In 1928, McKesson produced the Nargraft, Model H, a precision-built apparatus of extreme mechanical ingenuity which gave excellent results. By the 1930s, it began to be realized that a large flow of fresh gases—about 8 liters per minute in an adult—was essential to keeping the carbon dioxide content in the blood within normal limits. The practice of rebreathing was therefore abandoned. With the rise of carbon dioxide absorption methods, it was found that respiration was perfectly satisfactory, even though the carbon dioxide content of the inhaled atmosphere was virtually zero.

Cryobiology and Cryomedicine[4–9]

Cold is becoming more and more widely used as a preserver of life. Artificial insemination of cattle using bull semen preserved at liquid nitrogen temperatures and blood preservation by freezing are now standard. Banking of some human organs at liquid nitrogen temperature as spare parts for surgery and bone marrow for transplantation is nearly a reality. Cryosurgery is used successfully for some special forms of surgery.

In 1908 in Leiden, Paul Becquerel demonstrated that certain simple organisms such as seeds, algae, spores, and protozoa in the dried condition withstood freezing to the temperature of liquid nitrogen.

Basile J. Luyet, working originally at St. Louis University and later at the American Foundation for Experimental Biology at Madison, Wisconsin, was a pioneer in cryobiological research. In the United States in 1940 he and P. M. Gehenio showed that organisms could survive freezing if it took place extremely quickly. It was found

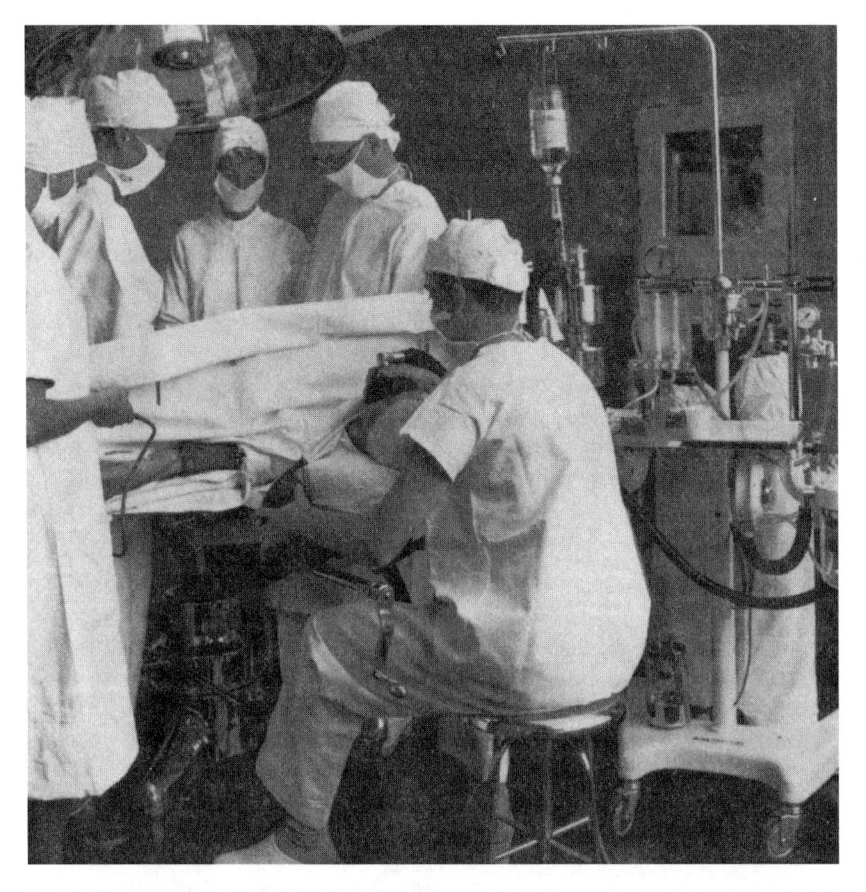

Figure 7.1.5. Anesthetic hospital equipment for surgery, ca. 1945. Courtesy of the AGA Historical Archive.

that quick-freezing prevented the development of ice crystals which otherwise would burst the cell membranes (Fig. 7.1.7). They pointed out that if the metabolic processes of living cells could be stopped by freezing without causing death, unlimited storage of cells and organisms would be possible.

In the 25 years following the publication of Luyet and Gehenio's work the feasibility of low-temperature preservation and the cause and nature of specimen damage during freezing were studied more in detail. During the 1940s American (Sheffner) and French (F. Rostrand) groups studied the effects of freezing on frog and fowl sperm down to −79°C. They observed that the cell could recover after thawing if it was imbedded in special substances during freezing.

Modern cryobiology started in 1949 with Alan Parkes, Audrey Smith, and Chris Polge at the MRC Laboratories in London. They found by chance that if sperm was frozen in glycerol, the glycerol protected the spermatozoa against the otherwise damaging effects of freezing and thawing. By systematic studies the group prevented the rapid deterioration of cells stored at −79°C by using liquid nitrogen to lower the storage temperature to −196°C. This formed the basis for the technique used today in

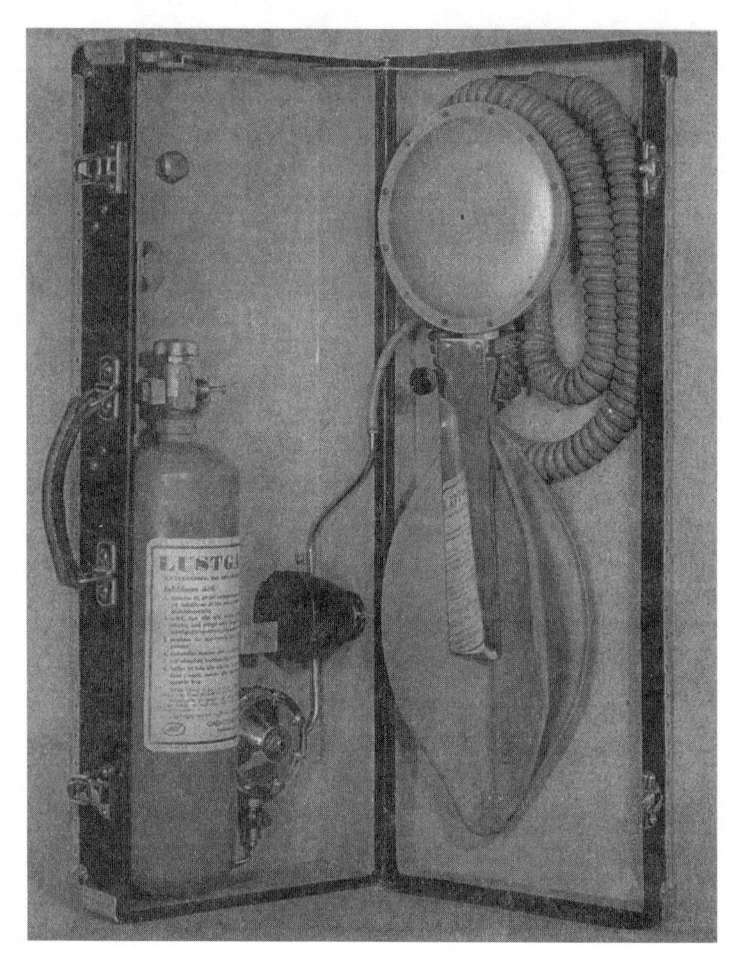

Figure 7.1.6. Portable nitrous oxide equipment used by midwives from 1940. Courtesy of the AGA Historical Picture Archive.

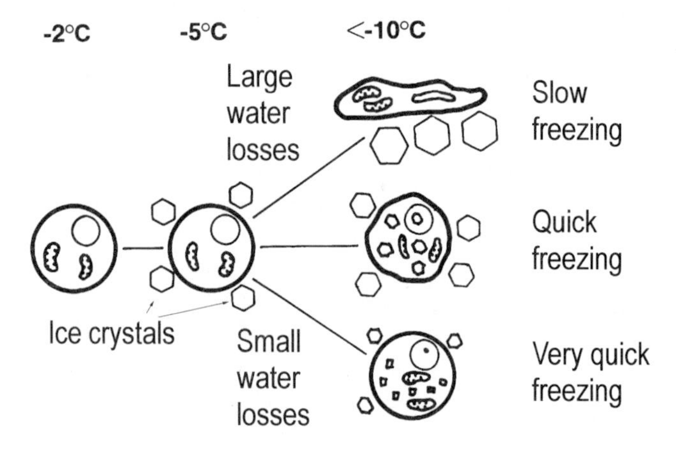

Figure 7.1.7. Cell damage by freezing. Rapid cooling with liquid nitrogen prevents the formation of large ice crystals. Based on a drawing in CIBA–Geigy (1977). Courtesy of Novartis International, Basel, Switzerland.

sperm and blood banks. In 1951 the method was tried for sperm from bulls and by 1955 artificial insemination centers having frozen sperm had been set up in 20 or so countries.

By the end of the 1960s attempts were being made to freeze complete organs. It soon became clear that different cells best survived the freezing and thawing process if first cooled to –25°C and held at that temperature for a few minutes before then being cooled to liquid nitrogen temperature. The reason for the success of this so-called two-stage cooling method was investigated during the 1970s. Cells shrink when being held at –25°C and a slow dehydration takes place. When they are then frozen down to the final storage temperature less internal damage occurs in the cells because water has disappeared.

In 1973 the first calf, Frosty, was born after its embryo had been cryogenically preserved for 1 year. Integrated groups of biologists, biochemists, biophysicists, and engineers have since been set up at various locations to study the basic phenomena of cryobiology and devise practical processes for freezing and storing all types of biological specimens. The availability of liquid nitrogen at low cost has done much to make these efforts fruitful.

Semen

Artificial insemination has long been known, having been used by the Arabs in the 14th century. It was not until 1920, however, that the breeding of farm animals

Figure 7.1.8. Cryogenic storage of biological specimens. From Koch (1993) with permission.

by such techniques became widespread in the Soviet Union. In the United States, improvement of dairy cattle by artificial insemination techniques was first practiced in New Jersey in 1938. The above-described work by Polge, Smith, and Parkes in 1949 gave rise to a new industry—the preservation of bull spermatozoa in the frozen state for the artificial insemination of dairy cattle. Today, the storage and transport of bull semen at cryogenic temperatures is used world-wide and consumes many tons of liquid nitrogen per day. By 1958 more than one-third of the dairy cattle population in the United States was bred by this means. Refrigerated sperm storage was first used in 1952, and today the preservation of bull semen is almost entirely a cryogenic operation.

Blood

Blood transfusions were successfully started in 1906 by direct delivery of blood from a donor to a patient. During the World War I a great demand for collection and storage of blood developed. The use of sodium citrate for preventing blood from clotting made storage for 24 hours possible.

Developments during World War II increased the storage time of chilled blood to 1 week. Later, with better anticoagulants, 3 weeks of storage was possible and large-scale blood banks were begun. When in the 1970s red and white blood cells and blood plasma could be separated, nationwide systems for blood storage were built with accurately controlled liquid nitrogen technology.

Bone Marrow

The function of bone marrow is to form blood cells. When bone marrow is destroyed by radiation or by massive chemotherapy, it no longer produces the required blood cells. In 1954 a team at the Radiobiological Lab at Harwell observed that bone marrow cells could be safely frozen and thawed back to life. Recently this observation has resulted in registers of bone marrow donors and banks of bone marrow and stem cells for treatment of leukemia and aplastic anemia.

Whole Organs

Different cell types require different cooling regimes. Therefore freezing of whole organs without damage is very difficult. From 1954 human corneas have been successfully frozen and transplanted. Research on cryogenic supplies of human "spare parts" is going on, but is not likely to be successful. Storage of microorganisms such as bacteria, yeast, and protozoa has been established.

Treatment with Liquid Nitrogen

Treatment of Skin Diseases

Liquid nitrogen can be used effectively in the treatment of skin diseases. Treatment of warts or of scarring caused by acne can be effected by application of liquid nitrogen using an ordinary cotton swab.

Cryosurgery

In cryosurgery, a probe cooled by liquid nitrogen is used partly to prevent bleeding and partly to kill certain types of malignant cells. It has been used for the destruction of local areas in the brain (such as the area involved in Parkinsonism or palsy) and of tumorous areas in other parts of the body.

Diving Applications: Deep-Sea Diving[10]

The first steps toward understanding the oxidative processes involved in metabolism were taken by Joseph Priestley in 1776. He studied the respiration of air and identified it as a phlogistic process. His investigations started developments which, among other things, led to practical applications such as general anesthesia and deep-sea diving apparatus.

Diving

Archaeological studies have found that early humans dived for food. Ancient Greeks dived for sponges and Alexander the Great is said to have been lowered into the sea at the Straits of the Bosphorus in a glass barrel during the 3rd century BC. More practical diving bell designs were used in the 16th century.

The next major step in the technology of diving was the development of a closed diving suit, supplied with air under pressure by a pump on the surface. This was to provide the basic technology of diving until the development of self-contained underwater breathing apparatus (SCUBA) equipment by Cousteau and Gagnan in 1943. Some years later two major industrial gas companies, Air Liquide and AGA, became involved in the development of light diving equipment.

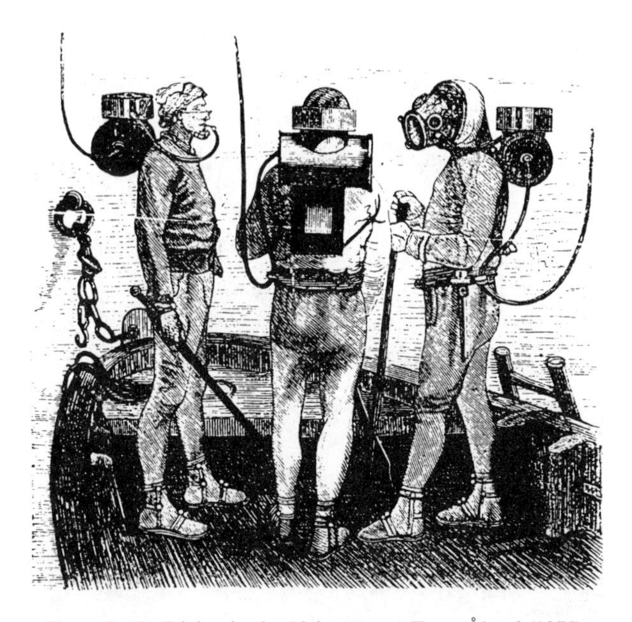

Figure 7.1.9. Diving in the 19th century. From Ålund (1875).

The pressure inside and outside the diver must be essentially equal, otherwise his or her lungs would collapse. Therefore the pressure of the air supply must be equal to the actual environmental pressure (pressure increases 1 bar per 10 meters depth below the surface). In deep diving, even with a reliable standard diving suit decompression sickness can occur. This problem was revealed in the 19th century when the air pump and pressurized chambers were first used in diving, but the cause took more than a century to understand and explain. When divers breathe gases under pressure, more gas dissolves in the body. If a diver returns to the surface too quickly, this excess gas can be released and can cause pain and even paralysis or death.

In the 1920s another problem arose, "inert gas narcosis" in deep-sea diving. Divers became confused and euphoric and eventually, at greater depths, could lose consciousness. At the end of the 1930s it was quite clear that this was due to the narcotic effect of the nitrogen in the air. This sets a limit on normal air diving to about 50 meters depth. In the late 1930s the U.S. Navy began to replace nitrogen in pressurized breathing mixtures with helium. General anesthesia can be induced by all nonreactive gases, but helium is much less narcotic.

Since pure oxygen has a number of adverse physiological effects at pressures of greater than about 2 bars, divers cannot breathe pure oxygen except in very shallow dives. Thus a gas mixture containing both oxygen and one or more dilutent gases is needed at these pressures. In recent years a sophisticated technology for saturation diving, in which divers remain at pressure for periods of many days or even weeks, has been developed.

The Pneumatic Medical Institute of Bristol[11–13]

In 1794 Thomas Beddoes (1760–1808) left a medical professorship at Oxford to work with gas medicine, that is, to treat diseases by the inhalation of gases, with the main goal of treating tuberculosis. In 1797, Beddoes proposed to the Lunar Society (see Chapter 3.2) that a Pneumatic Medical Institute should be established. The idea to use gases in this way was not exactly new, but Beddoes's concept was to work on a greater scale than before, establishing a joint laboratory and hospital where the possible curative effects of gases could be clinically tested.

Beddoes was supported financially by Lunar Society member Josiah Wedgwood and technically by James Watt, who helped design and make the apparatus for him. The Pneumatic Institution was opened in a suburb of Bristol in 1799. James Watt's 15-year-daughter had died in 1794 of consumption in the lungs. His son Gregory, born in 1777, suffered from tuberculosis, interrupted his academic career, and was sent to Cornwall to cure his disease. While there he also looked after the five steam engines which the company Boulton, Watt & Sons had delivered to local miners. In Cornwall he happened to lodge with Mrs. Davy, the mother of Humphry Davy, a boy of has own age.

Humphry Davy (1778–1829) (Fig. 7.1.10) was the son of a carpenter in Penzance, Cornwall. After his studies at the local grammar school, he was almost entirely self-educated by the time he met Watt, who taught him chemistry. Before the age of 20 Davy had been an apprentice surgeon and also studied metaphysics, ethics, and mathematics. He wrote a letter to Thomas Beddoes (Fig. 7.1.11) presenting new ideas for using gases in

Figure 7.1.10. Humphry Davy. From Eger (1985) with permission.

medicine. On a visit to Cornwall Beddoes met and decided to hire Humphry as superintendent of his new institute, the Pneumatic Institution.

Davy, like Priestley and the poet Coleridge, benefitted from Wedgwood's generosity, which supported Beddoes Pneumatic Institution in Bristol. There Davy, in his research on the oxides of nitrogen, and especially nitrous oxide, continued Priestley's tradition of experiments on gases. Davy's work, published in 1800, follows up Priestley's work on gases, which had been rather more qualitative than quantitative. He also unwisely experimented on

Figure 7.1.11. Thomas Beddoes. From Eger (1985) with permission.

Continued

himself, which had almost disastrous effects when he inhaled carburetted hydrogen gas in 1800.

He soon realized that the experiments with gases had failed: neither oxygen, carbon dioxide, carbon monoxide, nitrogen, ether, nor any or the other gases tried, in any combination, seemed to have an efficacious effect. The experiments which had greatest impact were those conducted with nitrous oxide, whose anesthetic and euphoric effects were widely tested and enjoyed.

In 1801 Davy was invited to become Professor of Chemistry at the Royal Institution in London. In 1807, he demonstrated the existence of sodium, potassium, calcium, magnesium, strontium, barium, and chlorine. He disagreed with Lavoisier's caloric concept, maintaining that heat was not matter, but a form of motion, and he decomposed by electrolysis some of the compounds Lavoisier had thought to be elements. He also defended Lavoisier's doctrine of acidity, demonstrating that the supposed generator of acidity, oxygen, was in fact a component of the strongest alkalis, but not of hydrogen chloride. In 1812 Davy hired as assistant Michael Faraday, a bookbinder's apprentice in the laboratory of the Royal Institution. Faraday soon became a skilled chemist and physicist. Before developing the science of electrochemistry and electromagnetism, he laid the ground for the liquefaction of gases. Like Davy, Faraday later became a famous professor and president of the Royal Society.

References

1. Eger II, E. I. (ed.). (1985). *Nitrous Oxide N_2O*. London: Edward Arnold and New York: Elsevier.
2. Gilbert, D. L. (1981). Perspective on the history of oxygen and life. In *Oxygen and living processes* (Carter, ed.), pp. 1–43. New York: Springer-Verlag.
3. Ross, J. A. S., Marr, I. L., and Tunstall, M. E. (1957). Entonox and its development. History of nitrous oxide and oxygen mixtures. In *Proceedings of the 8th BOC Priestley Conference*, pp. 27–41.
4. Luyet, D. J., and Gehenio, P. M. (1940). *Life and death at low temperatures*. Normandy, MO: Biodynamics.
5. Polge, C., Smith, A. V., and Parkes, A. S. (1949). *Nature* **164**:666.
6. Parkes, A. S. (1951). Preservation of spermatozoa, red blood-cells and endocrine tissue at low temperatures. In *Report of the Freezing and Drying Symposium*, pp. 99–106. London.
7. Westmore, A. (1984). First freeze-thaw baby is born. *New Scientist* **1984** April 12:3.
8. Sumida, S. (1986). Cryomedicine. In *IIR World Conference on Refrigeration in development, pp.* 210–241.
9. Rubinsky, B., and Onik, G. (1991) Cryosurgery: advances in the application of low temperatures to medicine *International Journal of Refrigeration* **14**:190–199.
10. Smith, E. B. (1986). On the science of deep-sea diving—observations on the respiration of different kinds of air. In *Proceedings of the 4th BOC Priestley Conference*, pp. 230–235.
11. Williams, L. P. (1960). Humphry Davy. *Scientific American* **1960**(6):106–116.
12. Rooney, R. P. (1986). Humphry Davy: The romantic chemist. *Journal of Chemical Education* **63**(9):739–740.
13. Smith, E. B. (1997). Humphry Davy, Thomas Beddoes and the introduction of nitrous oxide anaesthesia. In *Proceedings of the 8th BOC Priestley Conference*, pp. 155–162.

7.2. WELDING AND CUTTING

Prehistory: The Historical Development of Welding and Cutting[1]

Already in the Bronze Age it was known how to join metals by melting them together. In Chaldea and Egypt, for example, graves have been found containing hard-soldered and welded objects, which have been dated to a period long before our epoch. The only welding method used before the 19th century was so-called blacksmith welding. This consists in heating a piece of material over a coal fire until it is white hot and then pressing or hammering it together. The method was a precursor to several pressure welding methods, so-called resistance welding, which began to be used in production at the end of the 19th century. At that time it became possible to use new technical advances for the heating of metals. Energy from electrical resistance, an electric arc, or gas could be used as a heat source in welding. Around the same time gas cutting technology, a valuable complement to welding technology, was developed.

The basic principle of the oxygen cutting process was established by Antoine Lavoisier (1743–1794), who found in 1776 that iron at 1200°C could burn in oxygen. Because of the temperature rise caused by the combustion, the iron then began to melt.

In the first half of the 19th century scientists developed the oxyhydrogen torch as a laboratory tool. In 1838 the Frenchman Eugene Desbassayns de Richemont (1800–1859) patented a process of fusion welding using flexible tubes. By 1840 he had coined the name "soudure autogene" for his process. De Richemont's torch mixed air and hydrogen in a flexible rubber tube about 1 meter long and tipped with a metal "beak" (nozzle) from which the flame originated.

Robert Hare, Jr. (1781–1858), demonstrated an oxyhydrogen torch in Philadelphia in 1847. Oxygen and hydrogen were mixed and burned at the nozzle of his torch. Hare used a flexible air hose. It was difficult to make gas-tight flexible hoses at this time or contain gases without leakage. Leakage of combustible gas would have been dangerous. In the 1850s Henri Sainte-Claire Deville (1818–1881), Professor of Chemistry at the Ecole Normale Superieure in Paris, built oxyhydrogen torches for his work on the metallurgy of platinum. He mixed the gases very near the nozzle, which was regarded as a great safety innovation.

The Industrial Beginning of Welding and Cutting[2,3]

Arc Welding and Coated Electrodes

Two Russian scientists developed arc welding methods in the 1880s. Nikolais Benardos first patented a method where the heat source was an arc between a carbon electrode and the work piece. Filler material was fed from the side into the weld pool. The disadvantage of the method was the short life of the carbon electrode and the risk that the carbon would contaminate the final weld deposit and allow hardening to occur. These deficiencies were eliminated by Nikolai Slavianoff, who in 1892 developed metallic arc welding, that is, an arc between a melting metal electrode and the

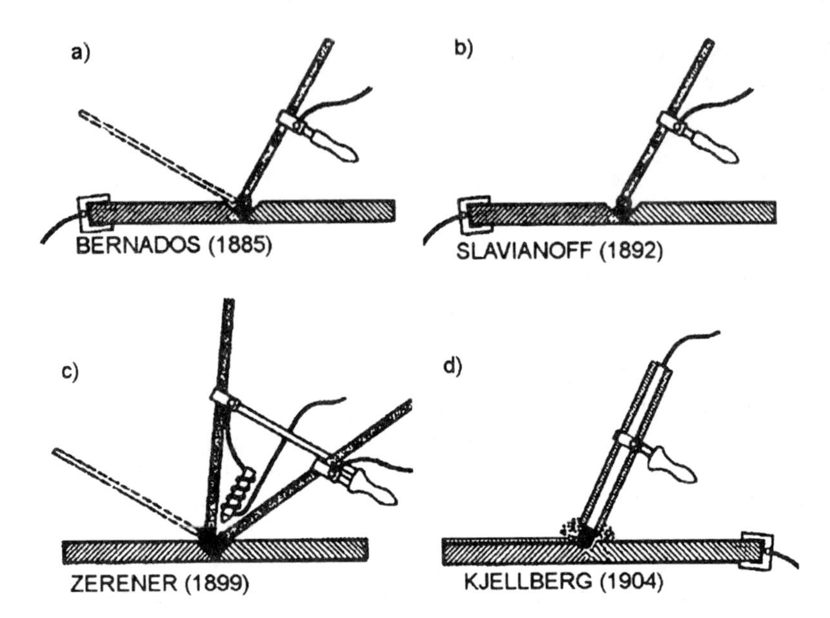

Figure 7.2.1. (a) Bernados's torch (1885), (b) Slavianoff's torch (1892) (c) Zerener's torch (1899), and (d) Kjellberg's coated electrode (1904).

work piece. In this way, filler material was supplied automatically. Both these arc welding methods, however, suffered from the disadvantage that the oxygen in the atmosphere freely entered the weld pool, thus causing a deterioration in the quality of the weld due to oxidation of molten material.

Twelve years later a method was discovered for excluding atmospheric oxygen from the weld puddle. The Swede Oscar Kjellberg (Fig. 7.2.2) solved the problem by

Figure 7.2.2. Oscar Kjellberg, inventor of the coated electrode and founder of ESAB. Courtesy of ESAB, Göteborg, Sweden.

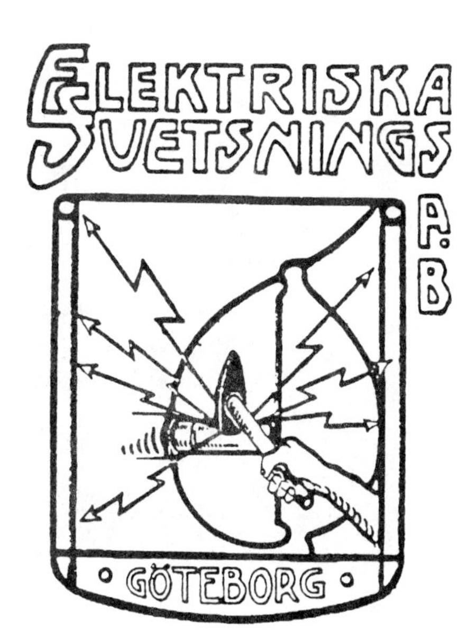

Figure 7.2.3. First ESAB logo. Courtesy of ESAB, Göteborg, Sweden.

coating welding electrodes with a deposit which melts at high temperature. The deposit then gives off a gas which shields the puddle. In 1904, he established ESAB (Fig. 7.2.3), now one of the leading world enterprises in the welding field. It was, however, gas welding and cutting that was first employed in industry.

When methods for the economical separation of pure oxygen from air were developed in the beginning of the 20th century practical welding and cutting methods were opened for a world-wide market.

Oxy-Fuel Cutting and Welding[4–6]

The 1888 oxy-fuel cutting torch of the English gas engineer Thomas Fletcher (1840–1903; Fig. 7.2.4) cut holes in an iron plate by fusion, not by oxygen cutting. He had become interested in the cheap oxygen provided by the newly established (1886) Brin's Oxygen Company. In his torch, coal gas entered the nozzle through the 3-millimeter annulus around the oxygen delivery pipe. He used oxygen to increase the flame temperature. Already in 1890 a progressive criminal demonstrated the efficacy of the cutting torch for robbing banks.

In 1901 a Belgian company, Société Oxyhydrique, began to market oxyhydrogen welding apparatus. Also in 1901 the German Ernst Menne (1869–1927; Fig. 7.2.5), a chemist at the Kreuztal Blast Furnace Plant of the Cöln-Müsener Bergwerks-Aktienverein, patented the oxygen lance. A French engineer, Charles Picard, succeeded in making the first usable oxyacetylene torch. In the next 5 years the design of oxyacetylene torches developed rapidly in France.

Figure 7.2.4. Thomas Flechter. From Hoffmann (1938).

Figure 7.2.5. Ernst Menne. From Hoffmann (1938).

The Oxygen Lance

Menne's oxygen lance was for a process to open the tapping hole of blast furnaces in steelworks. The oxygen lance was made of two concentric steel tubes. A stream of oxygen passed in the central tube under low pressure and the fuel gas passed in the space between the inner and the outer tube. The mixture of oxygen and fuel gas was lit at the end of the lance and the flame was directed upon the material blocking the tapping hole. Once the material was heated, the oxygen stream was raised to higher pressure. As a result the heated material burned violently and opened the tapping hole.

Oxyhydrogen Torches

The Société Oxhydrique was founded by the Belgian engineer Felix Jottrand (1863–1907; Fig. 7.2.6) in 1896 in Brussels (Molenbeek) to manufacture oxygen and hydrogen by electrolysis of water. In 1901 the company patented an oxyhydrogen torch and a system of oxyhydrogen welding. In 1904 Jottrand and Lulli designed an oxyhydrogen torch with an additional tube conveying a jet of oxygen and patented this first oxygen cutting torch. It employed two separate nozzles: one for a heating oxyhydrogen flame and another, a few millimeters away, for a jet of pure oxygen.

The main impulse for using this gas in processes based on torches was given by the German engineer Ernst Wiss in 1896 (Fig. 7.2.7; see also Fig. 5.3.1). He worked as an engineer in the maintenance service of the firm Chemische Fabrik Griesheim Elektron in Frankfurt, where he used an air–hydrogen torch for mounting and repairing lead tanks and pipes. Hydrogen was first produced on site by the

Figure 7.2.6. Felix Jottrand. From Hoffmann (1938).

interaction of zinc and diluted sulfuric acid. He found, however, that in another part of the factory large quantities of hydrogen were available as a by-product of the production of chlorine by electrolysis. He collected some of the hydrogen and compressed it into steel cylinders. In 1901 he designed and patented two types of torches, and also suggested the use of his equipment for the cutting of metals (Fig. 7.2.8).

Figure 7.2.7. Ernst Wiss welding (Griesheim). Courtesy of Messer Group, Frankfurt-am-Main, Germany.

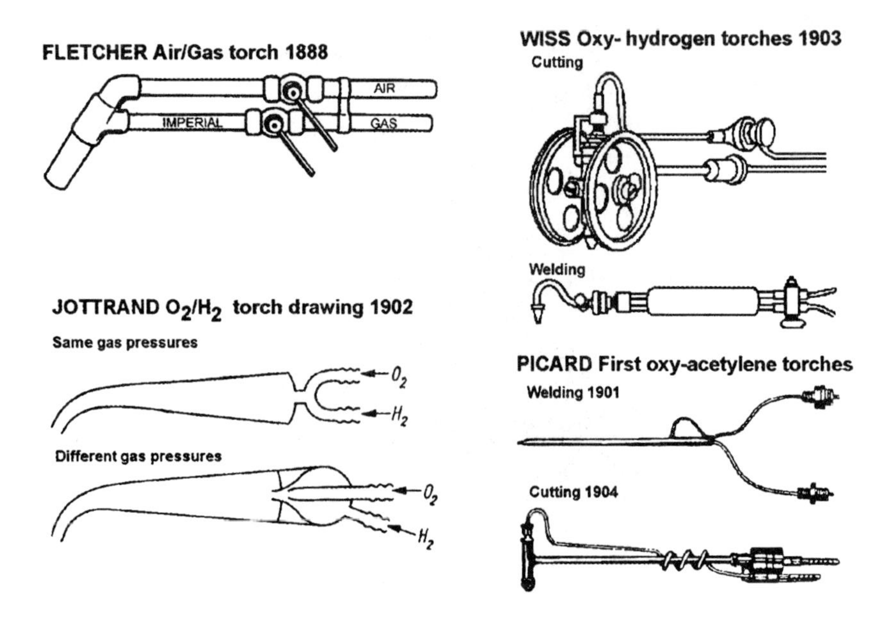

Figure 7.2.8. Fletcher's oxy-gas torch (1888), Jottrand's sketch of oxyhydrogen torches for different pressures (1902), Wiss's first oxyhydrogen torches (1903), and Picard's first oxyacetylene welding and cutting torches (1901, 1904). Based on Hoffmann (1938).

Oxyacetylene Welding and Cutting

After the discovery of acetylene in 1892, a market very soon developed for this gas as a very energy-rich light and heat source and it was soon used mixed with oxygen for welding. The disadvantage of acetylene was its tendency to explode, which led to tragic accidents. These problems were mainly solved during the first decade of the 20th century together with the development of new welding tools.

At first acetylene was produced on site in mobile generators by pouring water on calcium carbide. Very soon, however, attempts were made to compress it into steel cylinders, which was already common practice for hydrogen and oxygen. For acetylene, however, this solution was not as simple as for the two other gases because at a pressure above 1.5 bars it can decompose spontaneously with an explosion. The first step toward solving this problem was taken in 1896 by Georges Claude (1870–1960) and Albert Hess, who conceived the idea of filling the acetylene cylinders with a porous ceramic substance, which was then saturated with a precisely specified quantity of acetone. Thus they were able to dissolve and store a considerable quantity of acetylene in a normal-size cylinder at a pressure of up to 15 bars. A French company, Compagnie Française de l'Acétylène Dissous (CFAD), was set up to market this innovation.

Oxyacetylene Torches

Henri Le Chatelier, a French chemist, found in 1895 that he combustion of acetylene with oxygen produced a flame with a temperature far higher than that of any gas flame previously known. The primary combustion of acetylene with an equal volume of oxygen yields hydrogen and carbon monoxide. These two gases then burn, uniting with oxygen from the surrounding air to form water vapor and carbon dioxide, respectively.

In 1897 Charles Picard, one of Claude's colleagues at CFAD in Paris, was put in touch with Henri Le Chatelier by Henri's brother Louis, a board member of CFAD. Picard developed an oxyacetylene torch following Le Chatelier's discovery.

Picard's Work

The first attempt by Picard in 1900 failed. Le Chatelier then remembered the oxyhydrogen torches developed by Sainte-Claire Deville, who recommended that the oxygen and acetylene should be mixed at the nozzle. Picard took up that idea, encouraged by Edmond Fouché (Fig. 7.2.9), director of CFAD. This time he used a mixing chamber to promote homogeneous combustion. This development was successful and the torch burned quietly and regularly. In 1901 torches of a practical type were introduced, and by 1903 the process began to be used industrially. As experience was acquired, new techniques were developed for the welding of various metals. The CFAD company ran into economic difficulties and considering filing for bankruptcy, and the new torch seemed doomed. However, a patent was taken out and a firm set up to market the new technology, the Société des Applications de l'Acétylène.

Figure 7.2.9. Edmond Fouché. From Hoffmann (1938).

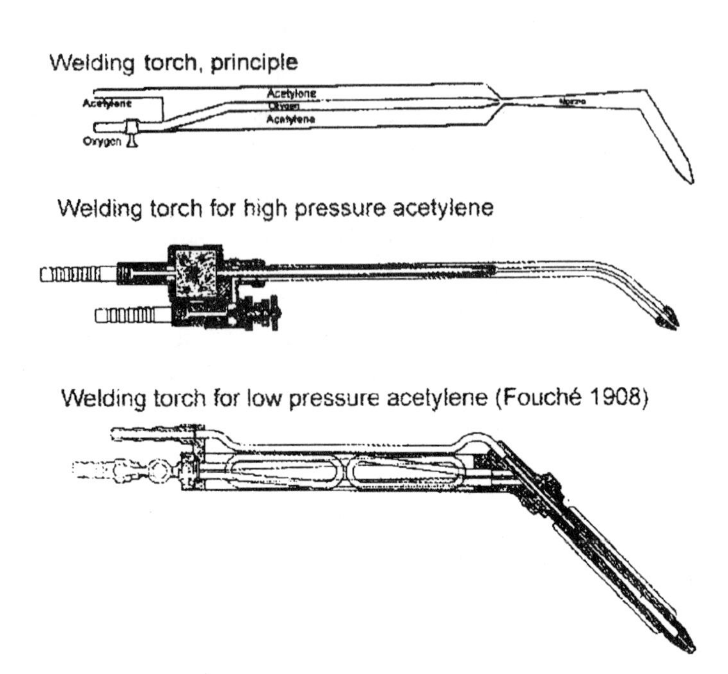

Figure 7.2.10. Oxyacetylene torch designs for different gas pressures. Based on Grandjon and Rosemberg (1912).

Fouché wanted to market a torch that could be used with the low-pressure acetylene from the acetylene generator, which was used far more than any source of dissolved acetylene. On the suggestion of the CFAD board member Arsène d'Arsonval in 1902, Picard built such a torch using a high-pressure oxygen injector to entrain the low-pressure acetylene. CFAD was not interested in Fouché's low-pressure torch. It would not sell dissolved acetylene. In October 1902 he patented the low-pressure acetylene torch, which jolted CFAD into patenting Picard's torch and caused a dispute between CFAD and Fouché. Fouché united with E. A. Javal of Paris in the production of a low-pressure torch.

New Developments and Applications

Because the high-pressure torch had serious defects and the low-pressure torch was not entirely satisfactory, depending as it did upon injection for its mixture of acetylene with oxygen, A. Boas, Rodrigues & Company of Paris undertook the production of a torch that would remedy both of these defects. They succeeded in producing a torch that would not only operate over the entire range of pressures, but could be arranged so that a number of tips of various capacities could be used in a single torch. Despite further improvements to the torch by Henri Le Chatelier's brother André, the reaction was still skeptical. Finally, though, the French Navy had to repair 80 of its ships, which had cracks in their hulls, by plug-welding, and training in the new technique was rapidly organized.

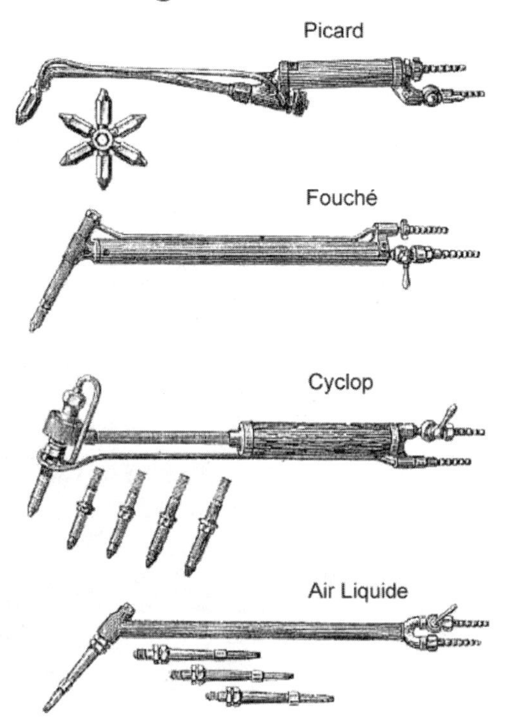

Figure 7.2.11. Different welding torch designs. Based on Grandjon and Rosemberg (1912).

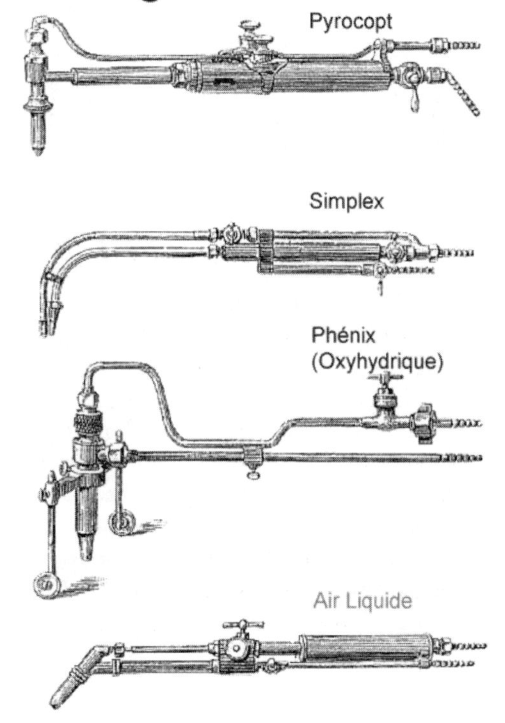

Figure 7.2.12. Different cutting torch designs. Based on Grandjon and Rosemberg (1912).

Figure 7.2.13. Steam dome welding. Fouché. From *Svetsning och skärning av metaller* (1909).

In 1909 the Soudure Autogène Francaise (SAF) was founded and obtained patent rights from the Swedish engineer Oscar Kjellberg, who had developed electrodes with a special coating designed to guide the arc and improve the quality of the weld puddle. Arc welding had made its entry onto the scene. The world wars greatly influenced the development of welding technology. Before World War I gas welding was mainly used for repair work, but during the war years it became an established tool of production in many countries. After the war, new arc welding methods were developed, but their use was limited by the lack of shielding gases.

During World War II there were new demands for welding various new metals and alloys. This greatly stimulated research and what has now become the most common arc welding methods, tungsten inert gas, metal inert gas, and metal active gas welding, were developed. At the same time, inert shielding gases such as argon and helium became available relatively cheaply through developments in industrial gas technology. In the postwar period several special advanced technological welding and cutting methods for new materials were developed, which use the energy from, for example, plasmas, electron beams, and lasers. The mechanization and automation of welding technology as well as the improvement of the welder's

Figure 7.2.14. AGA welding school started in 1916. Courtesy of AGA Historical Picture Archive.

working environment have a high priority in current development programs (Fig. 7.2.17).

Development of Welding in the United States[7]

From Europe to the United States: Fouché and Bournonville

In the United States acetylene was at first thought of primarily as a source of light. Augustine Davis and Eugene Bournonville both developed acetylene generators in 1896. Bournonville went to Paris and purchased the U.S. rights for the process of dissolving and compressing acetylene in acetone for the Commercial Acetylene Co. The compressed acetylene was at first used as a fuel for railway car acetylene lights. In 1903–1904 Bournonville received the first Fouché welding torches from France for his machine repair shop in New York and began working with the oxyacetylene flame. In 1899, a stove manufacturer in Cleveland, John Harris, was experimenting with the oxyfuel processes in his hobby work to produce synthetic rubies and sapphires. One day he heated aluminum oxide with an oxyhydrogen flame. Some of the free oxygen impinged on the steel plate which he used as a base for his experiments. When Harris studied this further, he found that the heat of the oxygen flame in some, then unknown, way actually cut the steel. This was the first time steel cutting was used in the United States. Harris later testified in a law suit against Linde patents that he had used oxygen for cutting steel before Jottrand. In 1905 Harris formed the Harris Calorific Co., now a division of Lincoln Electric Company, to make and sell oxy-acetylene welding and cutting apparatus of all types.

Figure 7.2.15. VSW advertisement for use of oxygen in welding and cutting, ca. 1910.

Around 1906 the Industrial Oxygen Co. of New York asked salesmen from France to come over to sell welding torches and also to introduce the compound known as Epurite from which to generate oxygen. The Epurite generators did not become successful and were soon replaced by the Oxygenite process. The early use of welding and the welding business in the United States were seriously retarded because the oxygen required had to be made in complicated and inefficient oxygen generators

Figure 7.2.16. Air Liquide welding advertisement, ca. 1912. From Granjon and Rosemberg (1912).

at high cost and low purity until the completion of the first commercial oxygen plant in 1907 at Buffalo, New York. The acetylene lighting industry continued to grow. Augustine Davis's firm, the Davis Acetylene Co. of Elkhart, Indiana, did very well, and early in 1907 Davis and Bouronville went to Paris and acquired the American rights to the Gauthier–Ely medium–pressure torch made by A. Boas, Rodrigues & Co. Back home, Bournonville made the first public demonstration of oxyacetylene cutting at the Williamsburg Bridge by cutting an iron stringer 5 feet wide. Some months later Bournonville left for Paris again, this time with Brown of the Ferrofix Co. to investigate an oxygen process because they had no adequate source of oxygen at that time. Because the barium process they were investigating proved to be imprac-tical, the Davis–Bournonville Co. started to manufacture oxygen by the potash

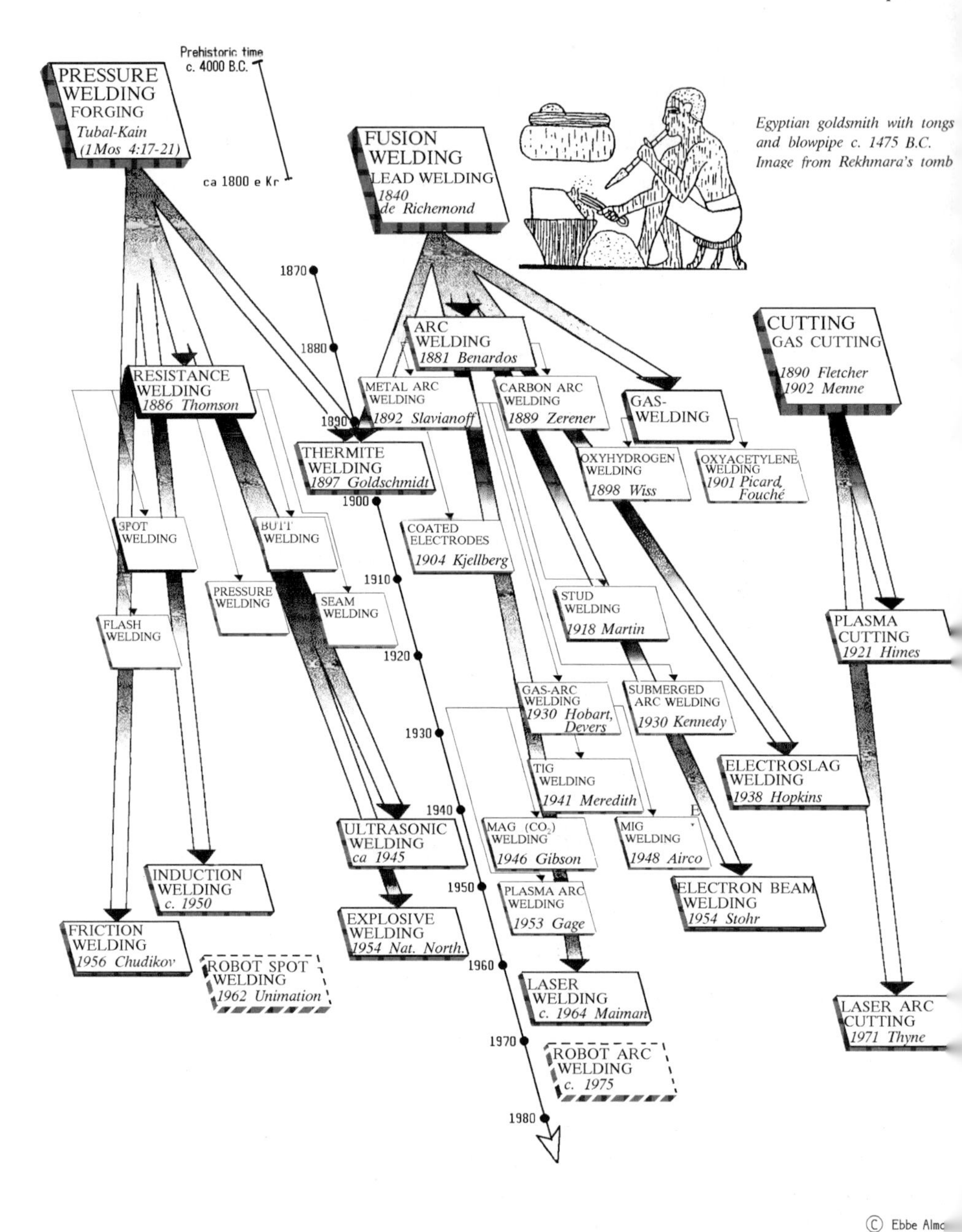

Figure 7.2.17. Overview of welding/cutting history.

chlorate process, in 1907 the most common oxygen production method in the United States.

In this period of the compressed gases industry, products were delivered exclusively in heavy steel cylinders. Distribution was a major problem and to cut distribution costs, oxygen producers began building small electrolytic oxygen plants. In 1908 F. W. Clifford set up the United Oxygen & Chemical Co. at Suspension Bridge, New York. In 1911 there were seven such plants installed in the United States: Chicago, Illinois (two units); Cleveland, Ohio; Pittsburgh, Pennsylvania; Elizabeth, New Jersey; and Jersey City, New Jersey. Oxyacetylene welding was put to work by small companies. Hundreds of gas welding manufacturers sprang up all over the country. In 1911 Modern Engineering Co. sold a non-backfiring torch.

Another successful torch was the Rego torch invented by Keith Dunham and sold by Bastian-Blessing Co. of Chicago. Elmer Smith developed a cutting torch in 1912. About 1914, shortly after Smith formed the Vulcan Process Co. to make and sell oxyacetylene welding and cutting equipment, the Linde Co., which owned the French Jottrand patent on cutting, started a law suit to protect its patent rights. To guard the interests of U.S. manufacturers, the Oxygen Protective Association was formed.

Oxyacetylene welding developed very fast in the United States. With regard to the volume of work being done by this process in the world in 1914, the United States is estimated to have been first, Germany second, France third, and England a poor fourth. In technology development, however, Germany was generally conceded the lead over all others. The war created a huge demand for scrap metal for use in the manufacture of the steel needed for ships, guns, and other war materials. The oxyacetylene cutting torch made possible the cutting up of old boilers, ships, tanks, and other forms of steel scrap that would otherwise not be reusable.

Linde Air Products—Union Carbide

Because of the great demand from U.S. industry for oxyacetylene welding and cutting, there was urgent need for low-cost, high-purity oxygen. Until late 1907 oxygen was only produced by chemical or electrolytic processes at high cost. At that time Cecil Lightfoot began operating the first Linde plant to produce oxygen from the atmosphere in Buffalo, New York. Linde Air Products Co. was formed in 1906 to exploit the American rights to the liquid air process developed by Carl von Linde.

In 1912 the Union Carbide Co. acquired part of Linde Air Products Co. For the manufacture and development of the oxyacetylene apparatus a newly formed organization, the Oxweld Acetylene Co., was formed. The Linde firm had earlier introduced torches of the Fouché type as well as other oxyacetylene apparatus. The Oxweld organization went on to develop a complete line of acetylene generators and welding and cutting items. At that time the Davis Oxygen Co. discontinued the manufacture of oxygen for the Davis–Bournonville Co. and arranged for the Linde company to supply Davis–Bouronville's customers with oxygen.

Air Reduction Company

Around 1905 the Albany banker Robert Pruyn became convinced that there was a great future in oxyacetylene welding if oxygen could be produced economically. He interested Percy Rockefeller, nephew of John D. Rockefeller, in the project. In 1908 they hired Herman Van Fleet, a young sales engineer with welding experience, to work with the American Oxygen Co., one of their enterprises. This company produced oxygen from potassium chlorate and manganese dioxide. Van Fleet tried to find a commercially satisfactory way of making oxygen and started initial experiments on different chemical methods at a small plant in Camden, New Jersey, but found that all the chemical methods were too costly. In 1911 Van Fleet started experiments on a method of producing oxygen from the air. He began construction of a small oxygen unit and by 1914 a plant producing about 300 cubic meters of oxygen per day had been put in operation.

At the same time Air Liquide sent W. A. Hollingsworth to the United States to explore the possibility of starting a liquid air business there. Hollingsworth opened negotiations with Rockefeller and Pruyn, which resulted in an exchange of patent rights and the formation of Air Reduction Co. in 1915.

Mechanization and Automation of Welding and Cutting

Although the oxyacetylene welding and cutting flame has been generally considered a repair and maintenance tool, development of automatic welding and cutting was started around 1910. Already in 1906 the first *screw-guided* and hand-driven cutting machines were in use for square-edge and straight-line cutting. In 1911

Figure 7.2.18. Early cutting machine, 1915. From Hoffmann (1938).

Figure 7.2.19. Welding machine for automatic joining of radiator elements, AGA ca. 1930. Courtesy of AGA Historical Picture Archive.

an automatic oxygen cutter called the **Oxygraph** was introduced by the Davis–Bournonville Welding Co. This firm then developed many new automatic oxyacetylene welding and cutting systems. In Europe the first machines with an electric motor and rheostatic speed control were built in 1913.

The first efforts at arc welding automation involved the use of bare or lightly fluxed wire. This occurred between 1925 and 1928, when the benefit of the shielded arc was just beginning to be recognized. One of the first firms to have an automatic wire feeder on the market was General Electric Co. During the 1930s efforts were made to find ways to automate all of the processes (gas, arc, and resistance) and adapt them to production situations.

Welding with a Gas Shield

Submerged-Arc Welding

The submerged-arc welding (SAW) process, introduced in the 1930s, was a step forward in automatic arc welding. It was much used in the shipbuilding industry. The basic idea behind it came from solving a problem in trials of welding enameled cast iron in the early 1920s. Harry E. Kennedy worked on the problem together with researchers at the University of California in Berkeley. The automatic "under-powder" welding process was developed by the National Tube Company for a new pipe mill at McKeesport, Pennsylvania, for welding longitudinal seams on pipes.

A solid or cored bare wire electrode is fed into the weld zone, where a layer of flux is placed on the seam. The heat of the arc melts the electrode and the flux to cover the weld pool. The flux refines the pool, shields it from the atmosphere, solidifies, and shapes the weld bead to form a clean weld. This process was patented by Robinoff in 1930, and was sold in the early 1930s to the Linde Air Products Co., who recognized its value. In 1935 they introduced their Unionmelt process commercially. Automatic submerged arc welding is much used in both shipyards and ordnance factories and is one of the most productive welding processes.

New Fuel Gases

During these early years there were many other gases besides acetylene used for cutting. Some of them, although possibly not as suitable for welding, afforded certain advantages in cutting, one of the most desirable being lower susceptibility to flashback and backfire when mixed with oxygen. Pure gases or gas mixtures with butane, propane, etc., are frequently used in cutting applications.

Shielding Gases

After the introduction of coated electrodes there was considerable interest in shielding the arc weld area by externally applied gases. During the 1920s it was realized that atmospheres of oxygen and nitrogen in contact with the molten weld metal caused brittle and sometimes porous welds. Therefore, research was started to develop gas-shielding techniques.

Peter Alexander and Irving Langmuir used hydrogen as a welding atmosphere. They utilized two electrodes, starting with carbon electrodes and later using tungsten electrodes. The hydrogen was dissociated to atomic hydrogen in the arc and then blown out of the arc forming an intensely hot flame. The method was called the atomic hydrogen welding process. It never became very popular and was used only for special welding applications during the 1930s and 1940s and later for welding of tool steels. In 1930 two American technicians, Henry M. Hobart and Paul K. Devers, patented a method using helium or argon as a gas shield for the weld. This process was not developed, for several reasons. The primary one was the cost of argon and helium at that time. General Electric, to whom the patent was assigned, felt that the welding industry was not yet ready for this process and torch equipment was not developed.

Various other companies experimented with argon and helium as shielding gases, but cost considerations were against the adoption of the process. During World War II, however, the problem of welding light-weight metals such as aluminum and magnesium became acute, especially in the aircraft industry. Argon was first used for shielding, but after some time alternative methods were developed using the much cheaper carbon dioxide, and the development of new welding processes and techniques was intensified.

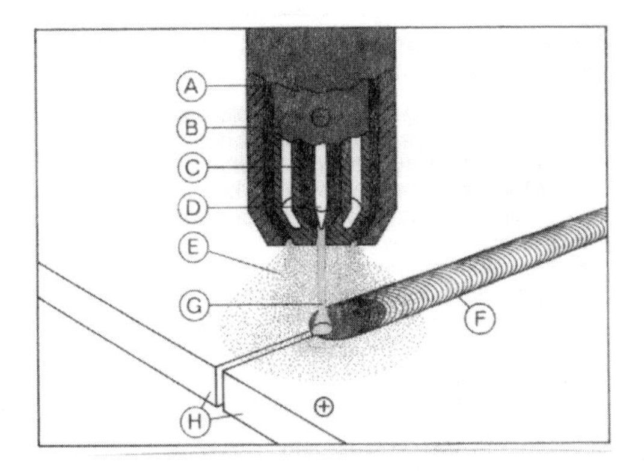

Figure 7.2.20. Arc welding with flux-coated electrode (principle). Courtesy of AvestaPolarit Welding, Avesta, Sweden.

Arc Welding Processes

Heliarc: Tungsten Inert Gas (TIG) Welding

In 1940, Russell Meredith, a welding engineer at Northrop Aircraft, was assigned the task of developing a process for welding an all-magnesium plane. Basing his investigations on the earlier work on argon and helium gases done by Hobart and Devers, he achieved success in less than 1 year.

At that time, all aircraft welding was done with oxy-fuel gas torches. After many preliminary trials Meredith tried hand-feeding magnesium wire through inert gas. This technique was not successful because of the high burnoff rate of smaller diameter wire and the low burnoff of large wire. Additional research indicated that a nonconsumable electrode of tungsten would be better, and a torch containing a simple electrode holder in inert atmosphere did the job. The process was patented by Meredith in October 1941 and soon after, Linde Air Products was granted a license to develop the TIG process under the trade name of Heliarc (Fig. 7.2.21). Later argon replaced helium as the shielding gas (argon-arc welding).

Sigma: Gas Metal Arc (GMA) or Metal Inert Gas (MIG) Welding

TIG welding was considered the way to join aluminum and magnesium, but at a material thickness above about 10 millimeters the process required a preheat. This was not feasible on larger weldments or in a production situation because the heat transfer rate of these metals is so great. Therefore new welding methods were sought that would supply sufficient control in welding thick and large pieces of these metals. Much work on the problem was done at Battelle Memorial Institute and the Air Reduction Co.'s laboratories in Murray Hill. A 3-year research project yielded an entirely new concept: an argon-shielded, consumable coiled electrode process where welding current was supplied by standard dc welding generators.

This metal inert gas (MIG) welding technique utilizes a reel of level-wound electrode wire, a wire drive mechanism, and a control panel to regulate the drive motor

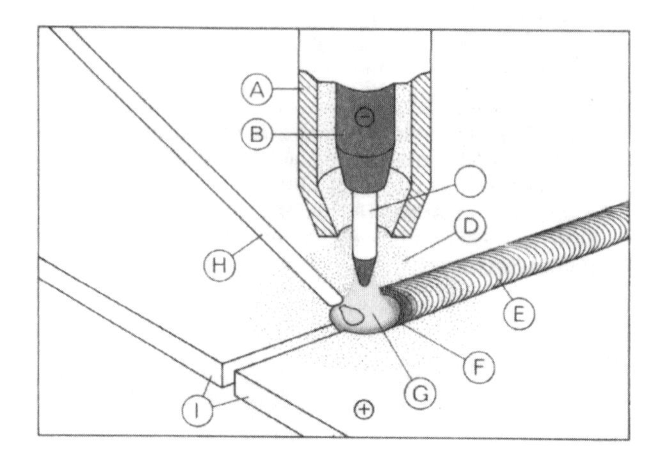

A Gas cup
B El ectrode holder
C Non consumeable tungsten electrode
D Shielding gas
E Weld
F Weld pool
G Electric arc
H Filler metal
I Base material

TIG Welding

A Gas cup
B El ectrode holder
C Welding wire
D Shielding gas
E Weld
F Weld pool
G Electric arc
H Base material

MIG Welding

Figure 7.2.21. TIG welding (principle). Courtesy of AvestaPolarit Welding, Avesta, Sweden.

speed (Fig. 7.2.22). Of the many innovations and improvements based on this basic method, one of the first was MIG spot welding, brought out by Linde Air Products in 1951 under the trade name of Sigma welding.

CO_2: Metal Active Gas (MAG) Welding

At first argon was generally favored for MIG welding and helium for TIG. There was also work on combining the various gases in order to stabilize the arc. When looking for cheaper shielding gases it was discovered that a mixture of argon and carbon dioxide was quite acceptable for welding stainless steel.

Researchers had been trying CO_2 for years because of its availability and low price. Much of the work was done between 1920 and 1930. P. Alexander of the General Electric Co. in 1926 reported trials with a consumable electrode shielded by carbon dioxide, but he stated that the weld was highly oxidized and brittle. In the period from 1920 to 1930, carbon monoxide (reducing) appeared suitable from the standpoint of weld ductility, but carbon dioxide (oxidizing) was declared unsuitable for producing ductile welds.

M. W. Mallet and P. J. Rieppel at the Battelle Memorial Institute found that stable arcs could be obtained by feeding carbon dioxide through a hollow steel

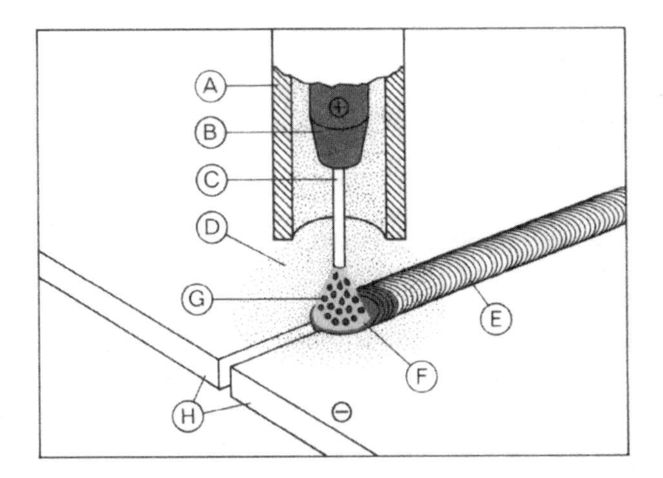

Figure 7.2.22. MIG welding (principle). Courtesy of AvestaPolarit Welding, Avesta, Sweden.

electrode. In 1946 G. J. Gibson reported the mechanical strength and elongation properties from weld metal deposited with a lightly coated electrode in an atmosphere of carbon dioxide. The MAG welding method began to be commercially applied in the United States in the early 1950s. Two European scientists, Lyubavskii and Novoshilov, reported the use of carbon dioxide shielding for welding carbon steel and stainless steel in 1953. They stated that the mechanical properties were satisfactory, provided the proper electrode composition was used. CO_2 welding became popular because it utilized equipment developed for inert gas metal arc welding, which now also could be used for economically welding steels.

Gouging and Scarfing

Both those techniques are derived from the oxygen cutting process and have been in use since 1930. Oxygen gouging has been supplemented by the compressed air arc gouging process with carbon electrodes since the development of the latter in the period about 1950.

Oxygen scarfing of ingots, blooms, and billets is in common use in modern steel production and is done with either hand scarfing torches or automatic oxygen scarfing machines.

High-Energy-Beam-Methods

Plasma Arc Cutting (PAC)[8]

Plasma-arc cutting was a spin-off from another investigation. The first industrial application of plasma was achieved in 1909 at Badische Anilin und

Sodafabrik (BASF) by the physicist Otto Schönherr in the manufacture of nitrogen dioxide (NO_2). Two years later, the American physicist Matters developed a plasma arc torch for heating a metal fusing furnace. In 1912, the American Langmuir gave the name "plasma" to a gas or gas mixture brought to such a high temperature that all diatomic molecules are dissociated and the atoms partially ionized and where also all monatomic gases are fully ionized. This plasma is electrically neutral, but a very good conductor of electricity. In 1921, the American Himes built a plasma-arc cutting torch with which it was possible to make fairly good cuts in metals.

In 1953 Robert M. Gage was working on the problem of arc melting titanium ingots at the Linde Research Laboratories in Tonawanda, New York. In his experiments with inert gases and arcs he observed the similarity of appearance between a long electric arc and an ordinary gas flame. It occurred to him that the intense heat and velocity of the arc might be better controlled and applied by putting it through a small nozzle. The result was a constricted arc, a plasma jet torch utilizing a plasma arc struck between a tungsten electrode (cathode) and the work piece (anode). The plasma gas was a mixture of argon and hydrogen. Now the most common plasma cutting gases used are argon, hydrogen, nitrogen, and mixtures of these as well as air and oxygen.

Experiments with plasma-arc cutting equipment of different metals were quickly successful. Before the invention of plasma-arc cutting, aluminum was cut slowly and expensively by mechanical means. Those who cut aluminum needed a cheaper, faster way to do it. The plasma-arc cutting torch, introduced in October 1955, filled this need.

The plasma-arc process proved excellent for cutting aluminum and led to entirely new concepts in both manual and automatic cutting of that metal. From 1955 to 1957 plasma arcs were sold mainly for cutting aluminum. In 1958, Linde developed the plasma arc to the point where it would cut stainless steel economically, its most common use at present. Since 1958 it has also been used on high-alloy steels, copper and copper alloys, and carbon steels, using oxygen or nitrogen or compressed air as plasma-arc gases.

Plasma-Arc Welding (PAW)

Plasma-arc welding, a process which is very similar to TIG welding, started to be used in the late 1950s after Gage's work. It uses a constricted arc, which creates an arc plasma that has a higher temperature than the tungsten arc (Fig. 7.2.23).

Electron Beam Welding (EBW)

The electron beam welding process, which uses a focused beam of electrons as a heat source in a vacuum chamber, was developed in Germany and France. The Zeiss Company and the French Atomic Energy Commission developed the process in the late 1940s. The first to publicly announce successful use of the electron beam to weld metal was Dr. J. A. Stohr of the French Atomic Energy Commission, who began work on the process in 1954. Electron beam welding is used for welding small details of special materials with very high precision.

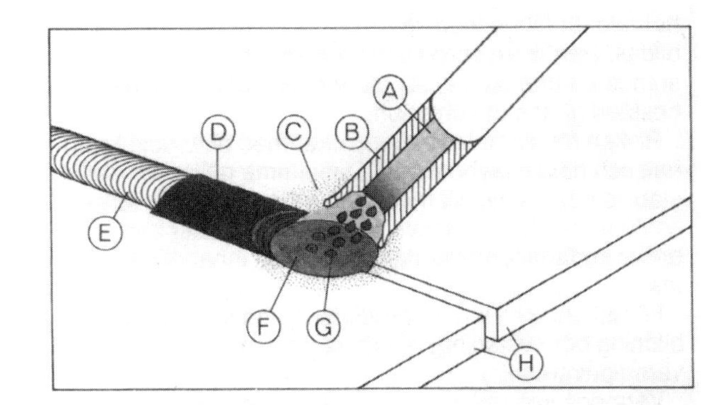

Figure 7.2.23. Plasma welding (principle). Courtesy of AvestaPolarit Welding, Avesta, Sweden.

The end of the 1960s saw attempts to take the work piece out of the normal shielding gas environment with argon or helium but still keep the welding torch in a vacuum chamber. This method remains at the experimental level of design.

Laser Beam Cutting (LBC) and Laser Beam Welding (LBW)

Laser welding and cutting use the combination of a focused laser beam and a process gas. The idea of using a light beam for the cutting and welding of materials has been studied for a long time. In 1960 the first ruby laser was developed by the American Maiman in the laboratories of the Hughes Aircraft Company. T. H. Maiman left Hughes and formed the Korad Corp. to exploit the laser process for welding. Korad later was bought out by the Linde Division of Union Carbide. Since then the laser has played an increasingly important role in welding and cutting (Fig. 7.2.24).

In 1959 Ali Javan and his team at the Bell Telephone Labs began studying laser action in gases. They constructed their first helium–neon laser early in 1961. This laser was able to operate continuously, whereas the ruby laser operated on a pulse mode. The rapid development of the CO_2 laser after 1965 resulted in systems with higher output power suitable for cutting and welding. In 1971, the Englishman Peter Thomas Houldcroft of British Welding introduced a supplementary axial gas jet around the laser beam, making it possible to do thermal laser cutting of metals more efficiently and narrower at higher cutting speeds and greater thickness.

In laser processes the purity of the gases is essential for an optimal result. Oxygen, argon, nitrogen, or combinations of these are used as the cutting gas, and in welding applications shielding gases such as argon, helium, helium–argon mixtures, carbon dioxide, and nitrogen are used.

Almost all of the new processes and equipment introduced during the past 20 years have been in the arc welding field or based on entirely new technology. Comparatively little of importance was developed for gas welding during this period.

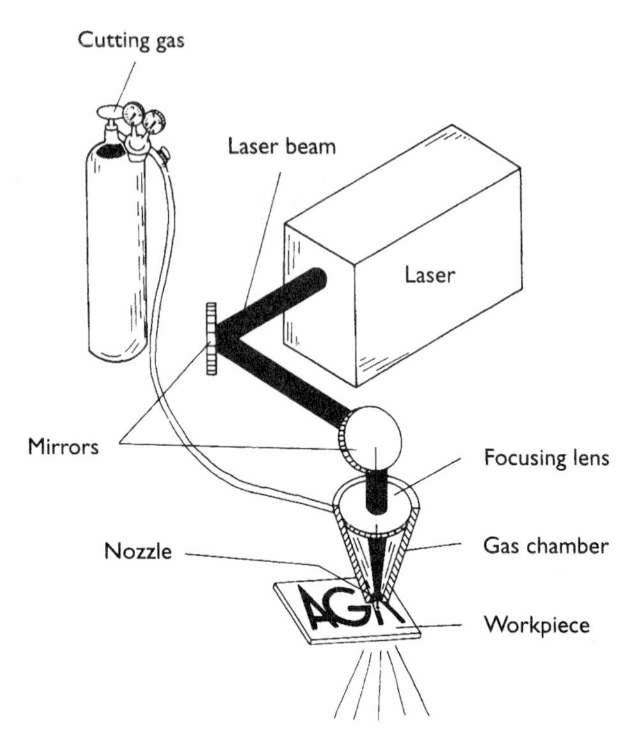

Figure 7.2.24. Laser cutting (principle). Courtesy of AGA Historical Picture Archive.

The First Uses of Oxygen Cutting (after Reference 5)

With the marketing of pure oxygen gas by the brothers Brin came ideas from many quarters. One was the aging of whisky in days instead of years, using oxygen. In Warrington, Thomas Fletcher, a gas engineer, had been studying ovens using a combination of hydrogen, oxygen, and coal gas in order to achieve high temperatures. He manufactured the forerunner to welding and cutting jets using a combination of oxygen and coal gas. On the 7 March 1888, the inventor demonstrated the efficiency of the jets on iron at a lecture to the Society of the Chemical Industry. He was subjected to unexpected criticism from the banking world. One critic said "This method can only be used for criminal purposes and should therefore be forbidden." The person who voiced those fears could say "I told you so" when, in 1901, a post office in London had its safe cut open and rifled.

It had happened in Germany somewhat earlier. A man calling himself Smith signed in at a hotel in Hannover on 22 December 1890. The hotel was situated above the offices of the Niedersächsisen Bank. A Mr. Brown, who had stayed at the hotel previously, reserved a room for himself and the adjacent rooms for his father and sister, who would be arriving later. In his baggage he had three small cylinders of oxygen from the Sauerstoff-Fabrik Dr. Elkan in Berlin and a Fletcher cutting torch. Mr. Brown first drilled a small hole in the floor, large enough to pass an umbrella through. Then he opened the umbrella in order to catch the pieces of plaster that fell as he made the hole large enough to get himself through. Using the rope ladder that he had with him, he climbed down into the room below. An iron spiral

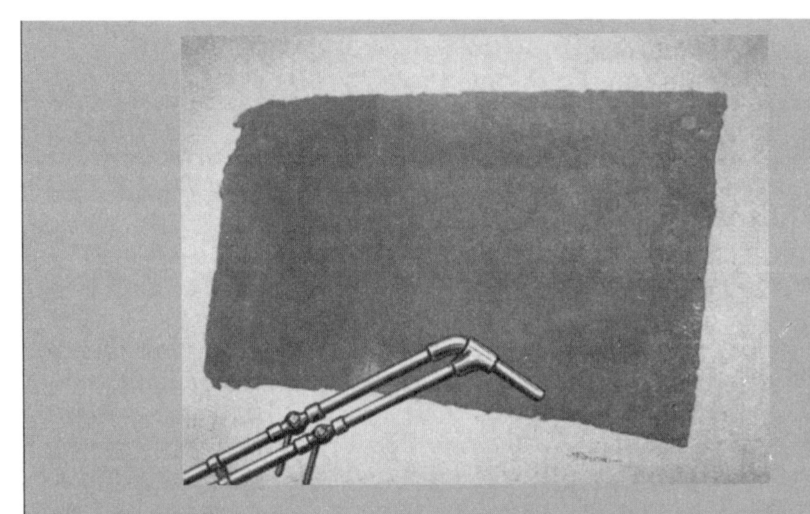

Figure 7.2.25. The first crime with the help of gas cutting. From Hoffmann (1938).

staircase led from there down to the vault. To avoid setting off the alarm, Brown unscrewed the staircase and insulated it with pieces of wood. Then he connected his cutting torch to the bank's gas supply and his own oxygen. All that separated Brown from 7 million marks was the double iron-plated vault wall. He cut a 30 × 50 centimeter hole in the first—and then his oxygen ran out! At 2 o'clock in the morning of December 26, Smith left the hotel on the excuse that urgent business had cropped up and he must take the express train to Cologne and he would be returning. But he never came back!

The equipment used by Brown and the results of his work can be seen in the Kriminal-Museum in Berlin (Fig. 7.2.25). The cutout piece of plate was to play an important part in the 1910s in a struggle for patent rights relating to the cutting method. Had the cutting method been used by Fletcher or had he just melted the iron? In 1914 it was decided that Fletcher had not invented the method for cutting. The honor was given instead to Ernst Menne, who patented the use of a hydrogen/oxygen flame to remove blockages of solidified iron in blast furnaces.

References

1. Bernsdorf, G. (1986). *Auf heissen Spuren vom Schmeltzen, Löten, Schweissen.* Düsseldorf, Germany: Deutscher Verlag für Schweisstechnik (DVS), and Leipzig, Germany: VEB Fachbuchverlag.
2. Granjon, R., and Rosemberg, P. (1912). *Manuel Pratique de Soudure Autogène*, pp. 44–47. Paris: Bibliotheque de l'Office Central de l'Acetylène.
3. Simonson, R. D. (1969). *The History of Welding*. Lake Zürich, IL: Monticello Books.
4. Nunes, A. C., Jr. (1977). Gas welding origins. *Welding Journal* **197**(6):5–23.
5. Hoffmann, P. (1938). *Aus Vier Jahrzehnten deutsche Autogentechnik Ein Rückblick*. Halle, Germany: Carl Marhold Verlag, 1938.
6. International Institute of Welding, Commision I. (1980). Some historical notes on thermal cutting processes. *Welding in the World* **18**(1/2):23–28.

7. Davis, A. (1911). The history and present status of the oxyacetylene process in America. In *Proceedings of the International Conference of the Acetylene Association*, pp. 109–117.
8. Hypertherm. (1990). *A history of plasma cutting*. Hanover, NH.

7.3. METALLURGY AND COMBUSTION

Background[1]

The use of gases within metallurgy is fundamental. Large volumes of oxygen are used for combustion, and gases are also used as reagents during reduction of iron ores and as oxidants during steelmaking. In ladle metallurgy, gases are used to accelerate refining reactions (e.g., desulfurization, deoxidization) and for a number of other purposes. Metals are often heat-treated in "controlled" or "protective" atmospheres. These include any gas or mixture of gases used to prevent or hinder the reactions that would take place during the heat treatment of the metal if it was surrounded by air instead of the synthetic atmosphere. Such reactions include oxidation (combination of oxygen with the surface layers of the metal), carburization (pickup of carbon by ferrous metals during heating), and decarburization (loss of carbon from surface layers of ferrous metals). Oxidation and decarburization are the commonest sources of trouble in the treatment of metals at elevated temperatures.

Some examples of use of industrial gases in metallurgy are as follows [after Brunner (1990); see reference list at the end of this section]:

Oxidizing gases	
O_2, O_3	Oxidant at heat generation
CO_2, H_2O	Decarburization
NO_X	Oxidation
Reducing gases	
CO, H_2	Fuel gases
NH_3, CH_4	Reduction
C_2H_2	Protective atmospheres
Inert gases	
Ar, He	Stirring, mixing
	Degassing
	Lowering of partial pressure
	"Vacuum effect"
Other	
N_2, SF_6,	Stirring
etc.	Nitriding
	Cryogenic cooling
	Fluorating

Historically, the large-scale gas industry emerged due to demands by the metallurgical industry. The first market for really huge quantities of oxygen started when in 1949 oxygen steelmaking was developed in Austria. The growth rate of the industrial gas market has since then been very great.

Figure 7.3.1. Blast furnaces for smelting copper–lead ores. From Agricola, *De re metallica*, 1556.

Early History

Wrought iron was the chief engineering metal until about 1860. Steel was used as a specialist material for tool making, derived from wrought iron by reintroducing carbon in the carburizing process and then melting the carburized iron in a crucible. The great step which opened the steel age occurred when Henry Bessemer (Fig. 7.3.2), searching for a way to make better materials for weapons, found in 1856 that air blown through liquid iron converted iron into steel. Another steelmaking process, the open-hearth process, was invented by William Siemens and Pierre Emile Martin and demonstrated by Martin in France in 1863. A difference between the open-hearth furnace and the Bessemer converter (Fig. 7.3.3) is that in the open hearth the fuel and air are turned over the top of the iron melt instead of being blown through the bottom of

Figure 7.3.2. Henry Bessemer. From *Berühmter Ingenieure* (1911).

Beſſemer-Stahlbereitung mittels des Birnapparates.

Figure 7.3.3. Early tests of the Bessemer method. From *Berühmter Ingenieure* (1911).

the furnace. Due to these new developments the price of steel fell about sevenfold and cast steel products, superior in every way to wrought iron, were available at one-fourth of the price. Very rapidly steel replaced wrought iron as the main engineering material.

The Bessemer and Siemens–Martin processes were later subdivided into acid and basic processes. The acid Bessemer process can use only molten iron low in phosphorus and sulfur. The basic process, which can use iron high in phosphorus, was developed in 1878 by Sydney Gilchrist Thomas, whose process is called either the basic Bessemer process or the Thomas process. A disadvantage of this process is that the steel produced has a high nitrogen content caused by air passing through the high-temperature liquid metal; this limits the type of steel that can be produced.

Use of Oxygen for Oxidation

The idea of using pure oxygen to convert molten iron into steel was proposed by Bessemer as one of several alternatives in his patent of 1856. At that time, however, there was no known way to produce the gas in quantity; pure oxygen was just a laboratory curiosity.

It was not until 1930, with the development of the Heylandt process for liquid oxygen and efficient heat exchangers by Fränkl, that the production of oxygen in tonnage quantities could make it available at prices sufficiently low for use in metallurgical operations. This encouraged experiments in oxygen use in electric arc and open-hearth furnaces. Oxygen or oxygen-enriched air was lanced below the metal surface through steel tubes to remove unwanted phosphorus, silicon, manganese, and carbon from the steel by oxidation.

Hearth furnaces had been used for centuries, but they could not attain temperature high enough to melt steel and retain it in a molten state for casting. The early furnaces melted the iron, and its impurities were removed by oxidizing through air let in to the furnace atmosphere. At appropriate intervals iron ore was added, which decomposed and supplied oxygen to the bath. Steel scrap was also added. These procedures dilute the impurities in the iron and also supply iron oxides such as scale. The same principles for supplying oxygen to the process continued to apply for almost 100 years; meanwhile, the open-hearth process became by far the largest producer of the world's steel requirements.

The Bessemer process using an acid lining and acid slag persisted on a very small scale. The Thomas process was much used in Europe. To overcome the problem of high nitrogen content, pure oxygen was tested as a substitute for air in the 1940s. Above a level of about 40 percent oxygen in the blowing air, however, the strongly exothermic oxidation process generates very high temperatures, which cause damage to equipment at the bottom of the converter.

As recently as 1962 over 50 percent of the total steel production of France, West Germany, and the Benelux countries was Thomas steel. In all other countries the open-hearth process developed rapidly, and eventually produced over 85 percent of world steel. These processes were replaced during the 1960s by the Linz–Donawitz (LD) and other oxygen steelmaking processes.

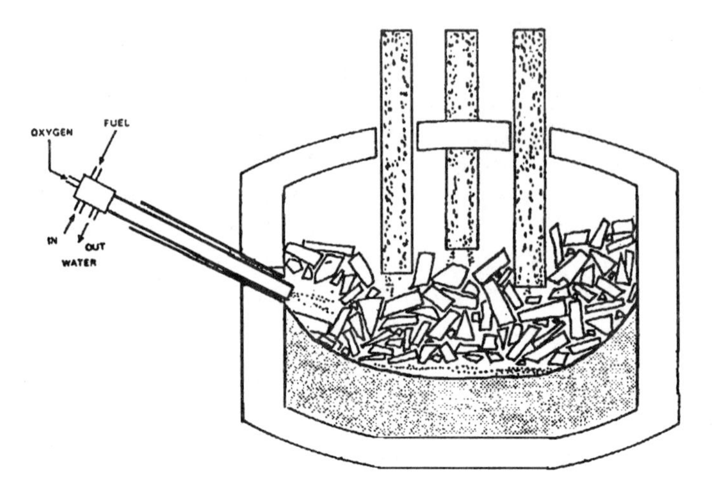

Figure 7.3.4. Oxy-fuel burner oxygen assisted melting of scrap in an electric furnace.

The very low nitrogen (VLN) process was used to produce steel on a large scale in the United Kingdom during the 1950s and 1960s using a mixture of oxygen and steam in a Thomas-like converter. The steam dissociated to oxygen and hydrogen, and steels made by this process tended to contain high levels of hydrogen, limiting the type of steel products which could be made by this method.

Modern Steelmaking Processes[2–9]

Three top-blown converter processes, the LD, Kaldo, and Rotor processes, using 99.5 percent pure oxygen blown at very high speed to penetrate the slag and underlying molten metal, were developed in Austria, Sweden, and Germany in the late 1940s to early 1950s. The LD process, which is the basis for present-day steelmaking, has been further developed in different parts of the world. The other two, the Swedish Kaldo process with a converter similar to an LD vessel in shape, but using an inclined rotating vessel, and the Oberhausen Rotor process, using a rotating horizontal cylindrical vessel, have faded into history as a result of the complex engineering required to provide rotation and because of excessive refractory wear.

The LD Process

The Swiss Robert Durrer, Professor of Metallurgy at the Technical University of Berlin-Charlottenburg, asked metallurgists to look into the possibilities of oxygen. In 1939 his colleague, Carl Valerian Schwarz, obtained the first patent on producing steel by letting a supersonic oxygen stream impinge upon an iron melt with such high kinetic energy that it penetrated into the melt.

The first experiments with oxygen were continued during and after World War II. In order to increase productivity and quality with lower investment and operation costs, attempts were made to blow mixtures of air and oxygen, instead of air alone, through the molten iron in the Thomas converter. These tests ran, however, into serious

Figure 7.3.5. Oxygen lancing at the Domnarfvet steelworks, Sweden, ca. 1970. Courtesy of AGA Historical Picture Archive.

trouble. Oxygen produced higher temperatures and accelerated the destruction of the equipment at the bottom of the furnace. Particularly active efforts were made at Linz in Austria after the war. Theodor Suess, who was the technical manager of the plant, cooperated with Robert Durrer and his staff to carry out a systematic program of studies on the Linz open-hearth steel furnaces. They began with top blowing of oxygen through a water-cooled lance into liquid iron contained in a vessel similar in shape to the Bessemer converter. Many attempts were made to develop new processes, but it was not until 1952 that a new method of steelmaking using high-purity oxygen came into commercial use. In most parts of the world the steel industry had invested heavily in conventional open-hearth furnaces and regarded oxygen as too expensive for large-scale utilization. Based on this work a 2-ton experimental converter was installed at the Vereinigte Östereichische Eisen und Stahlwerke (VOEST) at Linz, Austria, and the first cast of steel was successfully made in June 1949 (Fig. 7.3.6).

At Donawitz, further work was done in a 5-ton and a 10-ton vessel, and at Linz a 15-ton vessel was later used. In each case the plant was improvised and operated adjacent to an existing open-hearth melting shop.

Figure 7.3.6. LD process, first blasting at Linz, Austria, 1949. Courtesy of Voest-Alpine Stahl AG, Linz, Austria.

The operating principles of the LD process are still today very similar to those of the early production plants. Vessel size has, however, increased to 300-ton capacity. The rate of delivery of oxygen has also increased in proportion to capacity, so that the actual blowing has also increased in proportion to capacity and the actual blowing time of modern large vessels is almost the same as in the original production plants.

The Austrian experimenters succeeded in accelerating the conversion of iron to steel, but the increased flame temperature destroyed the roof of the furnace and the regenerators became clogged with dust. Attempts to feed oxygen into an electric furnace again failed because the heat ruined the electrode holders. The Linz engineers consulted Durrer, who, with his assistant Heinrich Hellbruegge, was experimenting in Switzerland using oxygen in a 2-ton converter. They were injecting a jet of oxygen into the molten iron through a water-cooled lance introduced just above the surface of the iron. The first trials of this system in experiments at Linz failed. The heat destroyed the lance; the stream of oxygen blown deep into the melt caused damage to the bottom and refractory materials of the vessel, and the treatment failed to remove enough of the phosphorus impurity from the iron.

By reducing the impact pressure of the oxygen and raising the lance farther from the surface of the melt the process worked splendidly. The bottom of the vessel was undamaged, the lance survived, carbon monoxide gas was effectively removed, and the experiment produced steel of good quality (Fig. 7.3.7).

The Linz researchers found that deep penetration of the oxygen jet had to be avoided. Since Bessemer's days researchers thought that refining by means of an

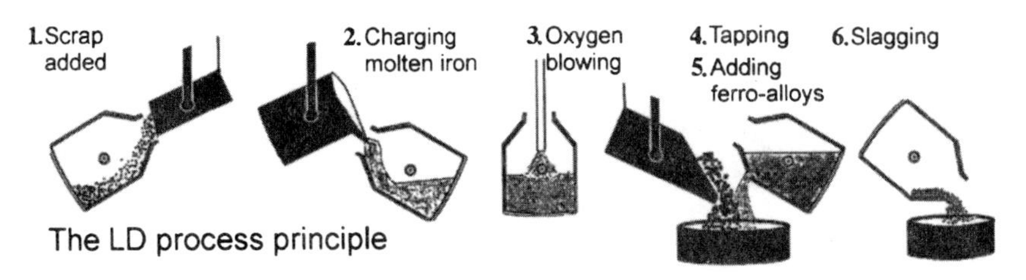

Figure 7.3.7. Principle of the LD process.

impinging oxygen stream required the oxygen and the iron melt to mix intensively. This was found to be wrong. In 1952 the system was installed on a commercial scale at Linz, using vessels of 35-ton capacity in the first LD plant in the world which produced steel on an industrial scale. Next year a second plant of the same kind went into operation at Donawitz. The system has since been called the LD process, although in the United States it is called BOS, which stands for basic oxygen steelmaking. The rate of introduction of the LD process has been phenomenal. In 1957 it accounted for about 1 percent of world steel production. By 1962 it had reached almost 20 million tons and in 1966 had exceeded 100 million tons per year, or 25 percent of world production. Typically, each ton of steel produced in the converter needs 55 cubic meters of oxygen of at least 99.5 percent purity to decarburize the molten pig iron and scrap charge.

The LDAC, OCP, and OLP Processes

Development work was also being done on the LD process in France, which eventually came to be called the LDAC process (named for Linz–Donawitz plus co-developers ARBED and CNRM). The LDAC process, which can use phosphoric molten iron (up to 2 percent phosphorus), is different from the LD process in the following respects: Powdered lime is injected through the oxygen lance, and a flush slag technique is used to continually remove phosphorus. The research organization CNRM (Centre National de Recherches Metallurgiques) in Liege, Belgium, developed a process using lime injection called the OCP process (oxygene, chaux, pulverisée). This was first brought to commercial success at the ARBED-Dudelange works in Luxembourg. A variation of the LDAC, the OLP process (oxygen, lime, powder), in which lime is injected with the oxygen stream, was developed in France by IRSID (L'Institute de Recherches de la Siderurgie) about the same time as the LDAC was developed.

The Kaldo Process

The Kaldo process was initiated and developed to production scale at the Domnarfvet works of Stora Kopparbergs Bergslags AB in Sweden. The process was mainly invented by Bo Kalling and his collaborators. The process can deal in an equally satisfactory way with both high- and low-phosphorus iron. Experiments commenced in 1948 and a production-scale plant, using a 30-ton vessel, was started at Domnarvet in 1954. It went on to commercial-scale production in 1956. A vessel with

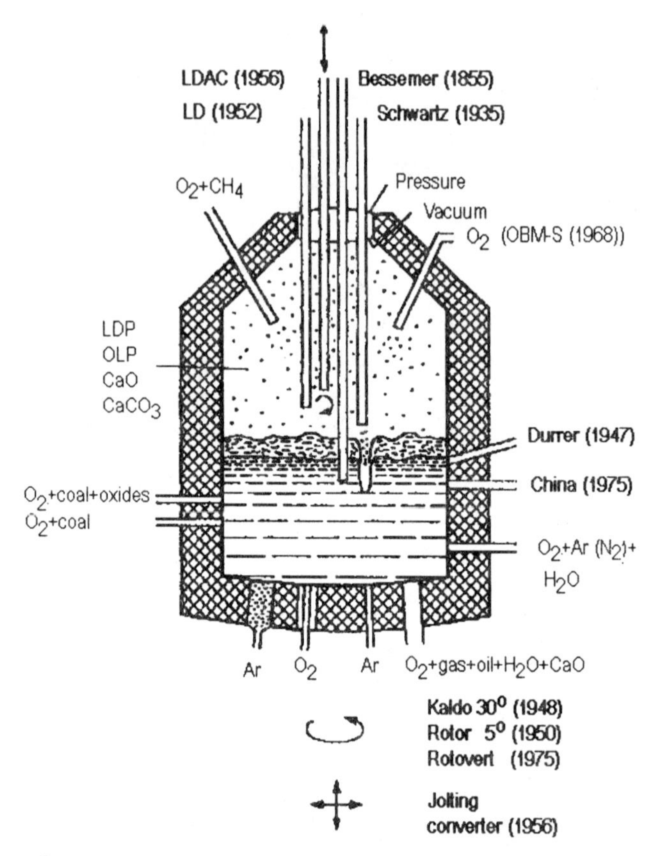

Figure 7.3.8. Summary of developments in oxygen steelmaking. From Eketorp (1981). Courtesy of S. Eketorp.

a nominal capacity of 125 tons commenced operations in France in 1960, followed by another of similar capacity in Sweden in 1961. Some years later the Kaldo process was used also in the United Kingdom, France, the United States, and Japan.

The Rotor Process

The first practical tests which led to the introduction of the Rotor process were made in April 1952 at Oberhausen, Germany, using a melting furnace in which the oil burner was replaced by an oxygen nozzle. Modern vessels are built on a turntable so that they may also be completely rotated in the horizontal plane, whereas for changing, the vessel can be tilted to an angle of 90°. Vessels were installed and operated in Germany, South Africa, and England. Two water-cooled lances are moved into the vessel opening during operations. The primary lance injects oxygen below the metal surface, thereby directly oxidizing the impurities. The objective of the secondary lance is to supply oxygen, oxygenated air, or air alone so as to consume a large proportion of this combustible gas in the furnace itself, thereby liberating heat.

The AOD Process

The AOD (argon oxygen decarburization) process (Fig. 7.3.9) for stainless steel production has allowed steelmakers to produce a higher quality product at a lower cost. The process was initially developed to reduce the cost of refining stainless steel, but a side effect of improving product quality with lower overall processing costs is one of the reasons for the rapid growth in stainless production and consumption.

The problem was to decarburize high-chromium raw steel. The oxygen supplied would oxidize carbon as well as chromium. The oxidation of carbon results in a gaseous end product, whereas the oxidation of chromium results in a solid oxide. The first reaction is desired, the second is harmful.

A mixture of argon and oxygen is blown into a molten charge, which reduces the partial pressure of carbon monoxide. The argon aids in the selective oxidation of the carbon and retards the oxidation of the chromium. In the AOD process, oxygen, argon, and nitrogen are injected into a molten bath through submerged, side-mounted tuyeres causing a shift in the decarburization thermodynamics that is afforded by injecting mixtures of oxygen and inert gas as opposed to pure oxygen.

The AOD process was invented by W. A. Krivsky in 1954 at Union Carbide. After considerable development work the first commercial vessel went into operation in 1965 at Joslyn Stainless Steel, Fort Wayne, Indiana. AOD is part of a duplex

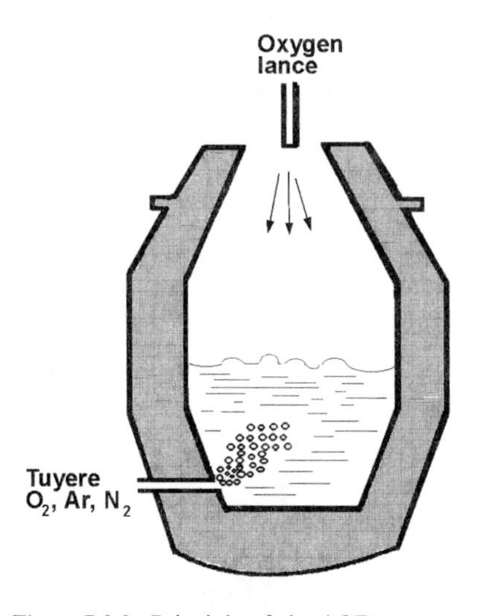

Figure 7.3.9. Principle of the AOD process.

process, with molten metal transferred from a separate melting furnace to the AOD refining unit. The AOD process consumes large amounts of argon. About 16 Nm^3 of argon gas per ton of steel is required. Over 75 percent of stainless steel production in the Western world is via the AOD process.

Hydrogen and Nitrogen in Steel

Hydrogen in Steel

When water vapor comes into contact with liquid steel, hydrogen is formed and immediately dissolves in the steel to some extent, depending on the water vapor pressure, the steel temperature, and the chemical composition. Hydrogen is a factor in causing fine cracks when forging and rolling the material.

Vacuum Degassing of Steel

Blowing bubbles of an inert gas such as nitrogen, argon, helium, carbon dioxide, or carbon monoxide through a bath of molten steel stirs the bath efficiently, aids in removal of dissolved hydrogen and other dissolved gases, and assists in the mixture of elements which may be added to the steel. It has also been developed to deoxidize unkilled steels by causing the contained oxygen and carbon to combine under lower pressures. The degassing principle was known during the late 19th century, but the first practical experiments were made in 1914.

Nitrogen in Steel

All steel contains some nitrogen. It tends to increase the yield strength, but in cold working it increases the rate of work hardening and strain aging; hence it is an undesirable element in steels that are to be heavily cold-worked without fracture. The air-blown Thomas process produces steel high in nitrogen. Efforts to reduce the nitrogen content led to the development of several modifications of the Thomas process.

Heat Treatment[10–12]

Heat treating is a way of changing the performance characteristics of a metal piece by running it through a cycle of heating, soaking, and cooling; the rate of cooling is particularly important for the properties of the microstructure. The gas surrounding the metal piece is also important for the properties of the end product. Since the 1960s ready-prepared mixtures of gases have often been used as protective and reactive gases in heat-treating furnaces containing combinations of nitrogen, hydrogen, carbon dioxide, carbon monoxide and hydrocarbons.

Machine parts, tools, and other metal parts usually need a wear-resistant, hard surface, but the part as a whole must remain elastic, that is, resistant to fracture. For centuries different substances such as salt baths and powders have been used for carburization, hardening, and annealing of iron and nonferrous metal products. In Germany coal gas was employed for hardening more than 150 years ago. Krupp-Werke used this gas for carburization of armor plating in the 1890s. Some years later

TREATMENT	nitrogen	nitrogen-hydrogen	nitrogen-hydrocarbon	nitrogen-methanol	hydrogen	argon	argon-hydrogen	helium			
Annealing, carbon steel	■	■	■	■							
Annealing, cast iron	■		■								
Annealing, alloy steel	■	■					■	■			
Annealing, stainless steel		■			■		■				
Annealing, non-ferrous metals	■	■			■						
Brazing		■			■	■	■				
Carburizing-carbonitriding				■							
Decarburizing		■									
Galvanizing		■									
Ion nitriding		+ CH4									
Neutral hardening	■	■	■	■							
Nitriding	+ NH3										
Quenching (austenite removal)	liquid										
Sintering	■	■	■	■	■		■				
Tempering	■	■	■								
Vacuum treatment	■					■		■			

Figure 7.3.10. Use of different industrial gases in the heat treatment of metals.

liquid petroleum and gasoline started to be used for this purpose. From about 1927 the first automatic case-hardening furnaces were used, producing so-called endogas (endothermic gas) by cracking gasoline and ammonia.

During the early 1970s, immediately following the energy crisis, the cost of natural gas and the lack of it in some countries became a major concern. As a result a range of nitrogen-based atmospheres to replace generated gases were developed by industrial gases companies. These atmospheres are created by mixing active gases with nitrogen to produce the desired chemical effects. The most common additives are

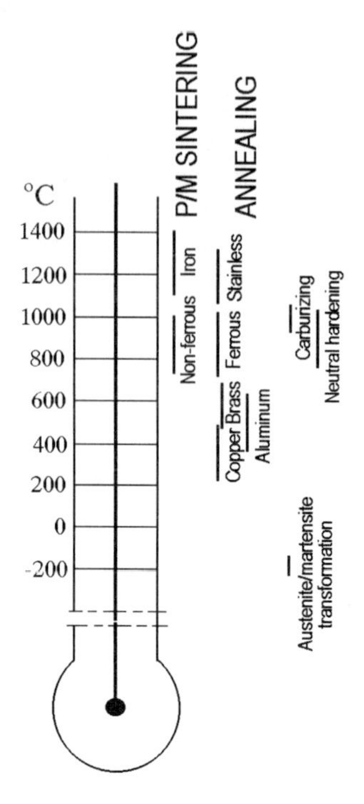

Figure 7.3.11. Typical heat treatment temperatures.

natural gas, hydrogen, methanol, and ammonia. Nitrogen-based atmospheres have replaced over half of the generated gases employed in metal working. In the aluminum industry nitrogen has become essential for degassing. The protective atmosphere keeps out the oxygen and prevents scaling, discoloration, and chemical changes, whereas reactive atmospheres add or remove chemical species by gas/solid chemical reactions. Usually a combination of nitrogen and methanol creates the base atmosphere.

In carburizing and carbonitriding, hydrocarbons are introduced into the furnace atmosphere in order to harden the surface of low-carbon parts. In neutral hardening, a neutral furnace atmosphere is used for hardening the surface of parts maintaining their carbon content. For powder metal sintering the atmosphere removes lubricants from the part, reduces oxides, prevents oxidation, and maintains or increases the carbon content.

Melting of Glass, Ceramics, and Non-Iron Minerals[13,14]

The glass, ceramics, and minerals industries are gradually making increased use of oxygen. For many years the glass container and tableware industries have used oxygen–hydrogen and oxygen–natural gas flames for smoothing or polishing the rims and edges of products by allowing by localized surface melting.

First studies at the Saint-Gobain glass works on the use of oxygen for combustion in glass furnaces date from 1948 and the first application was made in 1952 in the Escaupont plant. Efficient burners with water-cooled nozzles were used, which made possible to increase the production rate dramatically. The increase of the flame temperature and the diminution of the waste gas volume improved the combustion efficiency of the furnace. This experience proved that the principle was good, but Saint-Gobain did not develop new applications for a relatively long time because economic conditions were not favorable.

Further studies were made in April 1967 at the Lagnieu plant. The technology was the same as for Escaupont; an increase of production rate was obtained with a good quality of glass, but the energy aspect was not very good, and the large increase in chamber temperature showed that the flames were too long. The conclusion of this trial was that the technology had to be improved to obtain shorter flames. The real development of the use of oxygen was made at the beginning of the 1980s with the technology of oxy-fuel burners. The use of oxygen has became common.

For many years oxygen has also been used to increase the output from rotary kilns used for calcining dolomite, magnesite, lime, and Portland cement. Oxy-fuel burners or under-flame oxygen injection is used to intensify combustion in the kiln to produce a greater throughput of material.

References

1. Brunner, M. (1990). Use of gases within metallurgy. *Scand. J. Metallurgy* **19**:165–173.
2. Jackson, A. (1969). Oxygen steelmaking for steelmakers. London: Butterworth & Co.
3. Moore, C., and Marshall, R. I. (1980). Modern steelmaking methods. London: The Institution of Metallurgists and Chameleon Press.
4. Michaelis, E. (1977). The use of oxygen in the LD process for steel production. In *Proceedings of the 1st BOC Priestley Conference*, pp. 61–62.
5. Longden, I. P. (1980). Today's applications for industrial gases. *Steel Times* **1980**:796–801.
6. Anonymous. (1992a). BOS 40 years on.40 years of LD steelmaking. *Steel Times* **1992**:146–158.
7. Eketorp, S. (1981). Gedanken zur Stahlerzeugung in Jahre 2000. *Stahl und Eisen* **101**(13–14): 852–860.
8. Masterson, I. F., and Bury, R. P. (1990). Influence of AOD process on the production of stainless steel. *Steel Times* **1990**(1):38–39.
9. Sellers, N. (1980). Chemistry in steelmaking. *Journal of Chemical Education* **57**(2):139–142.
10. Matthews, M. J. (1979). The application of nitrogen to heat-treatment processes. *IOMA Broadcaster* **1979**(September–October):4–6.
11. Dale, J. (1983). Nitrogen-based atmospheres as the state-of-the-art. *IOMA Broadcaster* **1983** (November–December):13–15.
12. Anonymous. (1992b). ABCs of heat treating *Heat Treating* (Supplement) **1992** (December): ABC 1–7.
13. Miller, H. R., and Royds, K. (1973). The use of oxygen in glassmaking furnaces. *Glass Technology* **14**(6):171–180.
14. Janetta, G. (1988). The use of oxygen in glass furnaces. In *Proceedings of the Annual Meeting of the Scandinavian Society of Glass Technology*, pp. 3–4.

7.4. CHEMISTRY AND THE ENVIRONMENT

Chemical Processes

Many industrial chemical processes use molecular oxygen, traditionally in the form of air as oxidant. At present there are several fully developed chemical processes which use oxygen in preference to air so as to achieve a faster reaction, consequent smaller reactor, gas-handling, and purge volumes, and improvements in yield and environmental aspects.

Since the 1950s the need to build tonnage oxygen plants has been prompted mainly by demand from individual customers in the steel industry for large quantities of gas, a need which has been met by the direct piping of gaseous oxygen from a nearby plant to the user. Because the proportions of liquid oxygen needed to be produced, stored, and regasified are very small compared with those from traditional plants, changes in production technology have been introduced. The price of such oxygen has been lowered by a factor of at least three or four.

This also created a possibility for more economical use of oxygen in large-scale industrial chemistical processes. These include processes for acetaldehyde, acetylene from hydrocarbons, titanium dioxide by the chloride route, vinyl acetate, chlorine from hydrochloric acid by the Kellogg process, and others.

On a smaller scale, oxygen is used as an oxidizing agent in aqueous reactions. An example is its use as a substitute for air in reducing the BOD (basic oxygen demand) of waste water streams such as municipal sewage, and for bleaching in the pulp and paper industry.

Waste Treatment[1,2]

Industrial Wastes

Among the methods to destroy or reduce industrial wastes is incineration, which is employed for destroying industrial wastes containing at least 5 percent combustibles. The by-products of the process are mostly carbon dioxide and water, but there also some undesirable products: ash from the combustion and some poisonous gases.

In combustion processes where air blowers are used, oxygen can be utilized if the blower is operating at its maximum flow rate. The addition of oxygen in such cases must be very well regulated to avoid overheating, which could damage the incinerator's internal lining.

Waste Water Treatment

For removing organic material as well as nutrient from waste water (both industrial waste water and municipal waste water, i.e., sewage), biological methods are mostly used. In biological treatment of waste water, microorganisms (mostly bacteria) are used for removing the organic content of the water. The bacteria feed on the organic material, whereby they propagate, forming new cells (biological sludge), as well as convert some of the material to carbon dioxide and water. The produced cells can be removed from the water in the next stage after the biological stage, either aerobically or anaerobically; at present aerobic processes are predominant.

Aerobic Biological Treatment

In aerobic biological treatment of waste water the bacteria use oxygen for their respiration. The use of aerobic biological treatment can be traced back to the late 19th century. By the 1930s it was a standard method of waste water treatment. The source of oxygen can be either air or pure oxygen. The predominant source is air, but pure oxygen or pure oxygen-enriched air can be used for enhancement of the degradation capacity.

The oxygen dissolution rate (oxygenation capacity) increases almost proportionally with the partial pressure of oxygen in the gas phase. Increasing the pressure in biological tanks using air as oxygen source in order to increase the partial pressure of the oxygen is not feasible from economic and practical points of view. Besides capacity enhancement, the use of pure oxygen in the process gives a bacterial sludge that is easier to remove by sedimentation.

In biological nitrogen removal the use of pure oxygen can give a faster reaction rate in the nitrification stage. Different biological waste water treatment processes have been developed. The most used process is the activated sludge process, in which the waste water is mixed with a sludge containing bacteria (the activated sludge), which in combination with oxygen remove the contaminants from the water.

The use of pure oxygen in the activated sludge process can range from full-time use of pure oxygen alone to the use of pure oxygen only for peak shaving in periods of high load in treatment plants using air as normal oxygen source. Several processes for the use of pure oxygen in waste water treatment have been developed. The first and best known is the UNOX process developed by Union Carbide. The first full-scale operation of the UNOX system was installed at the Batavia municipal waste water treatment plant in New York, which was put into operation in 1968. The test continued for a 2½-year period and provided full-scale verification of the process. The UNOX system was commercialized and shortly thereafter Detroit, Michigan, in 1970, announced its commitment to install a UNOX system for the treatment of 300 million gallons per day of municipal waste (Fig. 7.4.1). In the beginning of 1980s, Linde AG

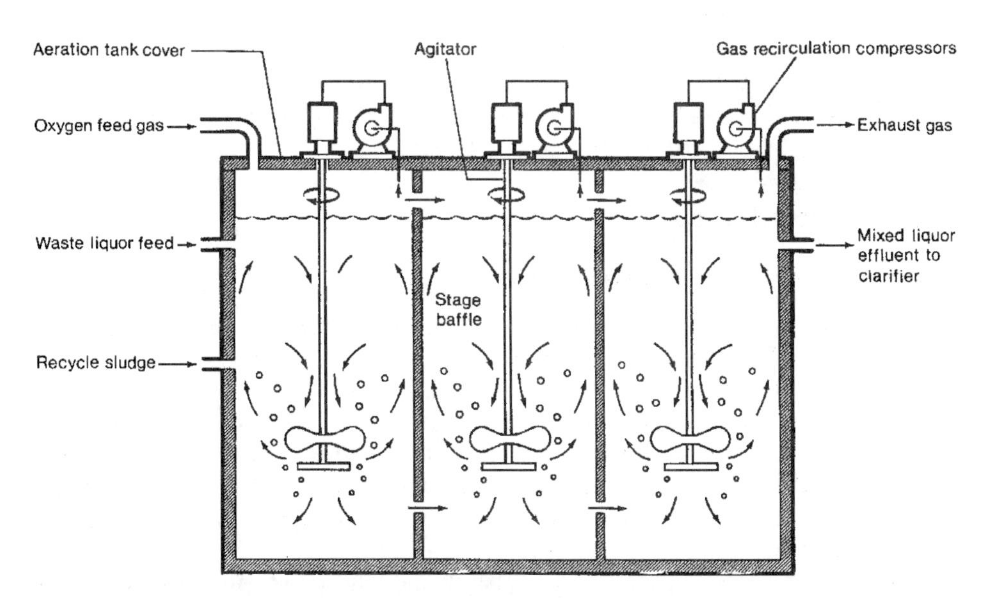

Figure 7.4.1. Typical waste treatment plant. The UNOX process, 1971. Courtesy of Praxair, Danbury, Connecticut.

acquired the UNOX business from Union Carbide and renamed it the LINDOX process.

Alternative pure oxygenation systems developed by other companies soon appeared on the market. In 1971, Air Products and Chemicals introduced the OASES system, and FMC Corporation marketed the MAROX system and Air Reduction Company presented the F^3O system. All were stimulated by the success of the UNOX system in using pure oxygen in waste water treatment.

A biofilm system called the Oxitron system using pure oxygen was developed at the end of the 1970s by the Dorr-Oliver company. This process uses a moving sand bed as carrier for the biofilm. This system has a very high volumetric capacity.

Today most gas companies have equipment for the use of pure oxygen in waste water treatment plants, for example:

- Linde: Solvox system
- BOC: Vitox system
- AGA: Aquaclean system
- Praxair: Surface aerators and the In Situ oxygenation system
- L'Air Liquide: Ventoxal system

Hydrogen Sulfide Abatement

In long sewage and waste water pipes the water becomes deficient in oxygen due to bacterial activity, creating septic conditions with emission of malodorous compounds (mainly hydrogen sulfide) as a consequence.

By keeping the waste water in the sewers aerobic, the emission of malodorous compounds can be prevented. Pure oxygen has been successfully used for maintaining aerobic conditions in sewers. The reason for using pure oxygen instead of air is that air contains 80 percent nitrogen, which strips the malodorous compounds from the water, thus giving a bad smell to the surrounding area. Furthermore, the nitrogen can cause gas pockets in high points of the sewer and thereby increase the pressure drop and decrease the sewage flow rate in the pipe.

Pulp and Paper Industry

Bleaching of Pulp with Oxygen and Ozone[3–6]

The kraft or sulfate process is the predominant process for the production of chemical pulp. The fiber raw material in the form of wood chips is digested at elevated temperature (150–170°C) in an aqueous solution (cooking liquor) containing sodium hydroxide and sodium sulfide. The objective of cooking is to dissolve the lignin and thereby bring about fiber separation. About 90 percent of the wood's lignin can be removed in the cooking stage. To achieve a high and stable level of brightness, the dark-colored lignin remaining in the pulp after cooking is removed in several bleaching stages.

Modern bleaching processes can be divided into two main types: totally chlorine-free (TCF), which is based on the use of oxygen, hydrogen peroxide, ozone, and per-acetic acids, and elemental chlorine-free (ECF), which uses chlorine dioxide together with non-chlorine bleaching chemicals. Classic chlorine bleaching, with high charges of elemental chlorine (chlorine gas), has to a large extent been phased out for environmental reasons. Bleaching of pulp with elemental chlorine produces chlorinated organic compounds in the effluent, which constitute an environmental hazard.

The kraft process was invented by the German chemist Carl F. Dahl in 1879. The bleaching power of chlorine was discovered in 1774 by Carl Wilhelm Scheele and continuous chlorine and hypochlorite bleaching was introduced in the 1930s. A decade later, in 1946, chlorine dioxide bleaching was introduced. The delignifying and bleaching properties of oxygen in alkaline solution was investigated in the early 1950s by the Russian chemists M. V. Nikitin and G. L. Akim, but it was not until the discovery of magnesium as a carbohydrate protector in France in 1964 by A. Robert and coworkers that oxygen could be industrially applied. Oxygen delignification was first introduced on an industrial scale in 1970 in South Africa and Sweden. Two groups developed the original process in parallel. Air Liquide (France), SAPPI (South Africa), and Kamyr (Sweden) formed one group and MoDo (Mo och Domsjö; Sweden), CIL (Canadian Industries Ltd.), and Sunds (Sweden) formed the other. Ozone, known since the 1830s, was introduced for bleaching of sulfite and kraft pulp in the early 1990s.

Oxygen Delignification

Oxygen bleaching or delignification is applied between cooking and final bleaching in the production of bleached chemical pulp. It is standard technology in modern bleached pulp mills and without it TCF bleaching would not be possible. The installed capacity for oxygen delignification exceeds 160,000 tons of pulp per day. Typical conditions for oxygen delignification are a temperature of 85–110°C, a pressure of about 0.5 megapascal, and a treatment time of 0.5–2 hours. Typical charges are 15–30 kilograms of oxygen and 20–40 kilograms of NaOH per ton of pulp. When oxygen delignification was introduced in the pulp industry in 1970 the process was operated at a high pulp consistency (dry solids content about 30 percent) in order to create a sufficient mass transfer rate between oxygen and the pulp fibers. With the advent of medium-consistency mixers at the end of the 1970s, operation at medium pulp consistency (10–15 percent) became possible (Fig. 7.4.2). The popularity of oxygen delignification grew because the investment costs were reduced and it was easier to operate than the high-consistency process. During the 1980s more stringent environmental regulations of emissions of chloroorganic compounds were introduced and in the early 1990s consumers entered the chlorine debate, which advanced the spread of the technology.

In the original systems it was possible to remove up to 50 percent of the remaining pulp lignin, whereas modern two-stage systems can remove up to 70 percent of the lignin at acceptable selectivity.

Ozone Bleaching

Ozone is very powerful oxidizing and bleaching agent and it can be applied to practically all types of pulp. It is applied after oxygen delignification to further

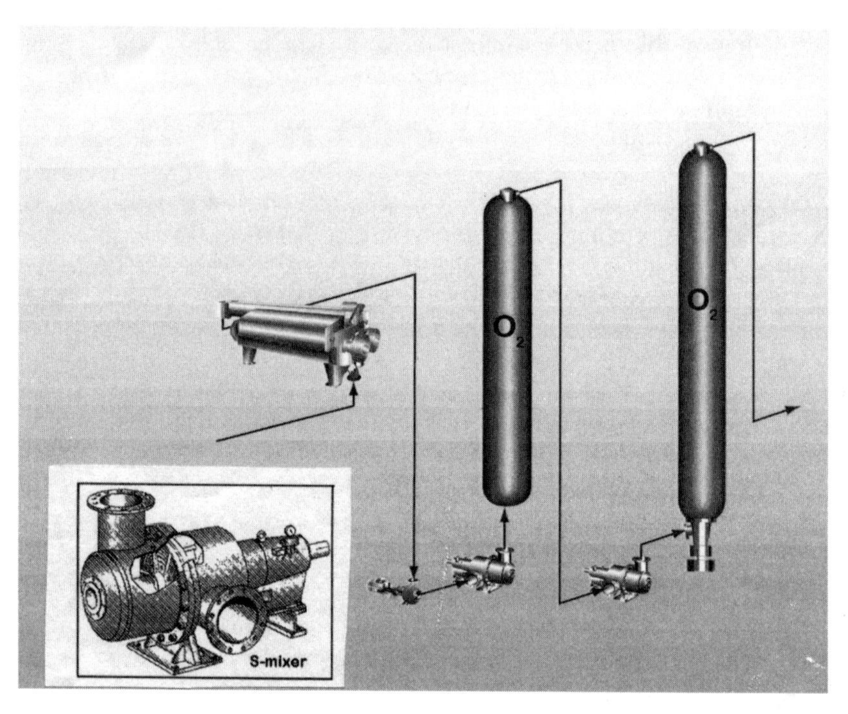

Figure 7.4.2. The OxyTrac system (Metso Paper) for oxygen delignification in two stages with efficient mixing between the stages. Courtesy of Metso Paper, Sundsvall, Sweden.

decrease the lignin content in the production of bleached chemical pulp. The main reason for the development of ozone bleaching technology was the need for alternatives to chlorine-based chemicals spurred by environmental legislation and consumer demands in the 1990s.

Ozone is produced on-site in special generators through silent electrical discharge in a gas stream containing oxygen. The ozone quantities required for pulp bleaching, 3–10 kilograms per ton of pulp, are most economically produced from oxygen. The upper technical ozone concentration in the ozone–oxygen gas mixture is about 12 percent by weight. The power consumption is relatively high, 10–15 kilowatt-hours per kilogram of ozone. At present the largest installations for the bleaching of pulp have an ozone generation capacity of 220 kilograms per hour.

Ozone bleaching can be carried out at either high pulp consistency (30 percent dry solids content) or medium consistency (10–15 percent). Conventional pulp bleaching equipment can be used, but the mixing of pulp and ozone is a critical operation due to the high reaction rate between pulp and ozone. Ozone can be pressurized up to about 0.5 megapascal. At high consistency the pulp is treated in the gaseous phase. Independent of pulp consistency, a pH of 2–3 is the optimum with regard to bleaching efficiency and the temperature must not exceed 50–55°C in order to avoid decomposition of ozone (Fig. 7.4.3).

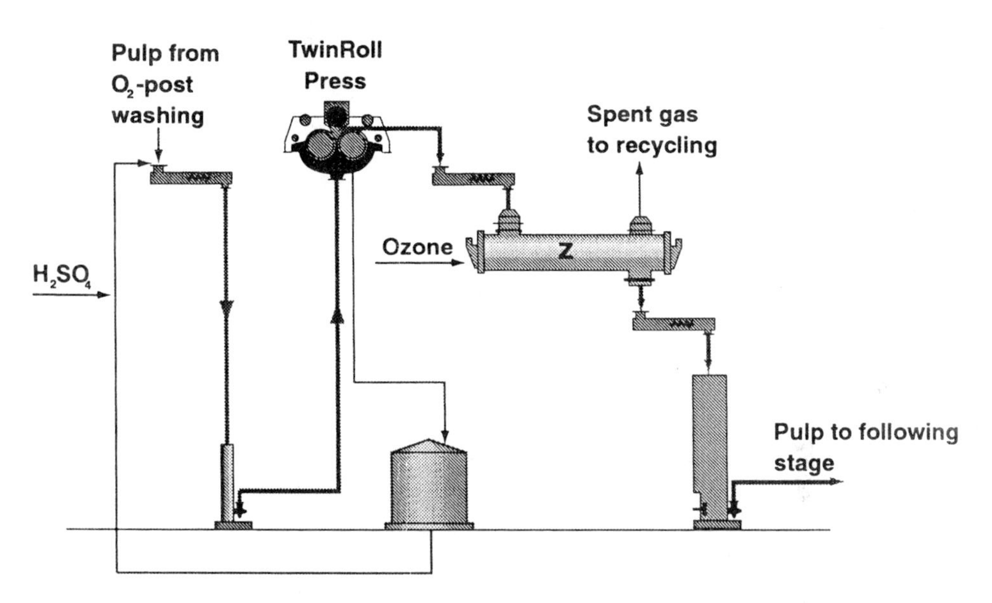

Figure 7.4.3. A high-consistency ozone stage for pulp bleaching (Metso Paper). Courtesy of Metso Paper, Sundsvall, Sweden.

Chemical Recovery in Kraft Pulping

The spent cooking liquor after kraft cooking contains dissolved wood substance and cooking chemicals. The liquor is recovered in the pulp-washing process and is then evaporated and incinerated in a recovery boiler, where the dissolved wood substance is the fuel. The spent cooking chemicals are regenerated from the smelt from the recovery boiler to fresh sodium hydroxide and sodium sulfide by means of causticizing with burned lime.

In recovery systems where the final evaporation of the black liquor takes place in open, so-called direct contact evaporators, mainly in North and South America, there is a risk of emissions of reduced sulfur compounds to the atmosphere. In order to decrease the reduced sulfur emissions, the black liquor can be treated with oxygen. The sulfides react rapidly and exothermally and are converted to thiosulfate. The heat of reaction reduces the need for steam for evaporation. About 1 ton of oxygen per ton of pulp is required to convert sodium sulfides to sodium thiosulfates.

Modern mills use so-called low-odor recovery boilers without direct contact evaporators. Furthermore, by increasing the dry solids content of the black liquor to 80 percent instead of the conventional 55–65 percent before combustion, the emissions of both sulfur oxides and hydrogen sulfide are decreased substantially.

Recycled Fibers

The driving force behind recycling in the paper industry has been economy—recycled fiber has been the cheapest alternative. Legislation and demands from "green" consumers have become even more powerful driving forces. Depending on the type of recycled paper and the demands on the final product, recycled fibers are deinked and bleached. Traditional bleaching chemicals are hydrogen peroxide, hydrosulfite (dithionite), and formamidine sulfidic acid (FAS), and more recently processes based on oxygen and ozone have been introduced. Oxygen and ozone are applied under process conditions similar to those applied in the bleaching of chemical pulps. Air Products has developed the Oxypro O_R system, with benefits that, in addition to brightening, include contaminant removal, improvement in pulp strength properties, and color stripping. Air Liquide's process is named Redoxal. It combines bleaching and dye-stripping capabilities and is particularly effective in reducing the content of fluorescent whitening agents.

References

1. Cook, G. A. (1980).The use of oxygen in the treatment of sewage. *Journal of Chemical Education* **57**(2):137–138.
2. Union Carbide. (1970). *UNOX System*. Union Carbide brochure.
3. Rydholm, S. A. (1967). *Pulping Processes.* New York: Interscience.
4. SUNDS. (1978). *No two pulping plants are exactly alike*. SUNDS brochure.
5. Dence, C. W., and Reeve, D. W. (eds.). (1996). *Pulp Bleaching-Principles and Practice*. Atlanta, GA: TAPPI Press.
6. Croon, I. (1992). Use of oxygen in the pulp and paper industry. In *Proceedings of the 6th BOC Priestley Conference*, pp. 257–264.

7.5. ENERGY AND FUEL

I suppose we shall soon travel by air-vessels; make air instead of sea voyages; and at length find our way to the moon, in spite of the want for atmosphere.

Lord Byron (1822)

Coal Gasification[1–3]

Coal gasification originated in the late 18th century and for about 150 years coal was a major source of town gas, industrial heating gas, and petrochemicals. The availability of natural gas at very low prices after World War II forced the closing of all but one coal gasification plant.

The conversion of coal to gas started in the late 18th century when William Murdoch gasified coal by heating it in the absence of air. The gas produced was called coal gas. The first coal gas companies started in London in 1812 and in Paris and Baltimore, Maryland, in 1816. The primary goal of coal-to-gas conversion was the production of gas which could be used for lighting purposes, especially street lights. In coal gas production about two-thirds of the original coal is not utilized, and the remaining carbon and ash residue is sold or burned in other plants. A gas similar in composition to coal gas is produced when coal is heated in the manufacture of coke. In coke oven gas production there is no low-priced by-product because the remaining solid carbon coke is used in the production of steel or cast iron.

A third process of coal gasification was developed in the middle of the 19th century. In this process dry air is blown through a bed of coke or coal of sufficient depth to produce a combustible gas consisting primarily of carbon monoxide and nitrogen at a ratio of about 1:2. In the late 19th century the blue gas process was developed in which steam is passed over red-hot carbon at a temperature from 940°C to 1190°C. The gas thus formed burns with a nonluminous bluish flame and for this reason is called "blue gas" or "blue water gas." Since the basic reaction used to produce the blue gas (mainly carbon monoxide and hydrogen in the proportion ca. 1:1) is endothermic, the temperature of the fuel bed has to be raised by blasting air through the carbon bed.

Processes for Coal Gasification (Fig. 7.5.1)

In the early 1920s the use of oxygen and steam was first introduced for coal gasification on a commercial scale in the Winkler process. It was superseded by more efficient processes because of its low carbon conversion and low-pressure operation. Two units were operated, however, by Rheinische Braunkohlenwerke in Wesseling, Germany, during 1956–1964. The high-temperature Winkler (HTW) process was developed in 1978 with a 35-ton per day pilot plant in Frechen, Germany. The second process to use oxygen and steam to produce gas from coal is the Lurgi process, commercially introduced in 1936. It was developed by Lurgi Kohle und Mineralöltechnik GmbH and AG Sachische Werke, who started a pilot plant in Hirschfelde, Germany, in the early 1930s. Several Lurgi plants are in operation. Best known is the Sasol plant of South African Coal, Oil and Gas Corp., which built 72 Lurgi gasifiers from the

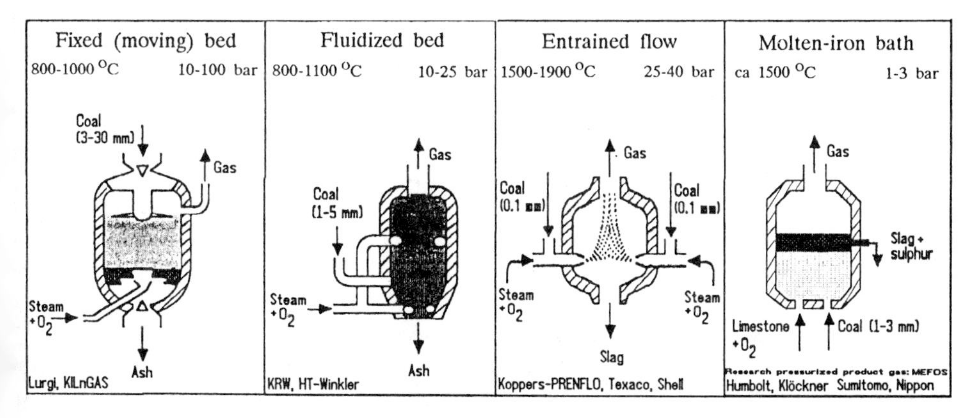

Figure 7.5.1. Different coal gasification processes.

1950s through the 1980s. The Lurgi countercurrent flow shaft gasifier operates at a pressure from 20 to 30 bars and ejects essentially a carbon-free ash.

A third process to use oxygen and steam for coal gasification is the Koppers Totzek process, which was developed by Friedrich Totzek in the late 1930s for Heinrich Koppers GmbH in Essen, Germany. The first commercial plant was built in France in 1949. In 1974 Koppers was acquired by Krupp International. The Koppers–Totzek reactor is a partial pulverized coal combustor operated at near-atmospheric conditions.

The Texaco synthesis gas generation process was first developed in the 1940s at the Texaco Montebello laboraties in California. A pilot plant delivering synthesis gas by partial oxidation was operated from 1946 to 1954. In 1977 Ruhrkohle/Ruhrchemie started up a 165-ton per day Texaco plant.

The Hygas gasification process has been in development since 1946. In the gasifier the hydrogen-rich gases and steam move countercurrently to coal through two stages.

Interest in coal gasification dropped in countries where domestic oil and natural gas resources were found during the late 1950s and the 1960s. Because these energy sources were potentially cheaper, the market for this technology was then countries with little or no petroleum sources, but with large amounts of coal. Interest was then renewed in 1973 due to the oil embargo, which drastically increased oil prices. Coal gas technology underwent great world-wide development in the 1970s and early 1980s.

LNG and Helium Recovery[4–7]

The first reported interest in liquefying natural gas occurred when the British government in 1917 proposed a liquid natural gas (LNG) facility to recover helium. At about the same time, Linde Air Products developed the separation of helium from natural gas. In 1920, Godfrey Cabot became interested in liquefying natural gas and transporting it in river barges and was issued a patent for the insulated containers to be used on the barges. The U.S. Bureau of Mines built a helium recovery plant at Forth Worth, Texas, in 1924 which incorporated the latest ideas in cryogenic separation. Development in this area of interest continued at Fort Worth and later at Amarillo, Texas.

By 1937 pure helium gas, obtained from helium-bearing natural gas, became commercially available in relatively large quantities under the regulation of the U.S. Bureau of Mines. The use of natural gas in liquid form had been tested in the Soviet Union and was now studied in detail in the United States. The concept was found feasible technically and economically because of its storage and transport advantages. In 1939 peak-shaving plants and transports were projected in United States.

Peak-Shaving Plants

The first U.S. facility for liquefying and storing LNG for later use was a pilot peak-shaving plant built by the Consolidated Natural Gas System through its subsidiary, Hope Natural Gas Company. In 1940 Hope built a pilot liquefaction plant at their Charleston compressor station in West Virginia (liquid capacity 8000 cubic meters per day, storage capacity 50 cubic meters). In 1941 the East Ohio Gas Company, also a subsidiary of Consolidated Natural Gas System, built a peak-shaving plant in Cleveland, Ohio, along the lines developed by Hope (liquid capacity 113,000 cubic meters per day, storage capacity ca. 8000 cubic meters). After 2½ years a new large storage tank was added. This tank failed in 1944, resulting in a disastrous fire with the loss of about 200 lives. Peak-load LNG plants were not built again in the United States until around 1964, when a liquefaction plant was built by the Wisconsin Natural Gas Co. in Oak Creek, Wisconsin. In quick order other states built LNG plants with industrial gas companies as contractors or consultants, for example, Alabama (Air Products), California (Messer), and Tennessee (Air Liquide).

In the Soviet Union a peak-shaving plant was built in 1947 near Moscow, with engineering aid from Dressers Industries Ltd. of Texas. In 1959 a test receiving facility was built in Canvey Island, United Kingdom, and test transports of LNG began. In France Gaz de France began studying LNG storage technology for the harbor in Le Havre.

LNG Ship Transport

The transfer of gas from North Africa to the United Kingdom had its inception in 1961 when the British government approved a plan for liquefying natural gas at Arzew, Algeria, and shipping it in two tankers to Canvey Island, England, where it was regasified, blended, and fed into English gas mains. In 1964 the tankers *Methane Progress* and *Methane Princess*, each with a capacity of 12,000 tons of LNG, began to make about 56 trips per year and supplied about 10 percent of Britain's yearly gas demand. From 1965 liquefied gas from Arzew has also been shipped by the LNG tanker *Jules Verne* to Gaz de France facilities at Le Havre, France (Figs. 7.5.2 and 7.5.3).

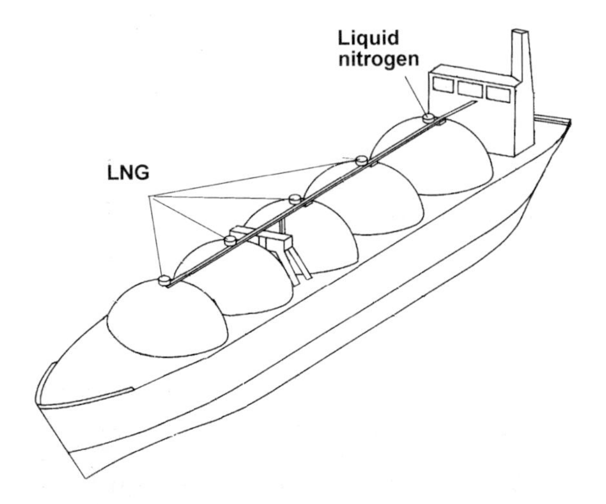

Figure 7.5.2. LNG ship. Courtesy of Teknisk Illustration, Umeå, Sweden.

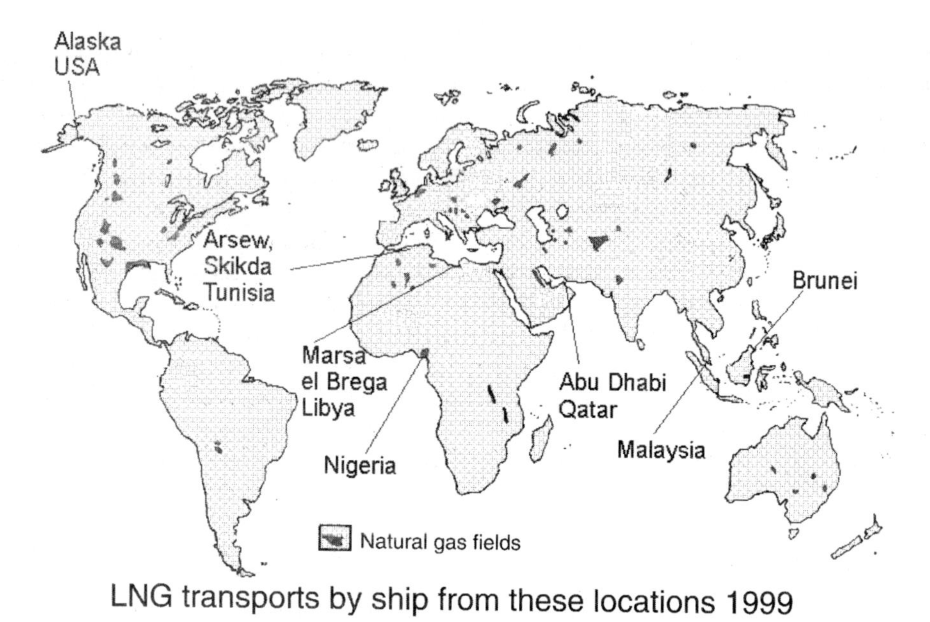

Figure 7.5.3. Natural gas fields and harbors for world-wide ship transport of LNG.

Underground Storage

There are basically four types of LNG storage:

1. Metal tanks of double-wall type with cork, mineral wool, perlite, etc., as insulants
2. Prestressed concrete tanks with a metal liner
3. Frozen-ground tanks in a frozen-ground pit
4. Vacuum-insulated tanks (for small capacities)

At first LNG storage in double-wall tanks was the only method available. In 1960 the American Gas Association and the Battelle Institute began studies of prestressed concrete for LNG storage. This method was employed in Texas in the mid-1960s. During 1960 and 1961, underground storage in caverns was widely discussed. Gaz de France began investigating the potential of storing large amounts of LNG in mined caverns as early as 1958. Theoretical work then began on frozen in-ground tanks, followed by laboratory work and experiments. Underground frozen-ground storage was used from 1967 at the Arzew LNG delivery station and in large storage facilities in the United States and Japan.

LNG Applications

LNG has approximately a 13 percent higher heat value than gasoline. In 1945 it was tested as a car fuel in England. In 1967 San Diego Gas and Electric began long-time testing running a pickup truck and a *Valiant Sedan* on LNG. The results of these tests started development of new LNG car fuel technology. At the end of the 1900s one million vehicles in the world were natural gas-powered, but only a few hundred of these, mostly in United States, were LNG-powered. In spite of advantages in energy density, low pollution emissions, and tank size and weight, technical and economic obstacles remain.

Hydrogen as Aircraft and Car Fuel and as General Power Source[8]

Aircraft and Car Fuel

During the 1950s the NACA Lewis Laboratory began experiments with liquid hydrogen–liquid oxygen and with liquid hydrogen–liquid fluorine. In their work they saw the potential of using liquid hydrogen for high-altitude aircraft. Such possibilities were reported in 1955 and a RB-57 twin-engine bomber was modified to run on either its normal jet fuel or liquid hydrogen.

These experiments demonstrated the feasibility of using liquid hydrogen in aircraft and apparently remain the only aircraft flight propulsion experiments with liquid hydrogen. Compared with conventional aviation fuel, liquid hydrogen for the same performance has three times less weight, but four times the volume. Tank insulation and control equipment further reduce the advantage, and safety aspects finally led to the decision to stop this development.

Figure 7.5.4. Solar-cell-produced hydrogen for car fueling; research project, Linde AG. Courtesy of Linde AG, Wiesbaden, Germany.

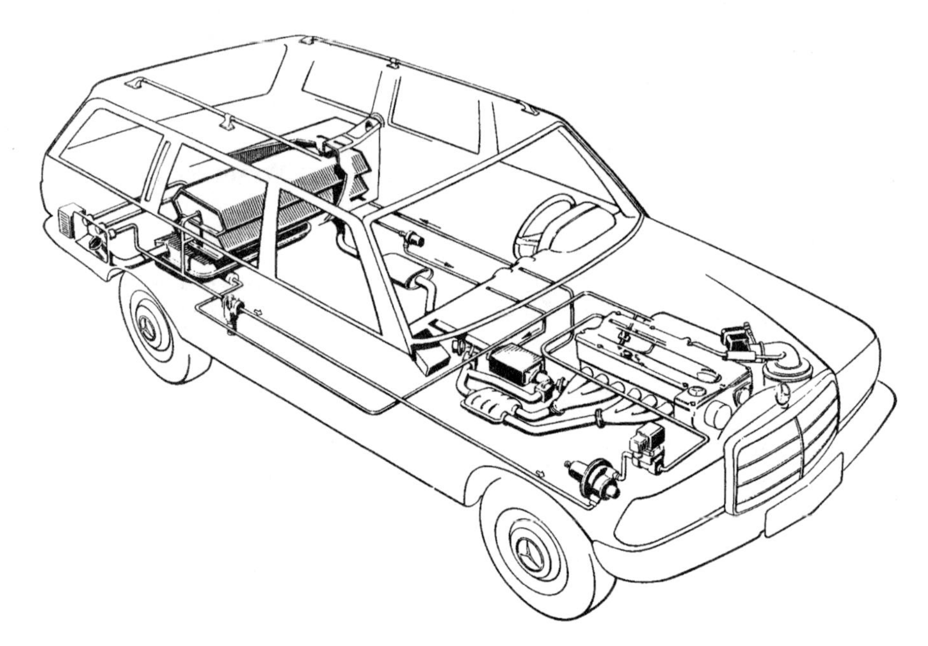

Figure 7.5.5. Mercedes driven alternatively by hydrogen hydride and gasoline fuel, 1989. Courtesy of Daimler Chrysler AG, Stuttgart, Germany.

The effort of the U.S. Air Force on hydrogen-fuelled aircraft was not in vain. August 1958 a program began for a hydrogen–oxygen upper stage for the Atlas rocket, called the Centaur. The engine development was given to Pratt & Whitney and they used the hydrogen liquefaction facilities adjacent to their Florida test facility to develop the first flight rocket engine to use liquid hydrogen–liquid oxygen, the RL-10.

As a fuel for cars, liquid hydrogen was first tested in California in 1960 by the Perris Company. Other attempts have been made since the beginning of the 1970s in the United States, Germany, and Japan.

Power Technology

After the power crisis of 1973 investigations began into a future society wherein hydrogen would form the dominant source of energy. These studies were based upon a cheap energy source producing liquid hydrogen, which is still the most realistic form for the large-scale distribution of hydrogen.

Rocket Fuels[9–11]

The concept of using liquid hydrogen for manned spaceflight was first proposed by the Russian Konstantin Tsiolkovskiy in 1903, but over 50 years elapsed before it was adopted for the Apollo moon landing mission. Although the use of liquid hydrogen is not essential for manned spaceflight, it was used in the development of the first

U.S. space vehicles, and has also been used as a fuel for other, unmanned planetary flights and for the space shuttle.

Tsiolkovskiy showed in 1903, only 5 years after Dewar first liquefied hydrogen, that hydrogen burned with oxidizers produces the highest exhaust velocity of all chemical fuels by a combination of high heat of reaction and low molecular mass of the exhaust gases (Fig. 7.5.6). Oxidizers tested included oxygen, fluorine, and ozone among others. Liquid hydrogen has the lowest density of all the fuels (1/14 the density of water), which is a disadvantage because more tankage volume and mass is required compared to that of denser fuels.

The first to experiment with a low-temperature liquefied gas in a rocket was Robert Goddard, who began using liquid oxygen in 1921. By 1923, Goddard had successfully operated a gasoline–liquid oxygen rocket. In 1926, he launched the world's first liquid-fuelled rocket at Auburn, Massachusetts (Fig. 7.5.7). It traveled about 54 meters to a height of 12 meters in $2\frac{1}{2}$ seconds. Goddard also calculated the amount of fuel necessary to send a craft into space (45 kilogram of liquid oxygen/liquid hydrogen per kilogram of load) and came up with the idea of multistage rockets.

In 1932 Walter Dornberger, a German army officer, organized a small rocket research station on the artillery proving grounds at Kummersdorf. The Germans were forbidden by the Treaty of Versailles in 1919 to do research on artillery and looked for other ways of ordnance development. During the 1930s they worked step by step to make military rockets based on Goddard's results. The know-how concerning liquid oxygen and much of the liquid gas equipment came from the rocket enthusiast and promoter of industrial liquid oxygen distribution Paul Heylandt.

The German A-4 (V-2) rocket using alcohol–liquid oxygen was the first practical application of a liquid-fuelled rocket and the first to be mass produced. In October 1942, the Germans launched their first flight of the 5000 kilogram A-4 (V-2) rocket at Peenemünde (Fig. 7.5.8). It traveled about 200 kilometers. During the war approximately 1100 V-2s were fired, which proved the feasibility of using liquefied gases as rocket propellants. Behind the successful German rocket designs were, among others, Walter Dornberger and Wernher von Braun. By the end of World War II the major properties of liquid hydrogen were well established.

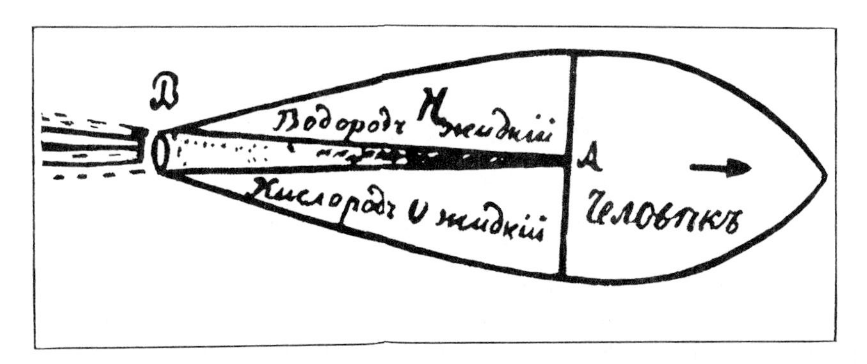

Figure 7.5.6. Tsiolkovskiy's drawing of his rocket principles, 1903.

Figure 7.5.7. Goddard with liquid oxygen/gasoline rocket, 1926. Courtesy of National Air and Space Museum, Smithsonian Institution, Washington, D.C.

In 1945 experiments with liquefied hydrogen and oxygen began in the United States at Ohio State University and Aerojet Engineering Corporation, sponsored respectively by the U.S. Air Force and the U.S. Navy.

After World War II about 2000 German rocket engineers from the Peenemünde and Nordhausen rocket bases were brought to the Soviet Union, where they started the development of military rockets and space rockets. Dornberger, von Braun, and about 500 of their technical staff members were taken to the United States after the war and later made contributions to rocket motor design for liquid hydrogen, which began to be developed in the United States in 1956. In 1954 Dornberger, who worked for Bell Aircraft Company, proposed a two-stage space shuttle for flights into space. Since 1963 liquid oxygen/liquid hydrogen has been used in U.S. space rockets.

Figure 7.5.8. V2 launch from Peenemünde (courtesy of Deutsches Museum, Munich).

Rockets for Satellites

In 1957, when Russia launched the first satellite, the United States had to speed up its space technology development. At that time its only launch vehicle was the small Vanguard, which still was under development. The space race made a rapid buildup of launch vehicle capability necessary. During the 1960s the giant Saturn V rocket was completed, which launched the Apollo lunar expeditions from 1968.

Already in 1958 and 1959 the decision was taken to use liquid hydrogen in the upper stages of the Centaur and Saturn vehicles. This decision was particularly bold, as engineers experienced in gas technology still remembered the *Hindenburg* disaster of 1937 and doubted the advisability of using hydrogen as a fuel. Other obstacles were the difficulty of liquefying a gas at temperatures so close to absolute zero and the problems of storage and handling.

In 1954 a breakthrough in nuclear weapons development reduced the payload requirement of intercontinental missiles to about 800 kilogram. A scaled-down Atlas rocket for this payload was built by Convair with a tank design separating fuel and oxidizer. The light-weight, pressure-stabilized tank concept of the Atlas was adopted for the Centaur, the first vehicle to use liquid hydrogen–oxygen, and was then used as the workhorse of the U.S. planetary programs. The first Atlas flew in 1957. The Saturn V rocket was used to put the first man on the Moon in 1969.

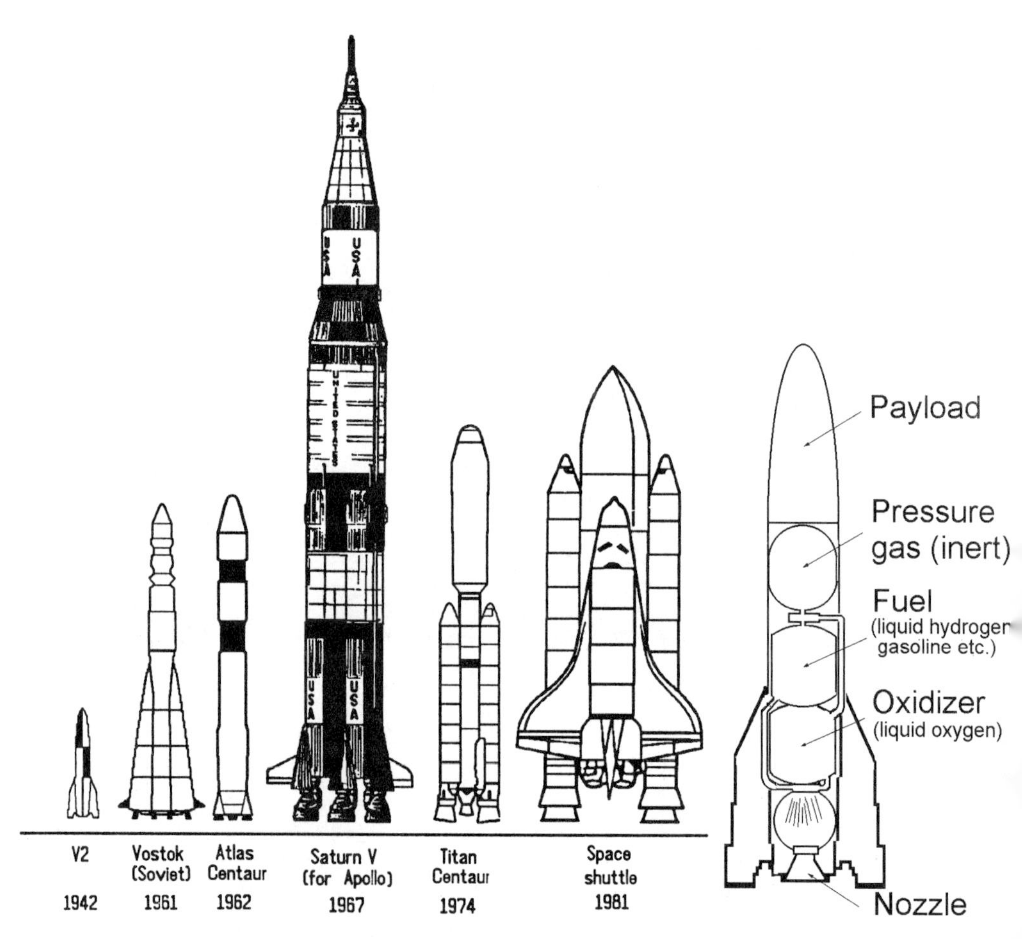

Figure 7.5.9. Size comparison, rockets fuelled with liquid gases. Right: Principle of liquid fuel rocket design.

Short Biographies[12,13]

Konstantin Eduardovich Tsiolkovskiy (1857–1935) was born of poor parents in a village 900 kilometers east of Moscow (Fig. 7.5.10). He lost his hearing at the age of 9, which caused problems in his contact with other children and normal social relationships. He became a reader and dreamer, educating himself and studying to become a schoolteacher in spite of his deafness. He became interested in winged flight and built an airplane model which his father proudly showed to guests.

As a teenager he made experiments with hydrogen and at 16 he was suddenly struck by a fascinating idea: Why not use centrifugal force to launch a spacecraft from Earth? This idea had a profound effect on his later activities. He dreamt of flying to the stars and became the first to develop the theory of rocket flight and the first to consider hydrogen–oxygen to propel rockets. But he did not attempt experiments.

Tsiolkovskiy's theory of gases earned him membership in the Petersburg Physico-Chemical Society. The Tsiolkovskiy equation states a key relationship for understanding the advantages and disadvantages of liquid hydrogen as a fuel. It shows that rocket vehicle velocity is directly proportional to the rocket exhaust jet velocity. The latter is essentially constant for a given rocket design, propellants, and operating conditions.

In a paper in 1903 Tsiolkovskiy described a manned rocket and discussed hydrogen–oxygen more than any other fuel. In an article in 1911, he referred to liquid hydrogen in the caption to a drawing. At that time liquid oxygen could be obtained cheaply, but not liquid hydrogen, and so he discussed replacing hydrogen by liquid or liquefied hydrocarbons such as ethylene, acetylene, etc. Tsiolkovskiy was aware of the high energy content of hydrogen, but a year before his death in 1935 he concluded that liquid hydrogen was unsuitable as a fuel because of its low density and difficulty of storage.

Figure 7.5.10. Konstantin Tsiolkovskiy. Courtesy of Smithsonian Institution, Washington, D.C.

Robert Hutchings Goddard (1882–1945) was born in Worcester, Massachusetts (Fig. 7.5.11). Poor health during his youth did not stop him in his passion for experimental work. At 16 his experiments with hydrogen led to a violent explosion. His parents tried to convince him to change his interests, but he did not stop such experiments.

Like Tsiolkovskiy, Goddard dreamt about spaceflight and from 1899 it became his greatest goal. He studied physics at Worcester Polytechnic Institute and Clark University, but had to choose other thesis subjects than rockets. He spent a year of postdoctoral research at Princeton University and continued studying rockets in his spare time; his initial experiments with powder rockets came during his undergraduate days.

In 1907, he calculated the lifting of scientific equipment to a great height by a combination of balloon and rocket. In 1908 he became Professor of Physics at Clark University. Like Tsiolkovskiy, he was attracted to the hydrogen–oxygen combination because of its high energy, but recognized the problems of obtaining and working with liquefied gases. In 1910 he wrote of a method for producing hydrogen and oxygen on the Moon by generating the gases from the electrolysis of water using solar energy.

In spite of Goddard's early interest in hydrogen–oxygen, he did not operate a rocket with either gaseous or liquid hydrogen as the fuel because of the handling problems. Among the rocket problems he solved was the theory for multistage rockets, and he pioneered the use of gyroscopes to control rocket flights.

On March 16, 1926 Goddard launched the world's first gasoline–liquid oxygen rocket. In the United States his rocket studies passed almost without attention, but they were noted by the Germans, who 16 years later launched the world's first military long-distance alcohol–liquid oxygen rocket (the V-2 rocket). Five years after this the experience from thousands of V-2 flights were taken over to the United States and the Soviet Union, where it laid the ground for the space age.

Hermann Oberth (1894–1989) was a German rocket pioneer of Romanian origin (Fig. 7.5.12). He became interested in space travel at the age of 11 after reading Jules Verne's '*From the Earth to the Moon*.' He analyzed several methods for achieving spaceflight and found reaction propulsion to be the only feasible method. He also saw major problems in the use and handling of the fuels then available.

Oberth's path through the classical school system and further academic studies was difficult. When he at last managed to present his thesis on rocket propulsion for his doctor's degree at Heidelberg University it was rejected, and a proposal in 1917 to the German War Department to build a long-range liquid propellant missile was also turned down. In 1923 Oberth published at his own expense a *Die Rakete zu den Planetenraumen* (*The Rocket into Interplanetary Space*), in which he proposed a rocket with a first stage using ethyl alcohol and oxygen and a second stage using liquid hydrogen–liquid oxygen. It was read by Goddard, but was largely ignored in Europe in the 1920s. After some years he was left disillusioned and broke. In 1929, a third edition of his book with refined and expanded calculations was published under the title *Wege zur Raumschiffahrt* (*Ways to Spaceflight*). After this Obert got some funding for rocket experimentation. In 1938, he began working on military rocket development for the Germans at Peenemünde until the time A-4 (V-2) development was completed. He

Figure 7.5.11. Robert Goddard. Courtesy of Smithsonian Institution, Washington, D.C.

Figure 7.5.12. Hermann Oberth. Courtesy of Smithsonian Institution, Washington, D.C.

Figure 7.5.13. Wernher von Braun. Courtesy of Deutsches Museum, Munich.

then worked mainly on solid fuels until the end of the war. When the Allies searched for rocket experts, they initially ignored Oberth. He spent 4 years on rocket research in Huntsville, Alabama, from 1955, and then returned to Germany. Oberth provided the theoretical basis for space rocket design and improved the old theories of rocket flight.

Wernher von Braun (1912–1977) was born in Wirsitz, Prussia; as a teenager he read Oberth's work on space rockets (Fig. 7.5.13). He decided to direct his education toward work on space research. In the late 1920s von Braun became a member of the German Rocket Society and was employed by the German Army Weapons Development Branch from 1932. He spent 13 years at the rocket research stations at Kummersdorf and Peenemünde, where he became chief designer of the V-2 rocket program. After the war, von Braun and a large team of Peenemünde engineers moved to the United States, where they continued their work on large military rockets. Among others they developed the first U.S. long-range missile, the launch vehicle for the first U.S. satellite, and the Saturn V rocket, which sent the first men to the Moon in 1969.

References

1. Perry, H. (1974). The gasification of coal. *Scientific American* **230**(3):19–25.
2. Electric Power Research Institute. (1983). Coal gasification systems: A guide to status, applications and economics. EPRI AP-3109. Final report. Palo Alto CA:
3. Malpus, B. (1984).The elegant use of oxygen. *BOC Group Magazine* 1984 (December):10–21.
4. Cady, H. P., and McFarland, D. F. (1907). *American Chemical Society*. 1907 29:1523.
5. American Gas Association. (1968). *Historical Background of LNG Development*. Arlington VA.
6. Haselden, G. G. (1992). The history of liquefied natural gas (LNG). In *History and Origins of Cryogenics* (R. G. Scurlock, ed.), (Chapter 17) pp. 599–649. New York: Oxford University Press.
7. Drake, E., and Reid, R. C. (1977). The importation of liquefied natural gas. *Scientific American* **236**(4):22–29.
8. Kleemenko, A. P. (1959) One flow cascade cycle. In *Proceedings of the10th International. Congress of Refrigeration*, Vol 1. pp. 34–39.
9. Tachtler, A., and Szyszka, S. (1993). Car fueling with liquid hydrogen. *Linde Reports on Science and Technology* **51**:31.
10. Sloop, J. L. (1978). *Liquid Hydrogen as a Propulsion Fuel 1949–59*. Washington DC: NASA.
11. Sloop, J. L. (1981). Liquid hydrogen for space flight: The long step from proposal to reality. *Journal of the Astronautical Sciences* **XXIX**(4):373–381.
12. Winter, F. H. (1990). *Rockets into space*. Cambridge MA: Harvard University Press.
13. Marsh, P. (1982). The shy eccentric who fathered the rocket. *New Scientist* **1982(**October 7):26–32.

7.6. INERTING AND COOLING

Background

In 1960, the industrial demand for nitrogen was less than one-fifth of the demand for oxygen. It was only some years later that inerting and freezing applications made nitrogen in the gaseous and the liquefied state an industrial commodity. By the 1970s its demand and value were similar to those of oxygen.

The main reasons for adding a gas with inerting properties to a process are to avoid unwanted reactions between different materials and the atmosphere, prevent fire and explosions, and use as a medium for drying, cleaning, and purging. Nitrogen is generally used, but in some special cases argon and helium may be used. The main application areas are chemistry, biology, food and materials processing, and various mechanical manufacturing processes.

Inerting Applications[1]

Chemistry

In the early 1960s nitrogen began to be used for inerting and blanketing in various chemical processes such as:

- Petroleum and petrochemicals processing and storage
- Polymer and synthetic fiber manufacture
- Reactive metals handling such as blanketing of the tin bath in float-glass manufacture
- Reactive organic chemicals handling

Inerting

Inerting is usually employed when packing a product or when there is a need to decrease fire and explosion risks in bulk storage of flammable liquids, flour, and granulates.

Blanketing

In blanketing or bulk-inerting, air is displaced by a nonreacting gas in the head-space of a liquid container (Fig. 7.6.1). Carbon dioxide, which is not strictly inert, may be used in some cases. It reacts with ammonia and some foodstuffs, but is sometimes used, for example, in breweries. The surface of the liquid is always covered with the inert gas, so that oxidation is minimized. The result depends very much on how low the oxygen content in the liquid itself is.

Sparging

Sparging is the injection of a gas into a liquid, and is often used the food industry (Fig. 7.6.2). It is used for many purposes, such as reducing the oxygen level,

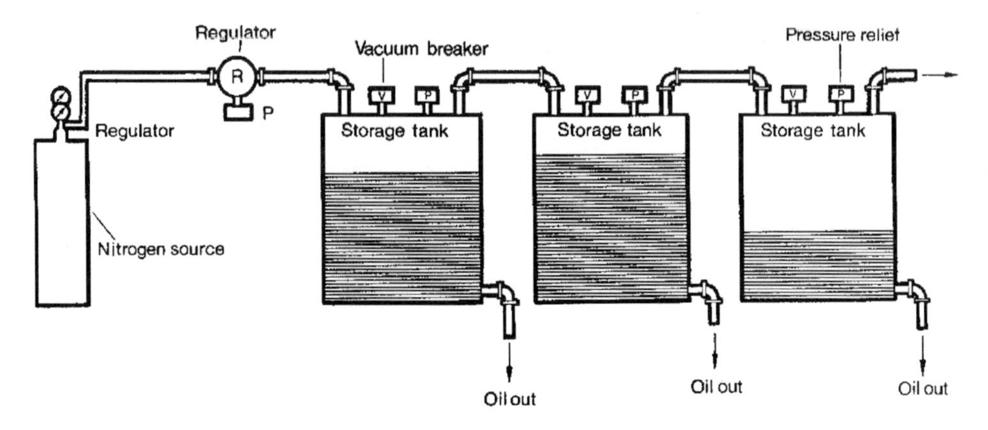

Figure 7.6.1. Typical blanketing setup. Courtesy of the AGA Historical Picture Archive.

controlling fermentation, carbonating beverages, changing pH, reducing water content, and achieving liquid agitation in a container.

Purging and Cleaning

Purging with inert gas has applications in industrial installations such as storage tanks, pipelines, processing equipment, or fixed or transportable containers. It is done to remove vapors from a vessel by displacement with an inert gas to reduce possible reactions or fire, or explosion risk where flammable or explosive materials are involved, or to avoid unwanted reactions between the material and the atmosphere.

The immediate requirement is to reduce the vapor to an acceptable level of reactive gas (such as atmospheric oxygen) in the system. The actual level depends on the nature of the industrial process and the final aim of the purging. If the intention is

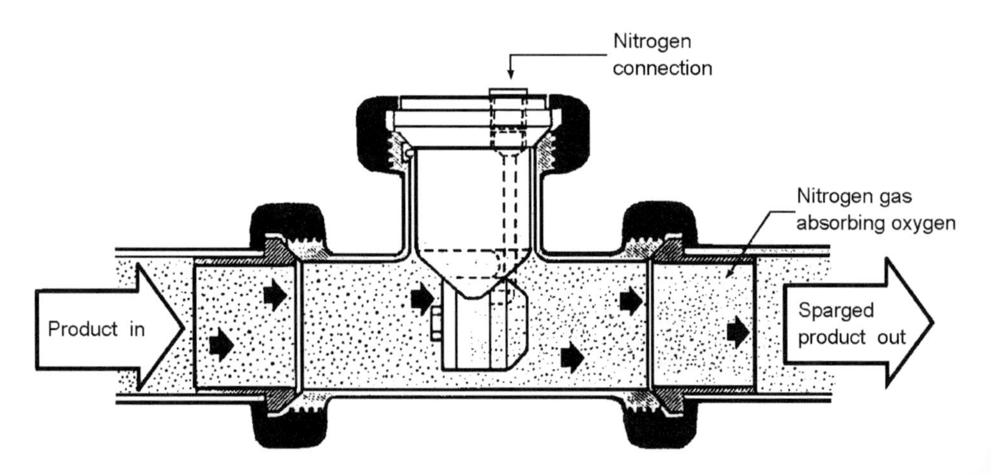

Figure 7.6.2. Scheme for sparging. Courtesy of the AGA Historical Picture Archive.

to eliminate the risk of fire on admitting a combustible gas to the system, it is necessary to reduce the concentration of atmospheric oxygen to a level just below the lower level of flammability.

Pressure Transfer

Any gas under pressure can be used to move a liquid from one vessel to another. Some materials (flammable, explosive, toxic, corrosive, etc.) require the use of an inert gas for transfer.

Biology

Blanketing

The preservation of such biological products as blood plasma, viruses, protein solutions, and the like in freeze-dried form requires inert atmospheres of very high purity. Even an oxygen content of 0.5 percent may be totally detrimental to the product.

Deoxygenation

Deoxygenation means the reduction or replacement of the oxygen content of air by a shielding or inert atmosphere in order to prevent oxidation, microbial growth, or ignition, which can cause fire or explosion. It is used to protect liquid or pulverized food products from becoming rancid or from fermentation processes.

Foodstuffs

Oxidative Deterioration

Nitrogen in food packaging is used in order to prevent oxidative deterioration of foods. Many foodstuffs, and especially those containing fats, are subject to such deterioration. Already in 1865 the first U.S. patent on preservation of foodstuffs was issued covering the use of carbon dioxide to preserve oysters or peaches. In 1948 Clarence Birdseye suggested the use of nitrogen or carbon dioxide to preserve fruits and meat products.

The purpose of inerting packages during filling is to reduce the oxygen content in the product as well as the headspace of the final package. In this way vitamin content, color, and taste will last for a longer time. Typical products where inerting is used are juices, beer, wine, edible oils, peanuts, milk powder, and other solid powders that are sensitive to oxygen.

Nitrogen, both in liquid and gaseous form, is the most common gas for this application (Fig. 7.6.3). In some rare situations argon is used because sometimes the high density of argon is desirable.

Protection against Mold Growth and Insect Infestation

If oxygen levels of 0.5–2.0 percent can be maintained by the use of nitrogen, considerable protection against mold growth is gained. These low oxygen levels not

Figure 7.6.3. Liquid nitrogen transport in a small vessel. Courtesy of the Messer Group, Frankfort-am-Main, Germany.

only destroy any insects present, but also destroy dormant eggs if the inert atmosphere is maintained for a few days.

Modified Atmosphere Packaging (MAP)

The ability of carbon dioxide to slow the rate of microbiological propagation was first found by Franklin Kidd and Cyril West in England in the early years of the 20th century. This was the origin of modified atmospheres and vacuum food preservation methods. Work on this method was carried on in the 1930s with meat shipments.

Dry ice was also employed to introduce elevated CO_2 levels in sealed transport packages. At the same time Robert Smock at Cornell University determined that elevation of the CO_2 concentration and reduction of O_2 in confined storage chambers of apples could prolong storage time to 6 months. More general practical uses of modified atmospheres started in the late 1950s for warehouse and distribution purposes and has been applied to the retail market since the mid-1970s.

During the 1940s nitrogen was first used in the packaging of food in cans. The filled can was passed through a chamber which was first evacuated and then filled with inert gas, after which the can was sealed, retaining the inert atmosphere.

When transparent film packages began to be used for foods in the late 1950s nitrogen packing became even more important. The use of nitrogen atmospheres in transparent packages also started the development of specialized machines for filling flexible containers or bags with nitrogen and sealing them (Fig. 7.6.4).

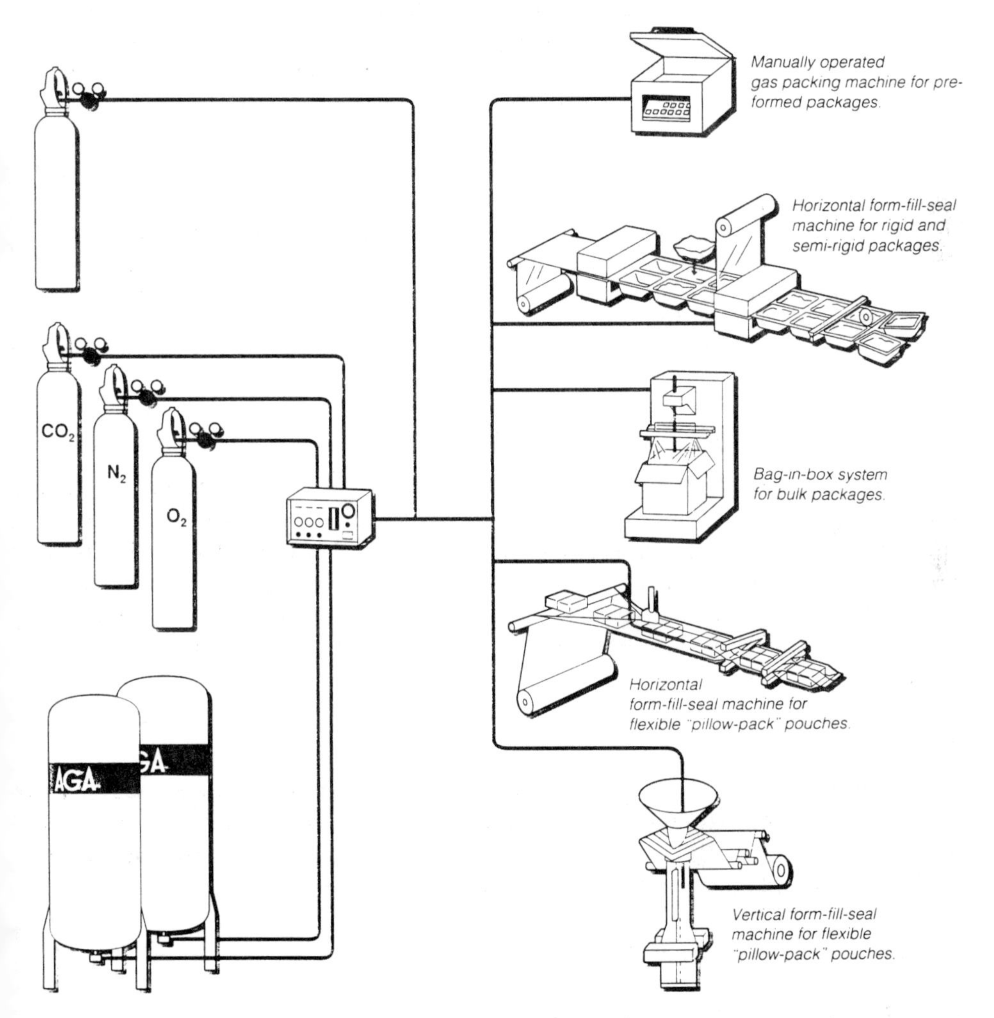

Figure 7.6.4. Gas packaging systems for modified atmospheres. Courtesy of the AGA Historical Picture Archive.

The type of gas and the gas concentration for MAP should be determined for each product, but for nearly all products a combination of carbon dioxide (CO_2), oxygen (O_2), and nitrogen (N_2) is used. Two- or three-component mixtures of gases are now common. Nitrogen has no inhibitory effect on microbial growth. Its indirect effect is due to its displacement of O_2. It is used as an inert filler to prevent the package from collapsing. From the microbiological point of view CO_2 is the most important gas used in MAP. This gas may kill, inhibit, have no effect on, or stimulate the growth of microorganisms. The ability of CO_2 to inhibit microorganisms is a complex phenomenon that depends on more than just the type of microorganisms present. The absorption of CO_2 in a package is highly dependent on the product. If excess CO_2 is absorbed, the package volume will be reduced and in some cases collapse. This is prevented by using N_2/CO_2 mixtures. Controlled concentrations of oxygen are used in MAP meat packing to maintain the myoglobin—the pigmentation agent—in its oxymyoglobin form. This keeps the fresh red color and slows down the creation of the brown color that arises due to natural aging. CO_2 and O_2 are also needed for fruit and vegetable respiration.

Other gases, such as carbon monoxide, ozone, ethylene oxide, nitrous oxide, helium, neon, argon, propylene oxide, ethanol, hydrogen, sulfur dioxide, and chlorine, have been used experimentally or on a limited commercial basis to extend the shelf life of a number of products. For example, carbon monoxide is very effective at maintaining the color of red meats, but its use in most countries is not permitted due to safety concerns.

Manufacturing

Lamps and Vacuum Tubes

Nitrogen is used for inerting at various stages of tube or lamp manufacture. Vacuum-tight bonds between ceramics and metals are made under pressure in a nitrogen atmosphere, and nitrogen may also be used as the lamp fill. Coiled tungsten wire filaments, heat-treated in hydrogen, are cooled in an atmosphere of nitrogen prior to use to prevent undesirable oxidation. Nitrogen is used as a carrier gas for iodine, which functions as a regenerative getter to prevent blackening of the lamp envelope.

Metals

In heat treatment, “controlled” or “protective” atmospheres are used. These include any gas or mixture of gases to prevent reactions that would take place during the heat treatment of the metal if it was surrounded by air. Such reactions include oxidation, carburization, and decarburization.

Liquid Nitrogen Applications[2–4]

Early Food Cooling and Freezing Applications

Use of compressed gases for the shipping of cooled foods was first primarily limited to carbon dioxide. Dry ice was used for shipment and storage of frozen foods, and the blast chilling of rail cars and trucks with liquid carbon dioxide became widely used, especially in the meat industry.

The food industry has long realized the vast potential of other liquefied gases such as nitrogen, nitrous oxide, and liquid air as refrigerants. The extremely cold temperatures of nitrogen and nitrous oxide made them desirable not only for providing cold, but also as immersion freezing media. During World War II the manufacture of liquid nitrogen and oxygen became common, which helped lead to a reduction in the cost of these gases. Prior to World War II most gases such as nitrogen or oxygen were sold in the gaseous state. During 1942–1945 the Germans used nitrous oxide as both a refrigerant and as an immersion, freezing medium.

After the war there was a resurgence of interest in both liquid nitrogen and nitrous oxide as immersion freezing media. The rapid advance in the marketing of frozen foods in the period 1946–1949 left little time for work on new and radical methods of freezing. The desire was to get frozen products onto the postwar market.

In 1955 there was a revitalization of interest in the freezing with liquid gases. This interest was primarily from in the seafood industry on the U.S. Pacific Coast. Once again the high cost of freezing was the factor that pushed this very worthwhile technique into the background. During the period 1954–1959 new large markets opened up for liquefied gases in the United States. Nitrogen became one of the major liquid gases and new production plants were built for production of both oxygen and nitrogen for electronic, missile, chemical, and other Industries.

Other arguments in favor of liquid nitrogen freezing than quality also arose. The industry began searching for new methods of shipping frozen products. The success of prepared and specialty frozen foods further increased this trend. Valuable improvements in quality and the increase in the frozen food market led to a new need for transport and storage under accurate temperature control, which inspired the use of liquid nitrogen for food transport.

Food Industry

From the early 1960s food freezing with liquid nitrogen developed very rapidly when it was found that food that is frozen cryogenically had superior quality to that frozen by mechanical freezing. Rapid freezing in liquid nitrogen prevents the formation of the large, long, needle-like ice crystals present in slow freezing, and less damage occurs to delicate food cells because the rupture of the cell walls during freezing is avoided. The inert atmosphere protection of foodstuffs with gaseous nitrogen also applies to liquid nitrogen, where the inert blanketing effect is valuable in the handling of frozen foods (Figs. 7.6.5 and 7.6.6).

In the early days ice-cooled refrigerator cars were use to transport refrigerated food by road and railway. Mechanical refrigeration then made it possible to accomplish large-scale controlled refrigeration of perishable and frozen foods by liquid nitrogen cooling.

In 1961 the Polarstream method was put on the market by the Linde Division of Union Carbide. This was also used by the British Oxygen Company, and Air Liquide used it under the name Cryogal. A liquid nitrogen cylinder mounted inside a truck body was equipped to discharge liquid nitrogen through a spray header, with

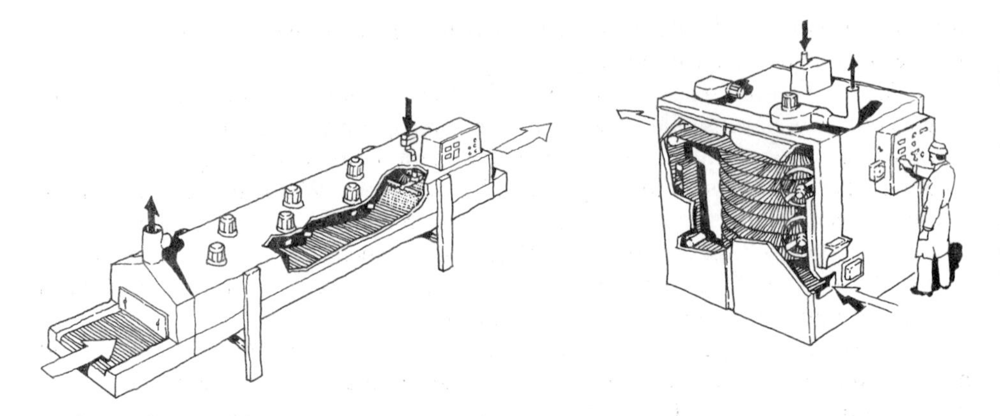

Figure 7.6.5. Cryogenic food freezing equipment. Courtesy of Teknisk Illustration, Umeå, Sweden.

the rate of cooling regulated by a thermometer (Figs. 7.6.7 and 7.6.8). Similar systems were developed by other companies, such as Cold Flow by Air Reduction, Superchill by Liquifreeze, Cryo-Guard by Air Products, and Cold Wall by Chemetron Corp.

In the early 1960s the food processing industry began to use the embrittlement of various food products by liquid nitrogen for cryogrinding in grinders such as pin mills or hammer mills. An important application was cryogrinding of spices, such as

Figure 7.6.6. A Cryo-Quick seven-line unit for McDonald's, 1970. From Butrica (1990) with permission.

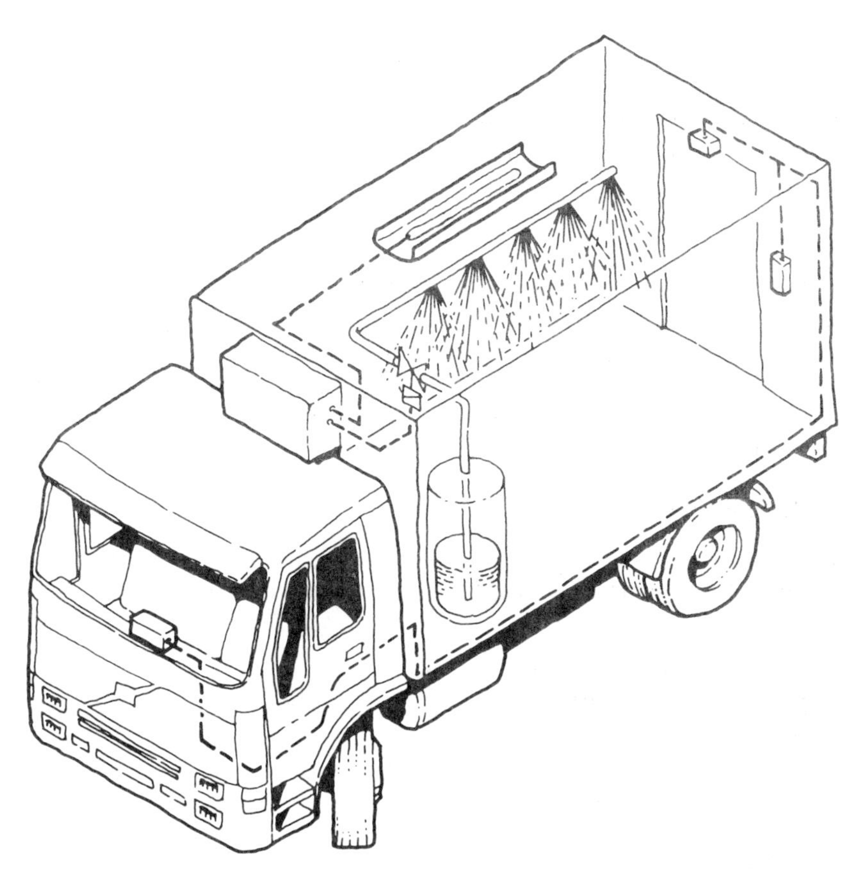

Figure 7.6.7. Cryogenic food transport. Courtesy of Teknisk Illustration, Umeå, Sweden.

Figure 7.6.8. The Polarstream (Praxair) transport cooler. Courtesy, of Praxair, Danbury, Connecticut.

cinnamon, pepper, or nutmeg, in order to balance the heat generated when they were ground in an abrasion or cutting mill. Heat causes oxidation and loss of the flavoring in the spices.

Medicine

Cold is becoming more and more widely used for preserving of living tissue. Artificial insemination of cattle using bull semen preserved at liquid nitrogen temperatures is now a standard commercial practice. Blood preservation by freezing is commonly used and successful cryosurgery has been reported. Banking of human organs at liquid nitrogen temperatures as spare parts for surgery is a hope for the future (see Chapter 7.1).

Other Industrial Uses[5,6]

Besides its use in cooling processes, liquid nitrogen has been used from the 1960s for a large number of applications where the effect of cold produces changes in materials:

- Liquefying substances in gas form
- Causing liquid material to solidify
- Shrinking material
- Making soft materials hard and brittle
- Reducing solubility
- Reducing reactivity
- Creating high vacuum
- Creating high pressure on evaporation
- Changing material properties

Since the 1970s many new of these industrial applications significantly increased the demand for liquid nitrogen. Newer developments are reviewed below.

Liquefaction of Gases: Recovery of Solvents

During the early 1980s Airco developed and perfected the Airco solvent recovery system (ASRS) for recycling solvents from industrial coating processes.

Solidifying of Liquids

Cooling in Extrusion Processes. Liquid nitrogen cooling is employed in both the metals and the polymer industries for faster solidification of materials. Besides an increased rate of production, a smoother surface finish is obtained.

Blow Molding of Glass and Plastic Bottles. Bottles are usually manufactured in a casting process in which an extruded sleeve of molten material is pressed outward against a die with compressed air. By cooling the compressed air with LIN the setting time is reduced and the production rate increased.

Ground Freezing. Ground freezing is used with foundation and tunnel work, where it is essential to prevent soil or water from penetrating into the area where work is

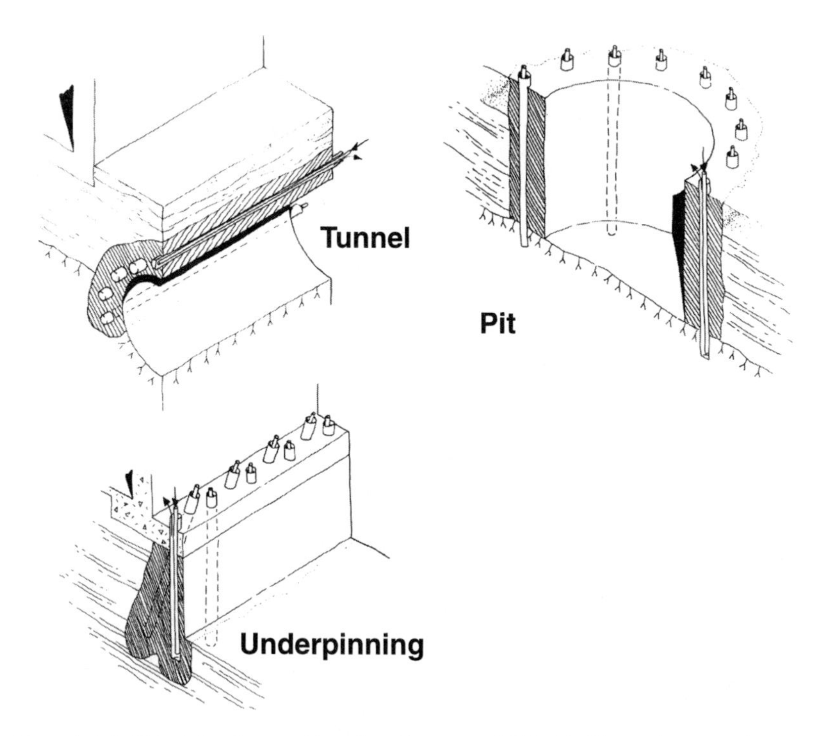

Figure 7.6.9. Principle of ground freezing. Courtesy of Teknisk Illustration, Umeå, Sweden.

taking place. Freezing with LIN is 10 times quicker than with the conventional cooling compressor technique (Fig. 7.6.9).

Pipe Freezing. Pipe freezing with LIN is used in the repair of pipe systems in heating plants, for example. Instead of draining down the whole system when a pipe must be opened, ice plugs are frozen on each side of the section under repair.

Shrinking

Shrink Fitting/Expansion Fitting. These methods are used for the retention of a wheel on a shaft, for example. The shaft pivot is shrunk by cooling with LIN and can then be fitted into the center hole of the wheel (Fig. 7.6.10). When the shaft has returned to the surrounding temperature the wheel is firmly fitted.

Paint Removal. During painting processes, quantities of paint land on the carrier of the object being painted. These layers of paint build up constantly and must be removed at regular intervals. Previous methods involved the use of solvents or burning off, both being detrimental to the environment. When a paint-coated object is dipped into LIN, the paint contracts more than the metal beneath. This causes the paint to crack and loosen from the surface underneath; it becomes brittle and can then be removed by mechanical means.

Figure 7.6.10. Shrink fitting. Courtesy of the AGA Historical Picture Archive.

Embrittling

Cryogenic Processing. Grinding at cryotemperature is used partly to make material brittle and partly to compensate for the heat generated during the grinding process. In 1974 "cold crushing" was first used for crushing such various materials as old automobiles, industrial waste of all sorts, and old tires, all materials which become brittle. In 1976 there were about 10 plants of this sort for metallic waste, in the United States, Germany, the United Kingdom, Japan, and Belgium. In Liege, Belgium, an installation handled 20 tons per hour of waste.

Cryogenic processing is now mostly used for plastics and rubber for recycling or for the production of powder. Due to their toughness, certain polymers cannot be ground to powder at room temperature. After cooling to below the glass point of the material, granules of down to 0.1 millimeters can be obtained.

Deflashing. When casting molded rubber components a flash ("whisker") or fin is formed at the mold joints. This has to be removed, and was previously done by hand with a knife. By cooling to the material's glass point the rubber becomes hard and brittle and the flash can be easily broken off by tumbling, shot blasting, or other mechanical means (Fig. 7.6.11).

Figure 7.6.11. Rubber gaskets before and after cryogenic deflashing. Courtesy of the AGA Historical Picture Archive.

Pressure Change

Pressure Raising. The use of drop injection during the can- or bottle-filling process began in the brewery industry in the 1980s. There was a problem arising from the introduction of aluminum cans. Because of the softness of the thin material, transporting and stacking the cans could not be done in the usual way. At the same time there was a need to extend the storage life of the product. By adding a few drops of LIN before the drink is poured in, the oxygen is driven out of the can, reducing the risk of fermentation. At the moment prior to the fitting of the lid, a few more drops are added, which vaporize (a volume increase of some 700 times). The expansion provides a pressure increase sufficient to make the can rigid.

Pressure Reduction. Cryopumping has become a major industrial application due to the advantage the cryopump has of quickly achieving ultrahigh vacuum with very high purity. The method is based on the fact that most gases liquefy at low vapor pressure when in contact with a very cold surface. A LIN cold trap is also used in diffusion pumps in order to liquefy the oil mist from the pump.

Changing the Properties of Materials: Cryohardening

Cryohardening of steel is used to achieve a harder material for knives and wear metals, etc., by deposition of martensite. Cryohardening is an example of a phase transformation which can be achieved by controlled rapid cooling.

References

1. Airco. (1961). The freezing and shipping of foods. In *Proceedings of the ASHRAE Symposium*.

2. Adams, R. J. (1947a). It's time to investigate nitrous oxide immersion freezing. Part I. *Frosted Food Field* **1947** (March).
3. Adams, R. J. (1947b). It's time to investigate nitrous oxide immersion freezing. Part II. *Frosted Food Field* **1947** (April).
4. Almqvist, E. (1998). Industriella kryoapplikationer [Industrial cryogenic applications]. *Kosmos* (Swedish Physical Society Annual) 1998:161–174 [in Swedish].
5. McFarland, A. I. (1945). Nitrous oxide meets low temperature refrigeration requirements. *Power* **1945** (June).
6. Scurlock, R. G. (1990). A matter of degrees: A brief history of cryogenics. *Cryogenics* **30**:483–500.
7. Sittig, M. (1965). *Nitrogen in Industry.* C Princeton, NJ: Van Nostrand.
8. Tressler, D. K., and Evers, C. F. (1968). *The Freezing Preservation of Foods.* Vol. 1. *Freezing Fresh Foods.* Westport, CT: AVI Publishing Co.
9. Butrica, A. J. (1990). Out of Thin Air: A History of Air Products and Chemicals, *Inc. 1940–1990.* New York: Praeger.

Further Reading

Barron, R. F. (1985). *Cryogenic systems.* Oxford: Clarendon Press.
Braton, N. R. (1980). *Cryogenic Recycling and Processing.* Boca Raton, FL: CRC Press.
Hands, B. A. (ed.). (1986). *Cryogenic Technology.* London: Academic Press.
Hausen, H., and Linde, H. (1985). *Tieftemperaturtechnik.* Berlin: Springer-Verlag.
Williamson, K. D., Jr., and Edeskuty, F. J. (1983). *Liquid Cryogens.* Boca Raton, FL: CRC Press.

7.7. SPECIALTY GASES. HIGH PURITY GASES[1]

Background

The specialty gas business emerged from the early use of air gases. The application of noble gases started a need for more efficient separation, liquefaction, and purification methods, improvements in cylinder technology, and advances in analytical techniques. Specialty gases and gas mixtures are defined as those of ultrahigh purity or special composition, sold in cylinder quantities.

Matheson Gas Products, established in 1927, is the world's oldest specialty gas producer. After becoming part of the Nippon Sanso Group in 1983, in addition to gas, they have marketed a product line which includes equipment and piping installation for arsine and phosphine. The other main industrial gas companies also organized specialty gases department in the 1960s. Although sales volumes were small, high unit selling prices made this business very profitable.

Noble Gases

Helium, argon, neon, krypton, and xenon were all discovered over a short period at the end of the 19th century (see Chapter 3.6).

In 1901 Cooper Hewitt in the United States used argon to improve the light of his mercury lamp. In 1913 the company Allgemeine Elektrizitätswerke used 99.5 percent pure argon to fill Osram tungsten filament lamps based on a patent by

Griesheim Elektron. Industrial large-scale helium separation from liquefied natural gas was first developed in 1917, based on the Linde–Claude principle.

Industrial manufacture of neon was first carried out by Georges Claude in 1917, employing the principle of fractional distillation of liquid air. Pure noble gases can also be obtained by low-temperature decomposition from ammonia synthesis.

The lighting industry's need for pure neon, krypton, and xenon in advertising signs and street lighting led to extensive work from the 1930s on pure gases and gas mixtures. The rare gas business at that time was characterized by the sale of small quantities of high-value products. In the 1940s pure argon found a large market as a shielding gas in welding. From the 1930s noble gases could be separated on an industrial scale from the atmosphere by liquefaction. A helium–neon-rich fraction from air separation plants is cooled under pressure in liquid nitrogen and then absorbed on passage through a charcoal bed in liquid nitrogen. On warming, helium is evolved, followed by a helium–neon mixture and finally pure neon.

After the end of World War II large growth in oxygen and nitrogen consumption occurred world-wide. Processing large volumes of atmospheric air enabled separation of larger quantities of the noble gases regardless of their very low content in air.

Before 1955, neon could only be marketed in small quantities in glass flasks, but by 1965 tube trailer volumes were being produced. Similarly, increased volumes of the other noble gases were sold.

Before the invention of the gas chromatograph, purity checks on noble gases were carried out by the use of a high-voltage, high-frequency discharge or by using a Teslar probe. During 1949–1950 the purity of argon produced by air separation reached a level sufficient for filling electric bulbs and for use as protective atmosphere in welding. This was achieved after the pressure purification system had been installed.

From about 1960 oxygen has been removed from raw argon by burning the mixture in hydrogen using a palladium catalyzer. The water produced is allowed to condense, and argon is dried in molecular sieves and led back to the distillation column. Excess hydrogen is separated and nitrogen present in the mixture is distilled off. Practically pure argon (99.7–99.9 percent Ar) is then compressed in membrane compressors and filled at 15°C and a filling overpressure of 4–15 bars into steel cylinders or at 15 bars into special transportation containers.

Pure gases are essential for use as protective inert atmospheres in metallurgy and in special steel production welding, lighting, analyzing procedures, and semiconductor technologies. The growth in their use caused an increasing demand for a high purity of noble gases and their mixtures. In particular, pure argon has been in demand in large quantities.

Gas Mixtures

Preparation of stable gas mixtures meeting the demands of customers was a major problem before data for the mixing properties of gases were available. Consumption of mixtures of various gases has steadily increased and there is a virtually unlimited need for the production of mixtures of various compositions. From starting as a "catalogue business" for a few customers, the specialty gas mixture business has become a market including many customers with very special requirements. In mixing, quantitative effects depend on many parameters, including the gas components used and the size and shape of the cylinder. As an example, a dramatic increase in the mixing rate occurs in a horizontal cylinder compared to a vertical one and still further improvements in mixing rates are achieved by rolling the cylinder.

Three basic methods are used when mixing gases:

- *Manometric*, corresponding to the partial pressures of individual components
- *Gravimetric*, in which individual components are weighed according to their densities
- *Volumetric*, where accurately determined volumes are mixed

The composition of mixtures must conform to technical requirements, chemical compatibility of the components, and corrosion properties in relation to the cylinder materials. In order to eliminate the effects of unidentified impurities present in the components it is necessary to produce the mixtures from high-purity gases.

Analysis of Gas Mixtures and High-Purity Gases

To manufacture various kinds of mixtures, it is necessary to have a high degree of accuracy in the concentrations of the components and in the stability of the system, including the mixture, gas components, and vessel. Optimizing a gas mixture puts high demands on the technical skill and analytical equipment of the manufacturer.

Examples of mixtures often used in different application areas are as follows.

Medicine, Biology, Environment

Applications in these areas include the following:

- Mixtures forming reducing atmosphere (N_2–H_2)
- Breathing gases for medicine (O_2–CO_2 mixtures, synthetic air)
- Mixtures for ripening fruit, for example, bananas (N_2–C_2H_4)
- Calibration gas, SO_2 in N_2, for checking air pollution
- CO in N_2 for checking exhaust gases
- CO_2 and O_2 in nitrogen, used in medical diagnostics

In medicine, specialty gases for anesthesia, breathing, and life-support dominate, and many diagnostic procedures depend on instrumentation that requires calibration with accurately prepared and analyzed gas mixtures.

Atmospheres for food packaging and fumigation and mixtures to provide suitable atmospheres for biological culture growth are other common applications.

Figure 7.7.1. Specialty gas filling in an aluminum cylinder Courtesy of the Messer Group, Frnakfurt-am-Main, Germany.

Analytical Instruments

Most methods of gas analysis require calibration standards. Examples are:

- Mixtures for counting techniques and measurement of ionizing radiation ($Ar–CH_4$, $He–C_4H_{10}$)
- C_2H_6, C_2H_2 and CH_4 in C_2H_4 for chromatographic control of ethylene manufacture
- CO_2 and O_2 in nitrogen for analyzers of flue gases

Calibration mixtures are required in a variety of analytical applications including production control in large chemical plants and multicomponent analysis of gas streams in the petrochemical industry.

Welding, Lighting, Lasers

Examples of these applications include:

- Protective gases for welding (Ar–CO_2 mixtures)
- Mixtures for laser technology (CO_2–N_2–He)
- Media for filling nuclear reactors and accelerators and for use in electronics (Ne–He)
- Mixtures for illumination purposes (Ne–Kr, Ar–HBr)

Process applications include the operation of lasers and, rather more unusually, mixtures with a radioactive component in switches for use in the lighting industry.

High-Purity Gases

High-purity products became part of the specialty gas industry because of the requirements for small quantities of especially purified industrial gases. Zero-grade gases originally devised for use in instrument calibration have impurities (frequently oxygen or moisture) reduced to levels which are either closely defined or are not detectable in the application envisaged. Requirements have been set not only for

Figure 7.7.2. Connection work in a high-purity room. Courtesy of the Messer Group, Frankfurt-am-Main, Germany.

very pure products, but also for the close definition of the presence of critical impurities.

There is no universally accepted definition of what is a "pure" gas. Very low concentrations of impurities, that is, undesirable gas components or particles, are commonly measured in parts per million. Gases that do not contain more than 100 parts per million of impurities are pure gases. The term "very pure" may be used for an impurity content down to parts per billion.

Gas purity is marked according to codes, for example, argon Ar 5.5 means a purity of 99.9995 volume percent, and He 5.6 means a 99.9996 volume percent content of helium, the content of impurities being a maximum of 4 parts per million by volume.

Semiconductors[2]

The transistor invented by Bardeen, Brittain, and Shockley at Bell Labs in the United States in 1948 was the start of the microelectric revolution. The development of microchips in the early 1960s rapidly increased the need for specialty gases, and this segment of the market continues to lead the demand for these gases and has broadened the specialty gases sector.

Apart from commodity gases such as nitrogen and argon, which are used for blanketing, the major gases used are helium, neon, krypton, and xenon. Products such as silane (SiH_4), dichlorosilane (SiH_2Cl_2), phosphine (PH_3), boron trichloride (BCl_3), tungsten hexafluoride (WF_6), and hydrogen fluoride (HF) are used in semiconductor

Figure 7.7.3. Semiconductor wafer. From Lockhart (1986) with permission.

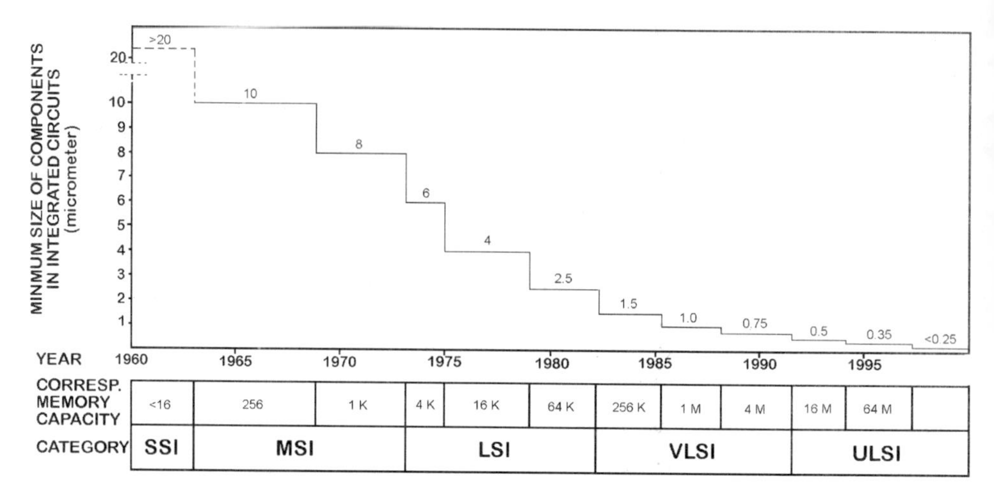

Figure 7.7.4. Minimum component size in integrated circuits versus year introduced.

processing. The use of these gases has been accelerated by the trend away from wet processing toward dry processing using gases.

Gas is supplied to this market from three sources: truck-delivered liquid (merchant) gas; cylinders carrying specialty gases from an on-site air separation plant as vaporized gas; and through a pipeline from a nearby delivery point. High purity must be ensured in the whole distribution chain, which means that special purity requirements are necessary for all equipment, piping, tanks, and cylinders. High-purity gases are also needed for other applications such as solar cells and fiber optics manufacturing.

Purity requirements for many specialty and electronic gases typically range from parts per million parts per billion, and the trend for almost all gases is toward low parts per billion levels. Purity is particularly critical in chip manufacturing because the decreasing size of integrated circuits has reached a level where individual atoms can affect devices (Fig. 7.7.4). At these low impurity levels, it has become very difficult to ensure that the gas quality meets customer requirements. This has led to a need for still more advanced analytical methods and equipment (Fig. 7.7.5).

Reactive Gases

Special handling techniques are required for some corrosive and reactive gases which are included in specialty gas mixtures used in laboratories for calibration, as reagents or dopants, for disinfecting and sterilization, etc. Some of these gaseous chemicals were known and used as far back as the alchemists and some are found in our natural environment. The development of modern chemistry and chemical engineering started the need for controlled and pure laboratory quantities of these substance for production control and quality testing. Therefore specialty gas departments of the industrial gas companies also market controlled gas mixtures containing hydrogen sulfide, ammonia, sulfur dioxide, carbon monoxide, ethylene, ethylene oxide, chlorine, fluorine and fluorine compounds, etc.

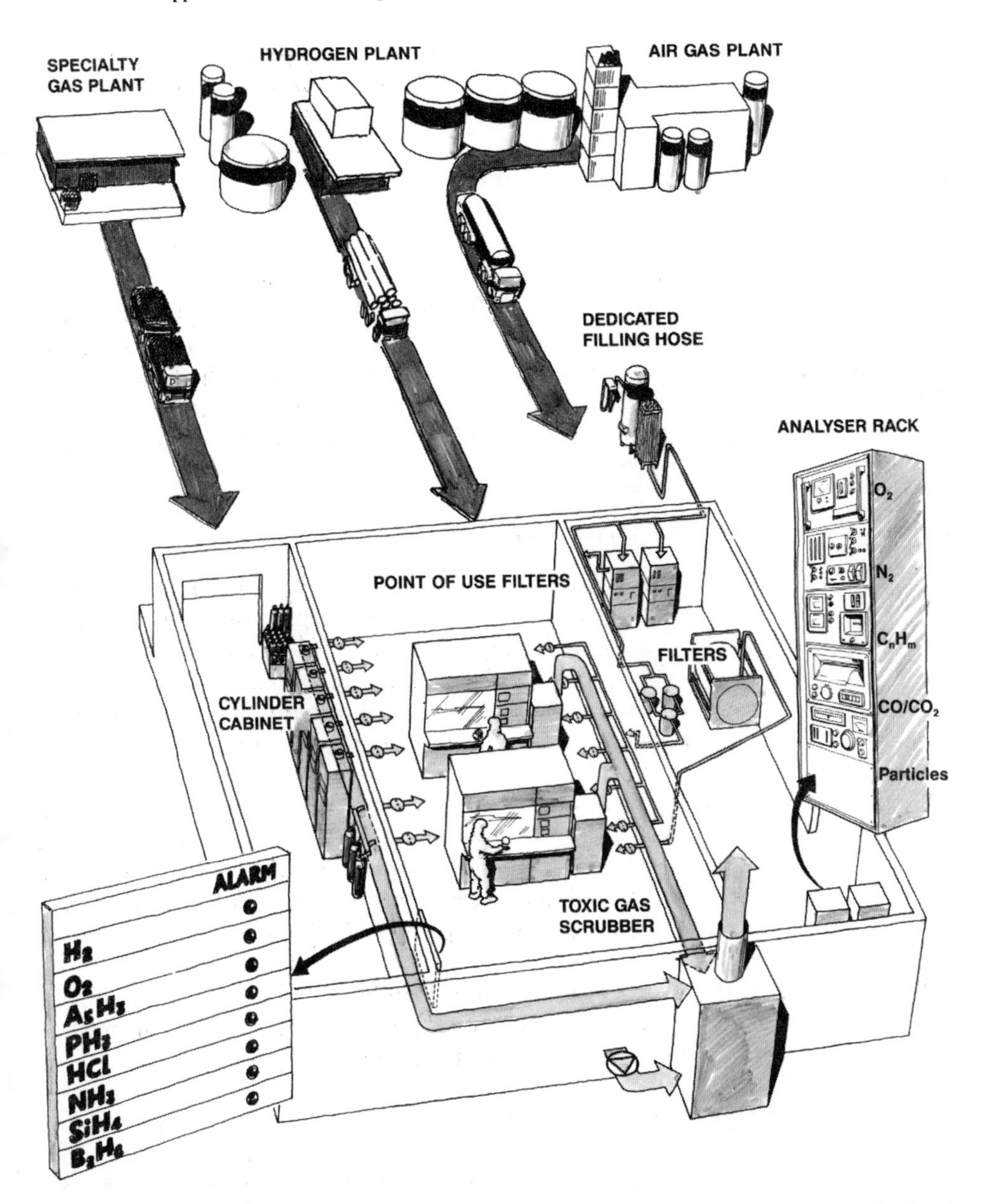

Figure 7.7.5. Supply of high-purity gases for the semiconductor industry. Courtesy of the AGA Historical Picture Archive.

Some decades ago it was found that the concentration in some mixtures containing low levels of reactive gases in inert balance gases delivered in steel cylinders could vary with time, temperature, and/or pressure. Therefore a world-wide requirement for stable, reliable calibration standards grew in the early 1970s. In 1975 Airco labs announced their development of a specially treated aluminum cylinder called Spectra-Seal, which guaranteed the long-term stability of mixtures of carbon monoxide, nitric oxide, nitrogen dioxide, and sulfur dioxide. Many years of experience with the Spectra-Seal process has confirmed the importance of this development in producing accurate calibration standards in cylinders.

Gas Analysis[3,4]

Gas analysis plays a key role in the application of the hundreds of gaseous chemical substances sold as specific products or mixtures and the vast amount of mixtures specifically produced to meet user needs. In the early years analytical measurement techniques used wet chemical methods as well as differences in physical and optical properties of the gas components. Analysis instrument methods available in the 1950s were inadequate, which led to the development and production of a range of specialized instrumentation which could be used both for production tests and establishment of user requirements.

Gas Impurity Detection and Measurement

The first patent on gas chromatography was granted to J. Janak in 1952, who used this method for accurate measurement of the gas components in air. The development of the first commercially available gas chromatograph in the late 1950s played a crucial role in the development of the business. It has enabled analysis to be carried out with greater speed, accuracy, sensitivity, and reproducibility than previously possible. It also brought new market opportunities for high-purity carrier gases such as helium and for calibration standards.

Gas chromatography is used with a wide variety of detection systems. Thermal conductivity detectors were used already in the 1930s. The flame ionization detector and the electronic capture detector arrived in 1958, and during the 1960s noble gas, ultrasound, and flame photometric detectors were developed. With gas chromatography impurities down to parts per billion levels or below can be quantified.

Mass spectrometry was originally used in 1919 by F. W. Aston, who used photographic detection. This method depends on differences in the molecular mass of different substances. Since 1940 various high-resolution instruments have been developed.

Infrared spectrophotometry was developed in the 1940s especially for the analysis of hydrocarbon fuels. High-resolution instruments are used for establishing impurity levels in nonreactive gases, and nondispersive instruments (NDIR) are widely used in the analysis of gases such as carbon monoxide and carbon dioxide in mixtures. Fourier transform infrared spectroscopy (FTIR), developed during the 1960s, increased the resolution of impurity detection in reactive gases.

Ultraviolet spectrophotometry has been used mainly in investigative work.

Chemiluminescent analysis for measurement of nitrogen oxides depends upon the chemiluminescence produced by the reactions of NO and O_3.

Dedicated instruments have also been designed for the analysis of moisture (capacitance, quartz resonance), oxygen (Hersch cell, phosphorus luminescence), argon, sulfur dioxide, etc,

Particle Measurement

Atomic absorption spectrophotometry, developed in the 1950s, has been frequently used for analysis of metal impurities in gases. It depends upon the absorption of light of a specific wavelength by metal atoms in an excited state. One application is in atmospheric monitoring, including the analysis of lead, mercury, and welding fumes in air. It is also used by the electronics and fiber optics industries for measurements of metallic impurities in gases and low-boiling liquids. Measurements are now being made routinely at the parts per billion level using flameless techniques.

Laser particle counting techniques employing light scattering are used to deal with particulate contamination problems at the submicrometer level in the microelectronics industry. They are often used for checking pipes and individual assemblies.

References

1. Davidson, J. M., and Kies, F. K. (1985). Microcontamination control for the electronics industry. Part 1. Particle reduction. *BOC Technology* **3**:3–13.
2. Hill, R. J. (1985). Advances in deposition techniques for microchip manufacture. *BOC Technology* **3**:14–19.
3. Lockhart, I. (1986). The development of the special gases business. *BOC Technology* **5**:16–25.
4. Riedel, E., and Schram, J. (1993). Gasanalytik—Damals und heute. *Gas Aktuell* **40**:4–10.

FIGURE REFERENCES

2.1.1 Lorens, Y. (ed.). (1922). *Bonniers Konversationslexikon* (in Swedish). Stockholm: Albert Bonniers.
2.1.2 Valentine, B. (1602). *Les Douze Chefs de la Philosophie*. Leipzig. (Drawing after engraving 1651).
2.1.3 Ercker, L. (1574). *Treatise on Ores and Assaying*.
2.1.4 Aberle, K. (1891). *Grabendenkmal, Schädel und Abbildungen des Theophrastus Paracelsus*. Salzburg, Austria.
2.1.5 Van Helmont J. (1662). *Oriatrike or Physik refined*.
2.1.6 Ålund, O.W. (ed.). (1875). *Uppfinningarnas bok* (in Swedish) Stockholm: L.J. Hiertas förlag.
2.1.7 Von Guericke, O. (1672). *Experimenta Nova Magdeburgica*.
2.1.8 Boyle, R. (1661). *The Sceptical chymist*.
2.1.9 Boyle, R. (1725). *Philosophical Works*.
2.1.10 Mayow, J. (1674). *Tractatus quinque medico physici*.
2.1.11 Hales, S. (1727). *Statical Essays containing Vegetable Staticks*.

2.2.1 Diderot, D., and d'Alembert, J. (1752). *Encyclopédie*.
2.2.2 Priestley, J. (1790). *Experiments and Observations on Different Kinds of Air*.
2.2.3 Scheele, C.W. (1777). *Chemische Abhandllung von der Luft und dem Feuer*.
2.2.4, 2.2.7 Lavoisier, A. (1789). *Traité élémentaire de chimie*.
2.2.5 Guillemin, A. (1884). *Le Monde Physique, IV*. Paris: Librarie Hachette et Cie.
2.2.6 Lavoisier, A. (1864). *Oeuvres de Lavoisier*.

2.3.1 Ålund, O.W, (ed.) (1875). *Uppfinningarnas bok* (in Swedish) Stockholm: L.J. Hiertas förlag.
2.3.2 Van Helmont J. (1662). *Oriatrike or Physik refined*.
2.3.3 Neudeck, G. (1905). *Das kleine Buch der Technik*. Stuttgart, Germany.
2.3.4–5, 2.3.16 Reference 2.3.1. Permission from the Journal of Chemical Education, copyright © 1935, Division of Chemical Education, Inc.
2.3.6–9, 2.3.11–15 Ramsay, W. (1905). *The gases of the atmosphere*. London: Macmillan.
2.3.10 Lomosonov, M. V. *Sovranje razhij sochinieniyl*Painting 1787 by L. Miropolsky.

3.1.1 Lana-Terzi F. (1670). *Prodromo dell'Arte Maestra*.
3.1.2, 3.1.5–6 Ålund, O.W. (ed.). (1875). *Uppfinningarnas bok* (in Swedish). Stockholm: L.J. Hiertas förlag.
3.1.3 Ganot, A. (1873) *Elementary Treatise on Physics Experimental and Applied*. (trans. E. Atkinson). London: Longman's Green.
3.1.4 Reference 3.1.1.
3.1.7 Reference 3.1–5.
3.1.9 Permission from Corbis Images, New York. Portrait: permission from Linde Gas, Höllriegelskreuth, Germany.
3.1.11 Permission from the Royal Institution, London UK/Bridgeman Art Library.
3.1.12–13 Permission from the Archive of Luftschiffbau Zeppelin GmbH, Friedrichshafen, Germany.

3.2.1 Permission from Edgar Fahs Smith Collection, University of Pennsylvania Library.
3.2.3–4, 3.2.12 Permission from the BOC Group, Windlesham, England.
3.2.5 Claude, G. (1926) *Air liquide, oxygène, azote, gas rares*. Paris: Dunod 1926.
3.2.6 Granjon, R., and Rosemberg, P. (1912) *Manuel Pratique de Soudure Autogène*, pp. 44–47. Paris: Bibliotheque de l'Office Central de l'Acetylène.
3.2.7 Brochure 1912.
3.2.8 Le Nature 1883.
3.2.10, 3.2.18 Ålund, O.W, (ed.) (1875). *Uppfinningarnas bok* (in Swedish) Stockholm: L.J. Hiertas förlag.
3.2.13 Permission from Linde Gas, Höllriegelskreuth, Germany.
3.2.14 Reference 3.2.11.
3.2.15 Reference 3.2.10.
3.2.16 Reference 3.2.13.
3.2.17 Reference 3.2.14.

3.3.1 Weeks, E. (1935). *The Discovery of the Elements*. Permission from the Journal of Chemical Education, copyright © 1935, Division of Chemical Education, Inc.
3.3.2 Westin, T. (1922). *Nordisk familjebok*. Stockholm: Nordisk Familjebok AB.
3.3.3 Permission from the Nobel Foundation, Stockholm, Sweden.
3.3.5–6 Ercker L. (1550). *Beschreibung Allerfürnehmisten mineralischen Ertzt und Bergwerksarten*. Prague.

3.4.1 Advertisement 1800s.
3.4.2–3 Permission from the Coca Cola Company.
3.4.4–5 *Le Monde Physique IV*, 1884.
3.4.6 O'Conor Slone, T. (1899). *Liquid air and the liquefaction of gases*. London: Sampson Low, Marston.
3.4.7 Reference 3.4.4.
3.4.8 Ålund, O.W, (ed.) (1875). *Uppfinningarnas bok* (in Swedish) Stockholm: L.J. Hiertas förlag.
3.4.10 Permission from Praxair, Inc., Danbury, Connecticut, USA.
3.4.13 Permission from Eric Snook, Bath, England.
3.4.15 Permission from the Nobel Foundation; Stockholm, Sweden.

3.5.1, 3.5.9 Permission from Linde Gas, Höllriegelskreuth, Germany.
3.5.2 Anonymous. (1898). *Calcium Carbide Fabrikation* **1898** No 20 (Jan 15), pp. 158–159.
3.5.3 Weeks, E. (1935). *The Discovery of the Elements*. Permission from the Journal of Chemical Education, copyright © 1935, Division of Chemical Education, Inc.
3.5.4 Anonymous. (1906). Dissolved Acetylen for the Lighting of Motor Vehicles. *Acetylene*, **1906** (Sept) 206–207.
3.5.5 Schmidt, J. (1913). Über Acetylenbeleuchtung und Acetylengaserzeuger. In *Zeitschrift für Beleuchtungswesen* (3) pp.29–33. Berlin: S. Fisher Verlag.
3.5.6 *Acetylene* 1906 (March) p.65.
3.5.7 *Acetylene*, March 1910.
3.5.8 *Acetylene* 1906.
3.5.11 De Szpcynski, S. (1898). Zur Geschichte der Acetylenbeleuchtung. *Zeitschrift für Calcium Carbide Fabrikation*. **1898** No 21 pp. 250-252.

3.6.1 Reference 3.6.1.
3.6.2–3 Weeks, E. (1935). *The Discovery of the Elements*. Permission from the Journal of Chemical Education, copyright © 1935, Division of Chemical Education, Inc.
3.6.5 Computer drawing from leaflet Goodyear Company 1927.
3.6.7 Permission from the University of Leiden.
3.6.8 Almqvist, E. (1988). Kryoteknikens historik [The history of cryotechnology]. *Kosmos* **1988**: 137–160.

4.1.1 Guillemin, A.(1884). *Le Monde Physique IV*. Paris: Librairie Hachette et Cie.
4.1.2 Girardin, M.J. (1880). *Lecon de chimie élémentaire*. Paris: G. Mason.
4.1.3 Reference 4.4.1.

4.1.4 *Le Monde Physique IV* 1884
4.1.5–7, 4.1.12–14 Reference 4.1.5.
4.1.8 Permission from Linde AG, Wiesbaden, Germany.
4.1.9–10 Permission from University of Leiden.
4.1.11 Reference 4.1.2.

4.2.1 Drawing by the author. Courtesy Teknisk Illustration, Umeå, Sweden.
4.2.2, 4.2.4 Permission from Torsten Lindqvist, Uppsala, Sweden. After Lindqvist T. (1970). Jakten mot absoluta nollpunkten (in Swedish) *Kosmos* **1970**:117–126.
4.2.3 Permission from the Nobel Foundation, Stockholm, Sweden.
4.2.5 Permission from University of Leiden.

4.3.1–5 Reference 4.3.1.
4.3.7 Reference 4.3.5.
4.3.9 Permission from the BOC Group, Windlesham, England.
4.3.10 Permission from Teknisk Illustration, Umeå, Sweden.

4.4.1 Ålund, O.W. (ed.) (1875). *Uppfinningarnas bok* (in Swedish) Stockholm: L.J. Hiertas förlag.
4.4.2 Permission from Praxair, Inc., Danbury, Connecticut, USA.
4.4.3 Reference 4.4.1.
4.4.4–5 Permission from Teknisk Illustration, Umeå, Sweden.
4.4.6 Reference 4.4.3.
4.4.7 Reference 4.4.5.
4.4.8 Permission from Linde AG, Wiesbaden, Germany.
4.4.9 Author drawing after Air Liquide brochure.
4.4.10 Permission from Linde Gas, Höllriegelskreuth, Germany.

4.5.2–3 Permission from Praxair, Inc., Danbury, Connecticut, USA.
4.5.4 Permission from Nippon Sanso Corporation, Tokyo, Japan.
4.5.5 Permission from Air Products and Chemicals Inc., Allentown, Pennsylvania, USA./Drawing by the author.

5.1.1–10 Reference 5.5.1. Permission from Praxair, Inc., Danbury, Connecticut, USA.

5.2.1–2 Claude, G. (1926). *Air liquide, oxygène, azote, gaz rares* (2nd ed). Paris: Dunod.
5.2.3–4 Reference 5.2. 6. Permission from Linde AG, Wiesbaden, Germany.
5.2.5–13 Permission from Linde AG, Wiesbaden, Germany.

5.3.1 Hoffmann, P. (1938). *Aus Vier Jahrzehnten deutsche Autogentechnik-Ein Rückblick*. Halle, Germany: Carl Marhold Verlag.
5.3.2, 5.3.4–6 Permission from the Messer Group, Frankfurt am Main, Germany.
5.3.3, 5.3.7–8, 5.3.10–11 From Reference 5.3.1. Permission from the Messer Group, Frankfurt am Main, Germany.

5.4.1–6, 5.4.8–12 Permission from L'Air Liquide, Paris, France.
5.4.7 Permission from Corbis Images, New York.

5.5.1–10 Permission from Linde Gas, Höllriegelskreuth Germany.

5.6.1–2, 5.6.4–7 From Reference 5.6.1. Permission from the BOC Group, Windlesham, England.
5.6.3 From Reference 5.6.2.
5.6.6–8 Permission from the BOC Group, Windlesham, England.

5.7.1–10 From Reference 5.7.1. Permission from Praxair, Inc., Danbury, Connecticut.
5.7.11 Permission from Praxair, Inc., Danbury, Connecticut.

5.8.1–9 Permission from Nippon Sanso Corporation, Tokyo, Japan.

5.9.1–5, 5.9.7 From Reference 5.9.1. Permission from Air Products and Chemicals Inc., Allentown, Pennsylvania, USA.
5.9.6–8 Permission from Air Products and Chemicals Inc., Allentown, Pennsylvania, USA.

6.2.2 Painting. Photo: E. Almqvist.
6.2.3 Advertisement ca 1906. Courtesy AGA Historical Archive.
6.2.4 View card Autogen Gasaccumulator, Brno plant (Czech republic) in 1927.

6.3.3–5 Permission from L'Air Liquide, Paris, France.

6.5.3–4 Permission from L'Air Liquide, Paris, France.
6.5.5 Permission from the BOC Group, Windlesham, England.

6.6.2 Technical note. In *Acetylene* 1910.
6.6.3–7 Permission from the BOC Group, Windlesham, England.

7.1.1–4, 7.1.10–11 Permission from Edmond Eger, San Francisco, CA.
7.1.5–6 Permission from Linde Gas, Höllriegelskreuth, Germany.
7.1.7 Drawing after sketch CIBA-Geigy 1977. Courtesy of Teknisk Illustration, Umeå, Sweden.
7.1.8 Permission from the Messer Group, Frankfurt am Main, Germany.
7.1.9 Ålund, O.W, (ed.). (1875). *Uppfinningarnas bok* (in Swedish) Stockholm: L.J. Hiertas förlag.

7.2.2–3 Permission from ESAB AB, Gothenburg, Sweden.
7.2.4–6, 7.2.8–9, 7.2.25 Reference 7.2.5.
7.2.7 Permission from the Messer Group, Frankfurt am Main, Germany.
7.2.10–12, 7.2.16, 7.2.18 Reference 7.2.2.
7.2.13 NSV. (1909). *Svetsning och skärning av metaller* (In Swedish). Stockholm: Nordiska Syrgasverken AB.
7.2.14, 7.2.19, 7.2.24 Permission from Linde Gas, Höllriegelskreuth, Germany.
7.2.15 Advertisement in *Acetylene* 1911.
7.2.20–23 Permission from AvestaPolarit Welding, Avesta, Sweden.

7.3.1 Agricola, G. (1556). *De re metallica*. Translation (1912). London: The Mining Magazine.
7.3.2–3 Hennig, R. (1911). *Buch Berühmter Ingenieure.* Leipzig: Verlag von Otto Spamer.
7.3.5 Permission from Linde Gas, Höllriegelskreuth, Germany.
7.3.6 Permission from Voest Alpine Stahl, Linz, Austria.
7.3.8 Eketorp, S. (1981). Gedanken zur Stahlerzeugung im Jahre 2000. *Stahl und Eisen* 101:852–859.

7.4.1 Permission from Praxair, Inc., Danbury, Connecticut, USA.
7.4.2 Permission from Metso Paper, Sundsvall, Sweden.
7.4.3 Permission from Metso Paper, Sundsvall, Sweden.

7.5.2 Permission from Teknisk Illustration, Umeå, Sweden.
7.5.4 Permission from Linde AG, Wiesbaden, Germany.
7.5.5 Permission from DaimlerChrysler AG, Stuttgart, Germany.
7.5.6 Tsiolkowsky, K. (1903, 1911). Reactive Flying Machines.
7.5.7 Permission from Smithsonian Institute, Washington DC, USA.
7.5.8, 7.5.13 (Left). Permission from Deutsches Museum, Munich, Germany.
7.5.10–12 From Reference 7.5.12. Permission from Frank Winter, Smithsonian Institute.

7.6.1–2, 7.6.4, 7.6.10–11 Permission from Linde Gas, Höllriegelskreuth, Germany.
7.6.3 Permission from the Messer Group, Frankfurt am Main, Germany.
7.6.5, 7.6.7, 7.6.9 Permission from Teknisk Illustration, Umeå, Sweden.

7.6.6 Permission from Air Products and Chemicals Inc., Allentown, Pennsylvania, USA.
7.6.8 Permission from Praxair, Inc., Danbury, Connecticut, USA.

7.7.1–2 Permission from the Messer Group, Frankfurt am Main, Germany.
7.7.3 Permission from the BOC Group, Windlesham, England.
7.7.5 Permission from Linde Gas, Höllriegelskreuth, Germany.

INSTITUTIONAL INDEX

NAME INDEX

SUBJECT INDEX